Optical biomimetics

Related titles:

Printed films: Materials science and applications in sensors, electronics and photonics
(ISBN 978-1-84569-988-8)
Printed films: Materials science and applications in sensors, electronics and photonics provides comprehensive coverage of the most significant recent developments in printed films and their applications. Materials and properties of printed films are the focus of Part I. Printed films as electrical components and silicon metallization for solar cells are discussed, as are conduction mechanisms in printed film resistors, and thick films in packaging and microelectronics. Part II goes on to review the varied applications of printed films in devices. Printed resistive sensors are considered, as is the role of printed films in capacitive, piezoelectric and pyroelectric sensors, mechanical micro-systems and gas sensors. The applications of printed films in biosensors, actuators, heater elements, varistors and polymer solar cells are then explored, followed by a review of screen printing for the fabrication of solid oxide fuel cells, and laser printed micro- and meso-scale power generating devices.

Quantum optics with semiconductor nanostructures
(ISBN 978-0-85709-232-8)
Quantum optics with semiconductor nanostructures is a key guide to the theory, application and potential of semiconductor nanostructures in the exploration of quantum optics. Part I begins with an overview of resonance fluorescence emission. Quantum optics with single quantum dots in photonic crystal and micro cavities are explored in detail, before Part II goes on to review nanolasers with quantum dot emitters. Light-matter interaction in semiconductor nanostructures, including photon statistics and photoluminescence, is the focus of Part III, whilst Part IV explores all-solid-state quantum optics, crystal nanobeam cavities and quantum-dot microcavity systems. Finally, Part V investigates ultrafast phenomena, including femtosecond quantum optics and coherent optoelectronics with quantum dots.

Laser growth and processing of photonic devices
(ISBN 978-1-84569-936-9)
Laser growth and processing of photonic devices is the first book to review this increasingly important field. Part I investigates laser-induced growth of materials and surface structures, with pulsed laser deposition techniques, the formation of nanocones and the fabrication of periodic photonic microstructures explored in detail. Laser-induced three-dimensional micro- and nano-structuring are the focus of Part II. Exploration of multiphoton lithography, processing and fabrication is followed by consideration of laser-based micro- and nano-fabrication, laser-induced soft matter organization and micro-structuring, and laser-assisted polymer joining methods. The book concludes in Part III with an investigation into laser fabrication and manipulation of photonic structures and devices. Laser seeding and thermal processing of glass with nanoscale resolution, laser-induced refractive index manipulation, and the thermal writing of photonic devices in glass and polymers are all considered.

Details of these and other Woodhead Publishing books can be obtained by:

- visiting our website at www.woodheadpublishing.com
- contacting Customer Services (e-mail: sales@woodheadpublishing.com; fax: +44 (0) 1223 832819; tel.: +44 (0) 1223 499140 ext. 130; address: Woodhead Publishing Limited, 80 High Street, Sawston, Cambridge CB22 3HJ, UK)
- contacting our US office (e-mail: usmarketing@woodheadpublishing.com; tel.: (215) 928 9112; address: Woodhead Publishing, 1518 Walnut Street, Suite 1100, Philadelphia, PA 19102-3406, USA)

If you would like e-versions of our content, please visit our online platform: www.woodheadpublishingonline.com. Please recommend it to your librarian so that everyone in your institution can benefit from the wealth of content on the site.

Woodhead Publishing Series in Electronic and Optical Materials:
Number 48

Optical biomimetics

Materials and applications

Edited by
Maryanne Large

WOODHEAD PUBLISHING

Oxford Cambridge Philadelphia New Delhi

Published by Woodhead Publishing Limited,
80 High Street, Sawston, Cambridge CB22 3HJ, UK
www.woodheadpublishing.com
www.woodheadpublishingonline.com

Woodhead Publishing, 1518 Walnut Street, Suite 1100, Philadelphia,
PA 19102-3406, USA

Woodhead Publishing India Private Limited, G-2, Vardaan House, 7/28 Ansari Road,
Daryaganj, New Delhi – 110002, India
www.woodheadpublishingindia.com

First published 2012, Woodhead Publishing Limited

British Library Cataloguing in Publication Data
A catalogue record for this book is available from the British Library.

Library of Congress Control Number: 2012911033

ISBN 978-1-84569-802-7 (print)
ISBN 978-0-85709-765-1 (online)
ISSN 2050-1501 Woodhead Publishing Series in Electronic and Optical Materials (print)
ISSN 2050-151X Woodhead Publishing Series in Electronic and Optical Materials (online)

The publisher's policy is to use permanent paper from mills that operate a sustainable forestry policy, and which has been manufactured from pulp which is processed using acid-free and elemental chlorine-free practices. Furthermore, the publisher ensures that the text paper and cover board used have met acceptable environmental accreditation standards.

Typeset by Toppan Best-set Premedia Limited
Printed by Publishers' Graphics LLC

Contents

Contributor contact details

(* = main contact)

Editor

Maryanne C. Large
Institute of Photonics and Optical Sciences
School of Physics
University of Sydney
NSW 2006
Australia

E-mail: maryanne.large@sydney.edu.au

Chapter 1

Andrew R. Parker
Department of Zoology
The Natural History Museum
Cromwell Road
London
SW7 5BD
UK

E-mail: a.parker@nhm.ac.uk

Chapter 2

M. J. Ranaghan* and N. L. Wagner
Department of Molecular and Cell Biology
91 North Eagleville Road
University of Connecticut
Storrs-Mansfield
CT 06269
USA

E-mail: mjr03009@gmail.com

M. N. Sandberg
Department of Chemistry
55 North Eagleville Road
University of Connecticut
Storrs-Mansfield
CT 06269
USA

Robert R. Birge
Department of Chemistry
University of Connecticut
Chemistry Building
55 North Eagleville Road
Storrs-Mansfield
Connecticut 06269
USA

E-mail: rbirge@uconn.edu

and

Department of Molecular and Cell Biology
91 North Eagleville Road
University of Connecticut
Storrs-Mansfield
CT 06269
USA

Chapter 3

Serge Berthier
Université Paris-Diderot
5 rue Thomas-Mann
75205 Paris
France

E-mail: serge.berthier@insp.jussieu.fr

and

Laboratoire de Physique des Solides
Facultés Universitaires Notre Dame de la Paix, Namur
61 rue de Bruxelles
B-5000 Namur
Belgium

Chapter 4

Surojit Chattopadhyay*
Institute of Biophotonics
National Yang Ming University
155, Sec-2, Li-Nong Street
Taipei-112
Taiwan

E-mail: sur@ym.edu.tw

Yi-Fan Huang
Institute of Biophotonics
National Yang Ming University
Taipei-112
Taiwan

and

Institute of Atomic and Molecular Sciences
Academia Sinica
Taipei-106
Taiwan

Kuei-Hsien Chen
Institute of Atomic and Molecular Sciences
Academia Sinica
Taipei-106
Taiwan

Li-Chyong Chen
Center for Condensed Matter Sciences
National Taiwan University
Taipei-106
Taiwan

E-mail: chenlc@ntu.edu.tw

Chapter 5

Shelley Wickham
Dana-Farber Cancer Institute
450 Brookline Avenue
Boston
MA 02115
USA

Leon Poladian
School of Mathematics and Statistics
University of Sydney
NSW 2006
Australia

Maryanne C. Large*
Institute of Photonics and Optical Sciences
School of Physics
University of Sydney
NSW 2006
Australia

E-mail: maryanne.large@sydney.edu.au

Peter Vukusic
Exeter University
Exeter
EX4 4QL
UK

Chapter 6

Hu Tao and David L. Kaplan
Department of Biomedical Engineering
Tufts University
4 Colby St.
Medford
MA 02155
USA

Fiorenzo G. Omenetto*
Department of Biomedical Engineering and Department of Physics
Tufts University
4 Colby St.
Medford
MA 02155
USA

E-mail: fiorenzo.omenetto@tufts.edu

Chapter 7

Anya Salih
Confocal Bio-Imaging Facility, School of Health and Science
University of Western Sydney
Richmond, NSW
Australia

E-mail: a.salih@uws.edu.au

Woodhead Publishing Series in Electronic and Optical Materials

1 **Circuit analysis**
J. E. Whitehouse

2 **Signal processing in electronic communications: For engineers and mathematicians**
M. J. Chapman, D. P. Goodall and N. C. Steele

3 **Pattern recognition and image processing**
D. Luo

4 **Digital filters and signal processing in electronic engineering: Theory, applications, architecture, code**
S. M. Bozic and R. J. Chance

5 **Cable engineering for local area networks**
B. J. Elliott

6 **Designing a structured cabling system to ISO 11801: Cross-referenced to European CENELEC and American Standards Second edition**
B. J. Elliott

7 **Microscopy techniques for materials science**
A. Clarke and C. Eberhardt

8 **Materials for energy conversion devices**
Edited by C. C. Sorrell, J. Nowotny and S. Sugihara

9 **Digital image processing: Mathematical and computational methods Second edition**
J. M. Blackledge

10 **Nanolithography and patterning techniques in microelectronics**
Edited by D. Bucknall

11 **Digital signal processing: Mathematical and computational methods, software development and applications**
Second edition
J. M. Blackledge

12 **Handbook of advanced dielectric, piezoelectric and ferroelectric materials: Synthesis, properties and applications**
Edited by Z.-G. Ye

13 **Materials for fuel cells**
Edited by M. Gasik

14 **Solid-state hydrogen storage: Materials and chemistry**
Edited by G. Walker

15 **Laser cooling of solids**
S. V. Petrushkin and V. V. Samartsev

16 **Polymer electrolytes: Fundamentals and applications**
Edited by C. A. C. Sequeira and D. A. F. Santos

17 **Advanced piezoelectric materials: Science and technology**
Edited by K. Uchino

18 **Optical switches: Materials and design**
Edited by S. J. Chua and B. Li

19 **Advanced adhesives in electronics: Materials, properties and applications**
Edited by M. O. Alam and C. Bailey

20 **Thin film growth: Physics, materials science and applications**
Edited by Z. Cao

21 **Electromigration in thin films and electronic devices: Materials and reliability**
Edited by C.-U. Kim

22 ***In situ* characterization of thin film growth**
Edited by G. Koster and G. Rijnders

23 **Silicon-germanium (SiGe) nanostructures: Production, properties and applications in electronics**
Edited by Y. Shiraki and N. Usami

24 **High-temperature superconductors**
Edited by X. G. Qiu

25 **Introduction to the physics of nanoelectronics**
S. G. Tan and M. B. A. Jalil

26 **Printed films: Materials science and applications in sensors, electronics and photonics**
Edited by M. Prudenziati and J. Hormadaly

27 **Laser growth and processing of photonic devices**
Edited by N. A. Vainos

28 **Quantum optics with semiconductor nanostructures**
Edited by F. Jahnke

29 **Ultrasonic transducers: Materials and design for sensors, actuators and medical applications**
Edited by K. Nakamura

30 **Waste electrical and electronic equipment (WEEE) handbook**
Edited by V. Goodship and A. Stevels

31 **Applications of ATILA FEM software to smart materials: Case studies in designing devices**
Edited by K. Uchino and J.-C. Debus

32 **MEMS for automotive and aerospace applications**
Edited by M. Kraft and N. M. White

33 **Semiconductor lasers: Fundamentals and applications**
Edited by A. Baranov and E. Tournie

34 **Handbook of terahertz technology for imaging, sensing and communications**
Edited by D. Saeedkia

35 **Handbook of solid-state lasers: Materials, systems and applications**
Edited by B. Denker and E. Shklovsky

36 **Organic light-emitting diodes and displays**
Edited by A. Buckley

37 **Lasers for medical applications: Diagnostics, therapy and surgery**
Edited by H. Jelínková

38 **Semiconductor gas sensors**
Edited by R. Jaaniso and O. K. Tan

39 **Handbook of organic materials for optical and optoelectronic devices: Properties and applications**
Edited by O. Ostroverkhova

40 **Metallic films for electronic, optical and magnetic applications: Structure, processing and properties**
Edited by K. Barmak and K. Coffey

41 **Handbook of laser welding technologies**
Edited by S. Katayama

42 **Nanolithography for fabricating nanoelectronics, nanophotonics and nanobiology devices and systems**
Edited by M. Feldman

43 **Laser spectroscopy for sensing: Fundamentals, techniques and applications**
Edited by M. Baudelet

44 **Chalcogenide glasses: Preparation, properties and applications**
Edited by J.-L. Adam and X. Zhang

45 **Handbook of MEMS for wireless and mobile applications**
Edited by D. Uttamchandani

46 **Subsea optics and imaging**
Edited by J. Watson and O. Zielinski

47 **Carbon nanotubes and graphene for photonic applications**
Edited by S. Yamashita, Y. Saito and J. H. Choi

48 **Optical biomimetics: Materials and applications**
Edited by M. Large

Introduction

M. C. J. LARGE, University of Sydney, Australia and
Canon Information Systems Research Australia, Australia

Perhaps the introduction of all academic texts should include some grandiose claims about the importance of the contents of the book. In this case, that is easily begun. The interaction of light with biological tissue is the basis of almost all life on the planet, and the dominant way that most animals receive information about the world. The scope of the structures and materials that have evolved in the context of biological interactions with light is therefore rich indeed. It includes photochemistry: the materials that are the basis of photosynthesis, bioluminescence, pigments and retinal sensors. Biomolecules often have very complex interactions with light: harvesting it, shifting wavelengths and using it transferring energy. The complexity of some of these interactions means that even in nature they are often used in more than one way. Chemistry, however, is only the start of the story. Physical structures have evolved in parallel which use the full range of optical effects: reflection, refraction, diffraction and interference. These effects have been used in eyes and other structures to produce colour, optimize absorption or manipulate polarization.

However, all of biological photochemistry and optics is probably too ambitious a task, even for a book as interesting as this one. We are aiming for something different. We aim to understand how we can *use* this huge variety of materials and structures. What can we learn from the optimization of millions of years of evolution? Can we make nature's tricks, our tricks? This more applied field itself is extensive, and has a surprising long history. As an example, graded index lenses were apparently first discovered by James Clerk Maxwell (he of the eponymous equations) while contemplating the eye of his breakfast kipper. Indeed many of the earliest examples of optical processes such as diffraction and thin film interference were first studied in nature, perhaps because it was not trivial to make structures of the right dimension and regularity. Although those structures are now well understood, the field remains surprisingly rich. Two recent high-profile papers, for example, have shown how jumping spiders can judge depth from a defocused image (Nagata *et al.*, 2012) and suggested how the iridescent structures found in the wing of the *Morpho* butterfly could be used to produce the next generation of thermal imaging sensors (Pris *et al.*, 2012).

Optical biomaterials are, in this respect, particularly compelling. Molecules of this complexity are hard to design and synthesize from scratch. Using naturally occurring biomolecules, however, has two profound benefits. The first is simple: the materials are generally biocompatible. For example, while many synthetic dyes are carcinogenic, those found in nature are non-toxic and can be used *in vivo*. This is even more appealing when combined with the second advantage. In many cases the genetic coding of biomolecules is understood, and can be manipulated to change the properties of the molecule. We are only beginning to exploit the benefits this brings.

In this book we will explore several families of important biomolecules: the retinal pigment Rhodopsin, GFP (green fluorescent pigment) found in a large number of marine animals (and the basis of the 2008 Nobel Prize in Chemistry) and, finally, a material that is not obviously optical at all: silk. Sometimes materials become 'optical materials' when you look at them in a new light. Silk and DNA have both been explored recently as optical materials. Apart from their high transparency, the attraction is their biocompatibility and eco-friendly processing. These allow us to envisage, for example, medical optical sensing in which the sensor can be safely integrated with biological tissue. In the chapters of this book, biological microstructures are discussed in the context of optical security devices, non-iridescent structural colour, ultra high absorption materials, photonic crystals and the manipulation of polarization.

We hope this book will stimulate and enable discussions between biologists, chemists, and physicists; doctors and engineers. Nature is no respecter of boundaries. There are surely many more interesting ideas to be borrowed from nature about how to use light in useful ways.

References

Nagata, T. *et al.*, Depth perception from image defocus in a jumping spider, *Science* 335, no. 6067 (January 27, 2012): 469–471.

Pris, A. D. *et al.*, Towards high-speed imaging of infrared photons with bio-inspired nanoarchitectures, *Nature Photonics*, doi:10.1038/nphoton.2011.355 (2012).

1

Natural photonic structures: an overview

A. R. PARKER, The Natural History Museum, UK

Abstract: This chapter reviews types of photonic structures found in nature and ways they can be exploited to produce photonic devices. It discusses both types of device, such as anti-reflective and iridescent surfaces, as well as techniques for their manufacture such as cell culture techniques to create butterfly scales, the use of diatoms and coccolithophores to build photonic devices as well as iridoviruses to create biophotonic crystals.

Key words: photonic structures, anti-reflective and iridescent surfaces, diatoms and coccolithophores, iridoviruses.

1.1 Introduction

Biomimetics is the extraction of good design or problem solving from nature. Its history can be taken back many centuries. Perhaps Leonardo da Vinci's sixteenth century helicopter was inspired by the corkscrew structure of some aerodynamic seeds, while Marc Isambard Brunel's tunnelling machine of 1818 was based on the waste-removing, boring system of the ship-worm. It is only now, however, that we are seeing concentrated efforts to convert some of the innovations in nature, uncovered during a century of biological research, into commercial products.

A few of evolution's innovations have already been successfully transferred to commerce. These include Velcro (the fastening device copied from the attachment of plant seed burs to animal fur) and a non-chemical adhesive tape inspired by the attachment mechanism of the gecko's foot. Yet there are not as many items in this list of successes as there should be. We all know the potential. The recent growth of an iridescent *Morpho* butterfly scale in the lab has raised the bar of potential for optical biomimetics. Animals and plants boast a variety of structural colours, or sub-micron photonic devices, comparable to the portfolio of physicists. The subject of photonics in nature began with the characterization of structural colours in animals and plants. Then came the manufacture of analogues of the natural photonic devices, employing modern, nano-fabrication techniques. Now we have the potential for a cell culture approach to commercial-scale productivity of optical devices. We can use living cells to make the devices for us, or at least learn from their engineering processes as they assemble their sub-micron structures under the microscope.

Why are animals and plants so well-adapted to their precise environmental conditions? Over millions of years, random genetic mutations have accumulated to code for structures that solve problems in a specific micro-environment, in the manner of trial and error. Structures that aid the survival of a species will be retained, or 'selected for', since the individuals that possess them will live longer and so pass on their genes more prevalently to future generations. The problems solved by their structures include how to appear bright and conspicuous in order to attract the eye of another individual. This and related optical problems have been solved in nature numerous times, resulting in the range of natural photonic structures that provide the foundation of *optical* biomimetics.

'In the country of the blind, the one-eyed man is king.' H.G. Wells's famous dictum tells us something that may seem self-evident: *sight matters*. Imagine for a moment, however, that the country of the blind is in fact the whole world, 522 million years ago. It is a world where life is primitive, and evolution is slow. Then something remarkable happens. Over the next million years, the process of evolution kicks into overdrive. For the first time, animals evolve hard external parts. Both hunters and prey develop armaments and defences in an event known as the Cambrian explosion. It is probable that this was instigated by the evolution of the first eye in a highly mobile, sizeable predator – an animal that could have had a profound effect on the entire community of animals. With the evolution of the predatory eye, the size, shape, colour and behaviour of animals was suddenly revealed for the first time. The animal kingdom exploded into life. Colour was amongst their main weapons. From that moment until today, colour on Earth has functioned to provide camouflage and warning colours as well as mating colours to attract the eye.[1]

1.2 Photonic structures found in nature

Today we find a diversity of photonic structures in nature that have become fine-tuned over 520 million years of 'trial and error'. For further information on the evolution of optical devices in nature, including those found in fossils, or when they first appeared on earth, see research by Parker.[2,3] Providing effects known as structural colours (in contrast to pigmentary colours), the simplest of these photonic structures are those where a light ray is (generally) reflected only once within the system (i.e. they adhere to the single scattering, or First Born, approximation, where the incident field is considered in place of the total field as the driving field at each point in the scatterer). These structures include:

- structures that cause random scattering;
- two-dimensional (2D) diffraction gratings;

- one-dimensional (1D) multilayer reflectors;
- three-dimensional (3D) liquid crystals.

Figure 1.1a shows a structure that causes random scattering. An irregular array of elements scatters incident light into random directions. The scattered (or reflected) rays do not superimpose. The gas spaces (>1 μm) within a liquid medium in the tentacles of some anemones can scatter sunlight to cause a white appearance by this means. Fig. 1.1b shows a 2D diffraction grating: a surface structure, from where light is diffracted into a spectrum or multiple spectra. Each corrugation is typically about 500 nm wide. Diffracted rays superimpose either constructively or destructively to provide a coloured appearance or not. The setae (halophores) of some ostracod (Crustacea) antennae are iridescent owing to diffraction gratings on their surfaces.

Figure 1.1c shows a 1D multilayer reflector known as a 'quarter wave stack', composed of thin (ca. 100 nm thick) layers of alternating refractive index, where light rays reflected from each interface in the system superimpose either constructively or destructively. Some degree of refraction occurs. The iridescent cuticles of *Ovalipes* crabs and jewel wasps contain such thin-layer stacks. Fig. 1.1d shows a 3D 'liquid crystal' composed of nano-fibres arranged in layers, where the nano-fibres of one layer lie parallel to each other yet are orientated slightly differently to those of adjacent layers. Hence spiral patterns can be distinguished within the structure. The height of the section shown here – one 'period' of the system – is around 200 nm. Many metallic-coloured (often green) beetles contain such cholesteric reflectors.

As well as these structures where each light ray is (generally) reflected once, there are structures where each light ray is more often reflected multiple times:

- 2D photonic crystals;
- 3D photonic crystals.

In 2001 the first photonic crystal was identified as such in animals[4] and, since then, the scientific effort in this subject has accelerated. Now we know of a variety of 2D and 3D photonic crystals in nature including some designs not encountered previously in physics. A 2D photonic crystal has periodicity in two planes. An example is shown in Fig. 1.1f from the hair of a sea mouse. The iridescence from these hairs acts as a warning colour, drawing attention to the protective spines nearby. A 3D photonic crystal has periodicity in three planes. An example is shown in Fig. 1.1e from the scale of a weevil. The iridescence of each scale contributes to the overall body pattern of the weevil, probably for recognition by its own species and/or to provide a disruptive pattern in the eyes of predators.

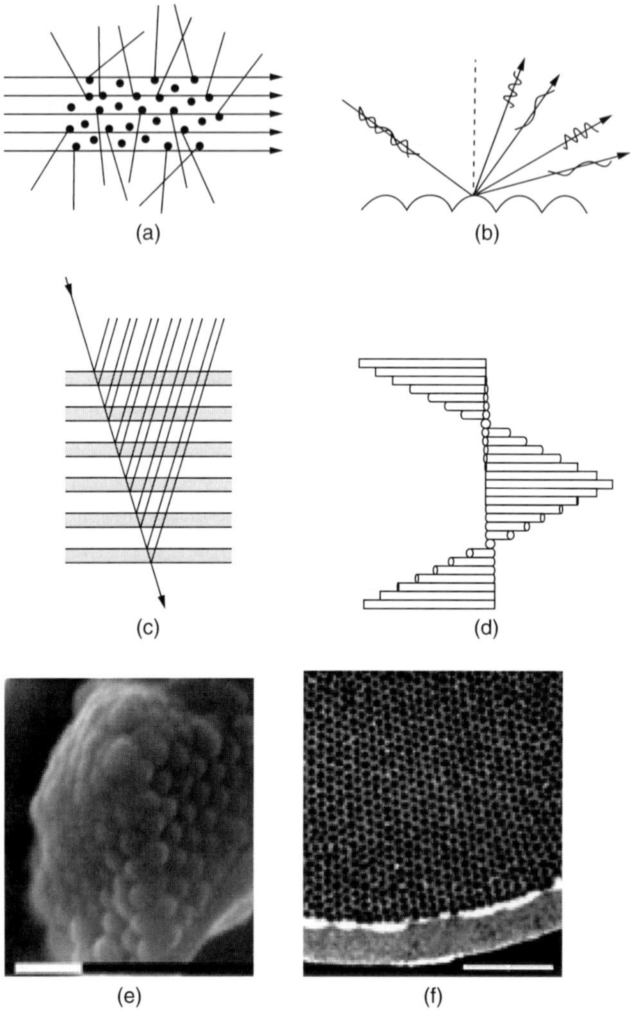

1.1 Summary of the main types of optical reflectors found in nature; (a–d) where each light ray can be considered as reflected once within the system, and (e) and (f) where each light ray is (generally) reflected multiple times within the system; (a) scattering system, (b) diffraction grating, (c) multilayer reflector, (d) cholesteric stack ('liquid crystal'). (e, f) Scanning electron micrographs of: (e) the 'opal' structure – a close-packed array of submicron spheres (a '3D photonic crystal') – found within a single scale of the weevil *Metapocyrtus* sp., scale bar = 1 μm; (f) a section through a hair (neuroseta) of the sea mouse *Aphrodita* sp. (Polychaeta), showing a cross-section through a stack of sub-micron tubes (a '2D photonic crystal'), scale bar = 5 μm.

Some photonic devices in nature are dynamic. Iridophores are cells that are packed with multilayer reflectors. Many have elastic cell walls and are under nervous and muscular control. Effectively, the reflectors can be made to turn on and off or to shift their reflectance wavelengths almost instantly, through the action of these microscopic muscles. Some beetles can alter their reflectors by interchanging one of the materials in a multilayered system (between water and air). In this way they can alternate between camouflage and conspicuousness. Other reflectors change thicknesses and therefore colour with changing temperature.

There are many biological functions for these photonic devices.[1] A few examples are given here. Multilayer reflectors in the corneas of some eyes are tuned to transmit waves of optimal detection by the retinal cells below (i.e. those most used for vision). Similar filters can also be found in plants, such as in leaves to transmit wavelengths employed in photosynthesis. Broadband multilayer reflectors are found in eyes and bioluminescent organs to focus light, while some fish possess a network of tubes lined with mirrors to guide light from a single source out of many exit points and into specific directions.

1.3 Examples of optical biomimetic devices

It has taken three centuries of research, beginning with Hooke and Newton, to reveal the diversity of natural optical devices at the sub-micron scale. Since 2000, we have begun to make them ourselves. Some examples are reviewed below.

1.3.1 Engineering of anti-reflectors

Some insects benefit from anti-reflective surfaces, either on their eyes to see under low-light conditions, or on their wings to reduce surface reflections in transparent (camouflaged) areas. Anti-reflective surfaces, therefore, occur on the corneas of moth and butterfly eyes[5] and on the transparent wings of hawkmoths.[6] These consist of nodules, with rounded tips, arranged in a hexagonal array with a periodicity of around 240 nm (Fig. 1.2b). Effectively they introduce a gradual refractive index profile at an interface between chitin (a polysaccharide, often embedded in a proteinaceous matrix; reflective index (n) 1.54) and air, and hence reduce reflectivity by a factor of ten.

This 'moth-eye structure' was first reproduced at its correct scale by crossing three gratings at 120° using lithographic techniques, and employed as anti-reflective surfaces on glass windows in Scandinavia.[7] Here, plastic sheets bearing the anti-reflector were attached to each interior surface of triple glazed windows using refractive-index-matching glue to provide a

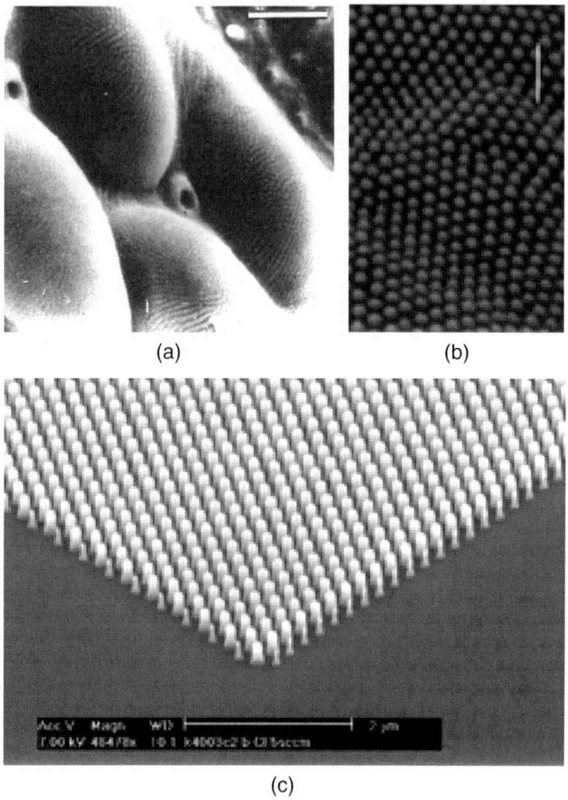

1.2 Scanning electron micrographs of anti-reflective surfaces. (a) Fly-eye anti-reflector (ridges on four facets) on a 45 million-year-old *dolichopodid* fly's eye. Micrograph by P. Mierzejewski, reproduced with permission and (b) moth-eye anti-reflective surfaces. (c) Moth-eye mimic fabricated using ion-beam etching. Micrograph by S.A. Boden and D.M. Bagnall, reproduced with permission. Scale bars = 3 μm (a), 1 μm (b), 2 μm (c).

significant difference in reflectivity. Today the moth-eye structure can be made extremely accurately using e-beam etching,[8] and is employed commercially on solid plastic and other lenses.

A different form of anti-reflective device, in the form of a sinusoidal grating of 250 nm periodicity, was discovered on the cornea of a 45 million-year-old fly preserved in amber[9] (Fig. 1.2a). This is particularly useful where light is incident at a range of angles (within a single plane, perpendicular to the grating grooves), as demonstrated by a model made in photoresist using lithographic methods.[9] Consequently it has been employed on the surfaces of solar panels, providing a 10% increase in energy capture through reducing the reflected portion of sunlight.[10] Again, this device is embossed onto plastic sheets using holographic techniques.

1.3.2 Engineering of iridescent devices

Many birds, insects (particularly butterflies and beetles), fishes and lesser-known marine animals display iridescent (changing colour with angle) and/or 'metallic' coloured effects resulting from photonic nanostructures. These appear comparatively brighter than the effects of pigments and often function in animals to attract the attention of a potential mate or to startle a predator. An obvious application for such visually attractive and optically sophisticated devices is within the anti-counterfeiting industry. For secrecy reasons, not all the work in this area can be described, although devices are sought at different levels of sophistication, from effects that are discernible by the eye to fine-scale optical characteristics (polarization and angular properties, for example) that can be read only by specialized detectors. However, new research aims to exploit these devices in the cosmetics, paint, printing/ink and clothing industries. They are even being tested in art to provide a sophisticated colour change effect.[11]

Original work on exploiting nature's reflectors involved copying the design but not the size, where reflectors were scaled-up to target longer wavelengths. For example, rapid prototyping was employed to manufacture a microwave analogue of a *Morpho* butterfly scale that is suitable for reflection in the 10–30 GHz region. Here layer thicknesses would be in the order of 1 mm rather than 100 nm as in the butterfly, but the device could be employed as an antenna with broad radiation characteristics, or as an anti-reflection coating for radar. However, today techniques are available to manufacture nature's reflectors at their true size.

Nanostructures causing iridescence include photonic crystal fibres (PCFs), opal and inverse opal, and unusually sculpted 3D architectures. Photonic crystals are ordered, often complex, sub-wavelength (nano) lattices that can control the propagation of light at the single wave scale in the manner that atomic crystals control electrons.[12] Examples include opal (a hexagonal or square array of 250 nm spheres) and inverse opal (a hexagonal array of similar sized holes in a solid matrix). Hummingbird feather barbs contain variation ultra-thin layers with variations in porosity that cause their iridescent effects, due to the alternating nanoporous/fully dense ultrastructure.[13] Such layers have been mimicked using aqueous-based layering techniques.[13] The greatest diversity of 3D architectures can be found in butterfly scales, which can include micro-ribs with nano-ridges, concave multi-layered pits, blazed gratings and randomly punctate nano-layers.[14,15] The cuticle of many beetles contain structurally chiral films that produce iridescent effects with circular or elliptical polarization properties.[16] These have been replicated in titania for specialized coatings,[16] where a mimetic sample can be compared with the model beetle and an accurate variation in spectra with angle is observed (Plate I – see colour section between pages 96 and 97). The titania mimic can be nanoengineered for a wide range of resonant wavelengths; the lowest so far is a pitch of 60 nm for a circular Bragg resonance at 220 nm in a Sc_2O_3 film (Ian Hodgkinson, personal communication).

Biomimetic work on the PCFs of the *Aphrodita* sea mouse is underway. The sea mouse contains spines (tubes) with walls packed with smaller tubes of 500 nm, with varying internal diameters (400–50 nm). These provide a band gap in the red region, and are to be manufactured by way of an extrusion technique. Larger glass tubes packed together in the proportion of the spine's nanotubes, will be heated and pulled through a drawing tower until they reach the correct dimensions. The sea mouse fibre mimics will be tested for standard PCF applications (e.g. in telecommunications) but also for anti-counterfeiting structures readable by a detector.

Analogues of the famous blue *Morpho* butterfly (Fig. 1.3a) scales have been manufactured.[17,18] Originally, corners were cut. Where the *Morpho* wing contained two layers of scales – one to generate colour (a quarter-wave stack) and another above it to scatter the light – the model copied only the principle.[17] The substrate was roughened at the nanoscale, and coated with 80 nm thick layers alternating in refractive index.[17] Therefore the device retained a quarter-wave stack centred in the blue region, but incorporated a degree of randomness to generate scattering. The engineered device closely matched the butterfly wing – the colour observed changed only slightly with changing angle over 180°, an effect difficult to achieve and useful for a broad-angle optical filter without dyes.

A new approach to making the 2D 'Christmas tree' structure (a vertical, elongated ridge with several layers of 70 nm-thick side branches; Fig. 1.3b) has been achieved using focused-ion-beam chemical-vapour-deposition (FIB-CVD).[18] By combining the lateral growth mode with ion beam scanning, the Christmas tree structures were made accurately (Fig. 1.3c). However, this method is not ideal for low-cost mass production of 2D and 3D nanostructures, and therefore the ion-beam-etched Christmas trees are currently limited to high-cost items including nano- or micron-sized filters (such as 'pixels' in a display screen or a filter). Recently further corners have been cut in manufacturing the complex nanostructures found in many butterfly scales, involving the replication of the scales in ZnO, using the scales themselves as templates[19] (Fig. 1.3d and 1.3e).

1.3.3 Cell culture techniques

Sometimes nature's optical nanostructures have such an elaborate architecture at such a small (nano) scale that we simply cannot copy them using current engineering techniques. Additionally, sometimes they can be made as individual reflectors (as for the *Morpho* structure) but the effort is so great that commercial-scale manufacture would never be cost-effective.

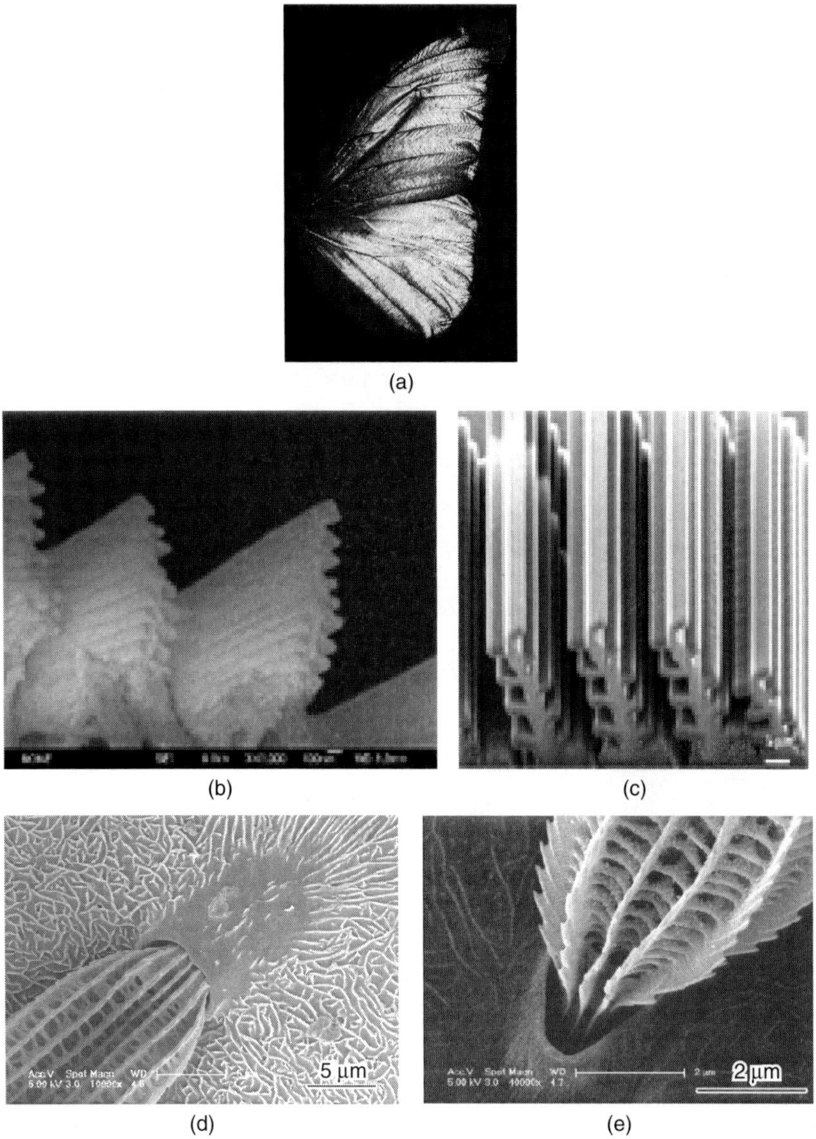

1.3 (a) A *Morpho* butterfly wing with (b) a scanning electron micrograph of the structure causing the blue reflector in its scales. Scale bar = 100 nm. (c) A scanning electron micrograph of the FIB-CVD fabricated mimic. A Ga^+ ion beam (beam diameter 7 nm at 0.4 pA; 30 kV), held perpendicular to the surface, was used to etch a precursor of phenanthrene ($C_{14}H_{10}$). Both give a wavelength peak at around 440 nm and at the same angle (30°). Scale bar = 100 nm. (d) Scanning electron micrograph of the base of a scale of the butterfly *Ideopsis similes*. (e) Scanning electron micrograph of a ZnO replica of the same part of the scale in (d). (a)–(c) by K. Watanabe, and (d) and (e) by W. Zhang, all reproduced with permission of the authors.

An alternative approach to making nature's reflectors is to exploit an aspect other than design – that the animals or plants can make them efficiently. Therefore we can let nature manufacture the devices for us via cell culture techniques. Animal cells are in the order of 10 μm in size and plant cells up to about 100 μm, and hence suitable for nanostructure production. The success of cell culture depends on the species and on the type of cell from that species. Insect cells, for instance, can be cultured at room temperature, whereas an incubator is required for mammalian cells. Cell culture is not a straightforward method, however, since a culture medium must be established to which the cells adhere, before they can be induced to develop to the stage where they make their photonic devices.

Current work in this area centres on butterfly scales. The cells that make the scales are identified in chrysalises, dissected and plated out. Then the individual cells are separated, kept alive in culture and prompted to manufacture scales through the addition of growth hormones. Currently we have cultured blue *Morpho* butterfly scales in the lab that have identical optical and structural characteristics to natural scales. The cultured scales could be embedded in a polymer or mixed into a paint, where they may float to the surface and self-align. Further work, however, is required to increase the level of scale production and to harvest the scales from laboratory equipment in appropriate ways. A far simpler task emerges where the iridescent organism is single-celled.

1.3.4 Diatoms and coccolithophores

Diatoms are unicellular photosynthetic micro-organisms. The cell wall is called the frustule and is made of the polysaccharide pectin impregnated with silica. The frustule contains pores (Fig. 1.4a–c) and slits which give the protoplasm access to the external environment. There are more than 100 000 different species of diatoms, generally 20–200 microns in diameter or length, but some can be up to 2 mm long. Diatoms have been proposed to build photonic devices directly in 3D.[20] The biological function of the optical property (Fig. 1.4d) is at present unknown, but may affect light collection by the diatom. This type of photonic device can be made in silicon using a deep photochemical etching technique (initially developed by Lehmann[21]) (e.g. Fig. 1.4e). However, there is a new potential here since diatoms carry the added advantage of exponential growth in numbers – each individual can give rise to 100 million descendants in a month.

Unlike most manufacturing processes, diatoms achieve a high degree of complexity and hierarchical structure under mild physiological conditions. Importantly, the size of the pores does not scale with the size of the cell, thus maintaining the pattern. Fuhrmann *et al.*[20] showed that the presence of these pores in the silica cell wall of the diatom *Coscinodiscus granii*

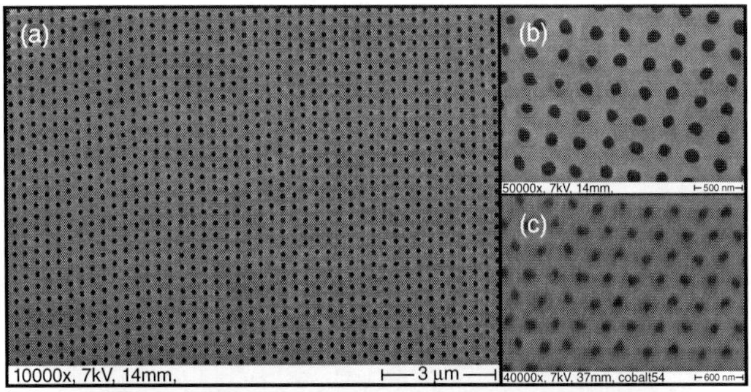

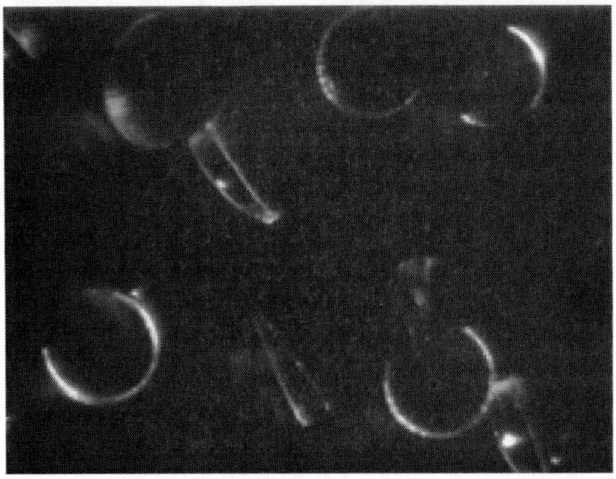

(d)

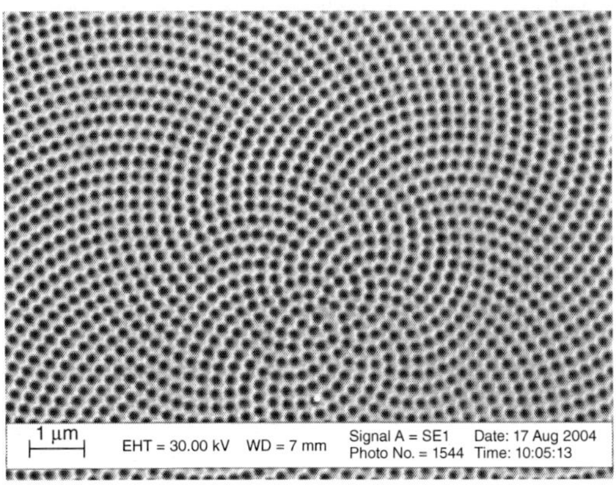

(e)

1.4 (a–c) Scanning electron micrographs of the intercalary band of the frustule from two species of diatoms, showing the square array of pores from *C. granii* ((a) and (b)) and the hexagonal arrays of pores from *C. wailesii* (c). These periodic arrays are proposed to act as photonic crystal waveguides. (d) Iridescence of the *C. granii* girdle bands. (e) Scanning electron micrograph of a Southampton University mimic of a diatom frustule (patented for photonic crystal applications) (by G. Parker, reproduced with permission).

means that the frustule can be regarded as a photonic crystal slab waveguide. Furthermore, they present models to show that light may be coupled into the waveguide and give photonic resonances in the visible spectral range.

The silica surface of the diatom is amenable to simple chemical functionalization (e.g. Fig. 1.5a–c). An interesting example of this uses a DNA-modified diatom template for the control of nanoparticle assembly.[24] Gold particles were coated with DNA complementary to that bound to the surface of the diatom. Subsequently, the gold particles were bound to the diatom surface via the sequence-specific DNA interaction. Using this method up to seven layers were added showing how a hierarchical structure could be built onto the template.

Porous silicon is known to luminesce in the visible region of the spectrum when irradiated with ultraviolet light.[25] This photoluminescence (PL) emission from the silica skeleton of diatoms was exploited by De Stafano *et al.*[26] in the production of an optical gas sensor. It was shown that the PL of *Thalassiosira rotula* is strongly dependent on the surrounding environment. Both the optical intensity and peaks are affected by gases and organic vapours. Depending on the electronegativity and polarizing ability, some substances quench the luminescence, while others effectively enhance it. In the presence of the gaseous substances NO_2, acetone and ethanol, the PL was quenched. This was because these substances attract electrons from the silica skeleton of the diatoms and hence quench the PL. Nucleophiles, such as xylene and pyridine, which donate electrons, had the opposite effect, and

1.5 Modification of natural photonic devices. (a)–(c) Diatom surface modification. The surface of the diatom was silanized, then treated with a heterobifunctional cross-linker, followed by attachment of an antibody via a primary amine group. (a)(i) Diatom exterior surface, (ii) APS, (iii) ANB-NOS, (iv) primary antibody, (v) secondary antibody with HRP conjugate. Diatoms treated with primary and secondary antibody with (b) no surface modification (c) after surface modification. (d), (e) Scanning electron micrographs showing the pore pattern of the diatom *C. wailesii* (d) and after growth in the presence of nickel sulphate (e). Note the enlargement of pores, and hence change in optical properties, in (e). (f) 'Photonic crystal' of the weevil *Metapocyrtus* sp., section through a scale, transmission electron micrograph; scale bar: 1 μm (see Parker[22]). (g) A comparatively enlarged diagrammatic example of cell membrane architecture: tubular christae in mitochondria from the chloride cell of sardine larvae (from Threadgold[23]). Evidence suggests that pre-existing internal cell structures play a role in the manufacture of natural nanostructures; if these can be altered then so will the nanostructure made by the cell.

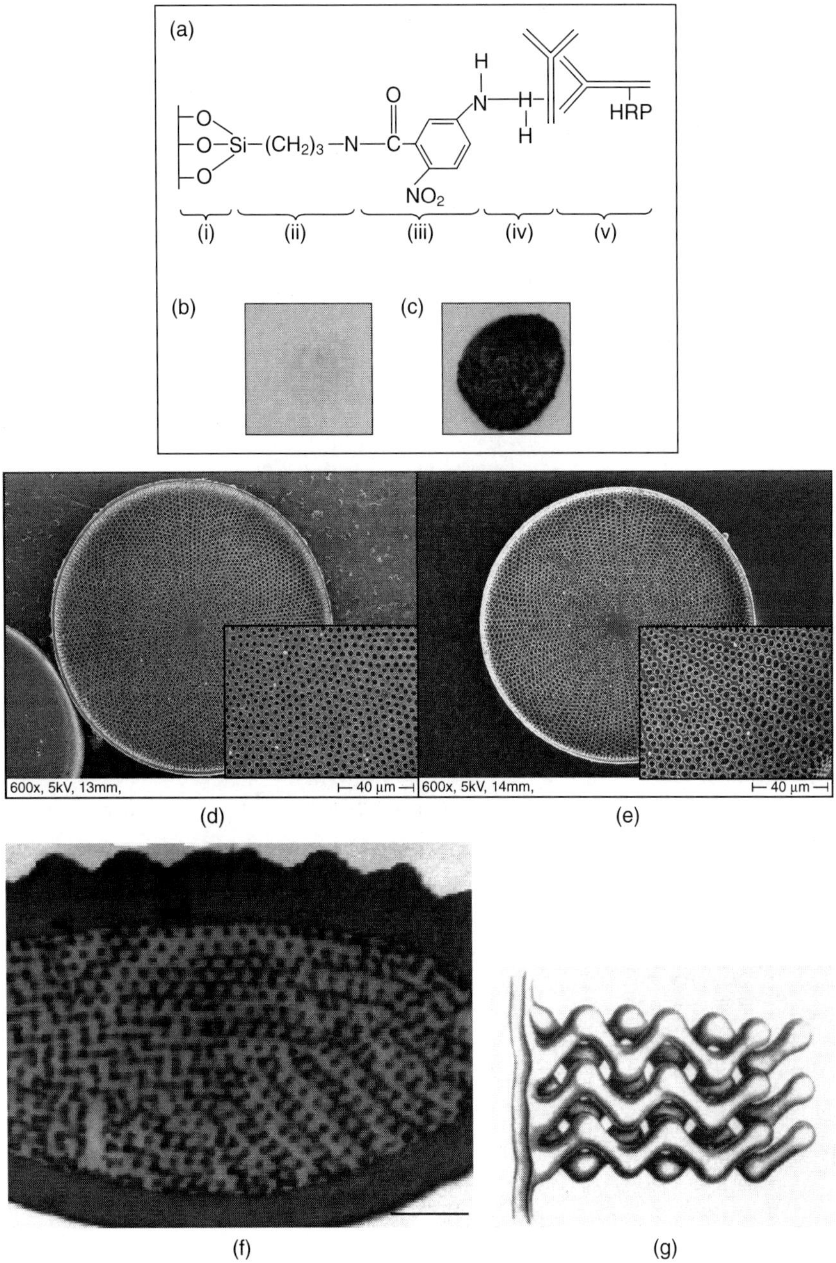
(a)
HRP
NO2
(i) (ii) (iii) (iv) (v)
(b)
(c)
600x, 5kV, 13mm,
40 μm
600x, 5kV, 14mm,
40 μm
(d)
(e)
(f)
(g)

increased PL intensity almost ten times. Both quenching and enhancements were reversible as soon as the atmosphere was replaced by air.

The silica inherent to diatoms does not provide the optimum chemistry/refractive index for many applications. Sandhage *et al.*[27] have devised an inorganic molecular conversion reaction that preserves the size, shape and morphology of the diatom while changing its composition. They perfected a gas/silica displacement reaction to convert biologically derived silica structures such as frustules into new compositions. Magnesium was shown to convert SiO_2 diatoms by a vapour phase reaction at 900°C to MgO of identical shape and structure, with a liquid Mg_2Si by-product. Similarly when diatoms were exposed to titanium fluoride gas the titanium displaced the silicon, yielding a diatom structure made up entirely of titanium dioxide; a material used in some commercial solar cells.

An alternative route to silica replacement hijacks that native route for silica deposition *in vivo*. Rorrer *et al.*[28] sought to incorporate elements such as germanium into the frustule; a semiconductor material that has interesting properties that could be of value in optoelectronics, photonics, thin film displays, solar cells and a wide range of electronic devices. Using a two-stage cultivation process the photosynthetic marine diatom *Nitzschia frustulum* was shown to assimilate soluble germanium and fabricate Si-Ge oxide nanostructured composite materials.

Porous glasses impregnated with organic dye molecules are promising solid media for tunable lasers and non-linear optical devices, luminescent solar concentrators, gas sensors and active wave-guides. Biogenic porous silica has an open sponge-like structure and its surface is naturally OH-terminated. Hildebrand and Palenik[29] have shown that Rhodamine B and 6G are able to stain diatom silica *in vivo*, and determined that the dye treatment could survive the harsh acid treatment needed to remove the surface organic layer from the silica frustule.

Now attention is beginning to turn additionally to coccolithophores – single-celled marine algae, also abundant in marine environments. Here, the cell secretes calcitic photonic crystal frustules which, like diatoms, can take a diversity of forms, including complex 3D architectures at the nano- and micro-scales.

1.3.5 Iridoviruses

Viruses are infectious particles made up of the viral genome packaged inside a protein capsid. The iridovirus family comprises a diverse array of large (120–300 nm in diameter) viruses with icosahedral symmetry. The viruses replicate in the cytoplasm of insect cells. Within the infected cell the virus particles produce a paracrystalline array that causes Bragg refraction of light. This property has largely been considered aesthetic to date but the research group of Vernon Ward (New Zealand), in collaboration with the Biomaterials laboratory at Wright-Patterson Air Force base, are using iridoviruses to create biophotonic crystals. These can be used for the control of light, with this laboratory undertaking large-scale virus production and purification as well as targeting manipulation of the surface of iridoviruses for altered crystal properties. These can provide a structural platform for a broad range of optical technologies, ranging from sensors to waveguides.

Virus nanoparticles, specifically *Chilo* and *Wiseana* invertebrate iridovirus, have been used as building blocks for iridescent nanoparticle assemblies. Here, virus particles were assembled *in vitro*, yielding films and monoliths with optical iridescence arising from multiple Bragg scattering from close-packed crystalline structures of the iridovirus. Bulk viral assemblies were prepared by centrifugation followed by the addition of glutaraldehyde, a cross-linking agent. Long-range assemblies were prepared by employing a cell design that forced virus assembly within a confined geometry followed by cross-linking. In addition, virus particles were used as core substrates in the fabrication of metallodielectric nanostructures. These comprise a dielectric core surrounded by a metallic shell. More specifically, a gold shell was assembled around the viral core by attaching small gold nanoparticles to the virus surface using inherent chemical functionality of the protein capsid.[30] These gold nanoparticles then acted as nucleation sites for electroless deposition of gold ions from solution. Such nano-shells could be manufactured in large quantities, and provide cores with a narrower size distribution and smaller diameters (below 80 nm) than currently used for silica. These investigations demonstrated that direct harvesting of biological structures, rather than biochemical modification of protein sequences, is a viable route to create unique, optically active materials.

1.4 Biomimetic approaches to fabrication of optical devices

Where cell culture is concerned it is enough to know that cells *do* make optical nanostructures, which can be farmed appropriately. However, in the future an alternative may be to emulate the natural engineering processes ourselves, through reacting the same concentrations of chemicals under the same environmental conditions, and possibly substituting analogous nano- or macro-machinery.

To date, the process best studied is the silica cell wall formation in diatoms. The valves are formed by the controlled precipitation of silica within a specialized membrane vesicle called the silica deposition vesicle (SDV). Once inside the SDV silicic acid is converted into silica particles, each measuring approximately 50 nm in diameter. These then aggregate to form larger blocks of material. Silica deposition is moulded into a pattern by the presence of organelles such as mitochondria spaced at regular intervals along the cytoplasmic side of the SDV.[31] These organelles are thought physically to restrict the targeting of silica from the cytoplasm, to ensure laying down of a correctly patterned structure. This process is very fast, presumably due to optimal reaction conditions for the synthesis of amorphous solid silica. Tight structural control results in the final species-specific, intricate exoskeleton morphology.

The mechanism whereby diatoms use intracellular components to dictate the final pattern of the frustule may provide a route for directed evolution. Alterations in the cytoplasmic morphology of *Skeletonema costatum* have been observed in cells grown in sublethal concentrations of mercury and zinc,[32] resulting in swollen organelles, dilated membranes, and vacuolated cytoplasm. Frustule abnormalities have also been reported in *Nitzschia liebethrutti* grown in the presence of mercury and tin.[33] Both metals resulted in a reduction in the length to width ratios of the diatoms, fused pores and a reduction in the number of pores per frustule. These abnormalities were thought to arise from enzyme disruption either at the silica deposition site or at the nuclear level. *C. wailesii* was grown in sublethal concentrations of nickel and observed an increase in the size of the pores (Fig. 1.5d,e), and a change in the phospholuminescent properties of the frustule. Here, the diatom can be 'made to measure' for distinct applications such as stimuli-specific sensors.

Further, *trans*-Golgi-derived vesicles are known to manufacture the coccolithophore 3D 'photonic crystals'.[34] The organelles within the cell therefore appear to have exact control of (photonic) crystal growth ($CaCO_2$ in the coccolithophores) and packing (SiO_3 in the diatoms).[35,36] Indeed, Ghiradella[11] suggested that the employment of pre-existing, intracellular structures lay behind the development of some butterfly scales and Overton[37] reported the action of microtubules and microfibrils during butterfly scale morphogenesis. Further evidence has been found to suggest that these mechanisms, involving the use of moulds and nano-machinery (e.g. Fig. 1.5f,g), reoccur with unrelated species, indicating that the basic 'eukaryote' (containing a nucleus) cell can make complex photonic nanostructures with minimal genetic mutation.[22] The diversity of butterfly scale architecture, at the submicron level,[38,39] could have evolved through mutations that caused the application of different cell structures, or changes to those structures, during scale development. The ultimate goal in the field of optical biomimetics, therefore, could be to replicate such machinery and provide conditions under which, if the correct ingredients are supplied, the optical nanostructures will self-assemble with precision.

1.5 Conclusion

The identification of new cases of photonic structures in nature has accelerated since the identification of photonic band gaps in an animal reflector in 2000. A variety of sub-micron architectures have since been revealed, many of which can be manufactured using current engineering techniques for the production of prototypes and/or commercial devices (made on an industrial scale). A greater contribution from biomimetics, however, would be the application of cell culture techniques to industry, whereby actual biological material could be harvested from living cell lines and incorporated into commercial products. Alternatively, the engineering processes of living cells that are used to make photonic devices could be copied or partially copied by industry in order to make new devices and/or reduce an industry's environmental footprint, since cells make their optical devices at relatively low temperatures and pressures (requiring a low energy input).

1.6 Acknowledgements

This work was funded by The Royal Society (University Research Fellowship), The Australian Research Council, European Union Framework 6 grant and a Research Councils UK Basic Technology grant.

1.7 References

1. Parker, A.R. 515 Million years of structural colour. *J. Opt. A* **2**, R15–28 (2000).
2. Parker, A.R. *In the Blink of an Eye*. Simon & Schuster (London)/Perseus Press Cambridge, USA, 316 pp. (2003).
3. Parker, A.R. A geological history of reflecting optics. *J. R. Soc. Lond. Interface* **2**, 1–17 (2005).
4. Parker, A.R., McPhedran, R.C., McKenzie, D.R., Botten, L.C. and Nicorovici, N.-A.P. Aphrodite's iridescence. *Nature* **409**, 36–37 (2001).
5. Miller, W.H., Moller, A.R. and Bernhard, C.G. The corneal nipple array. In: *The functional organisation of the compound eye* (ed. C.G. Bernhard). 21–33. Pergamon Press, Oxford (1966).
6. Yoshida, A., Motoyama, M., Kosaku, A. and Miyamoto, K. Antireflective nanoprotuberance array in the transparent wing of a hawkmoth *Cephanodes hylas*. *Zool. Sci.* **14**, 737–741 (1997).
7. Gale, M. Diffraction, beauty and commerce. *Phys. World* **2**, 24–28 (1989).
8. Boden, S.A. and Bagnall, D.M. Biomimetic subwavelength surfaces for near-zero reflection sunrise to sunset. *Proc. 4th World Conference on Photovoltaic Energy, Conversion, Hawaii* (2006).
9. Parker, A.R., Hegedus, Z. and Watts, R.A. Solar-absorber type antireflector on the eye of an Eocene fly (45Ma). *Proc. R. Soc. Lond. B* **265**, 811–815 (1998).
10. Beale, B. Fly eye on the prize. *The Bulletin*, 46–48 (May 25, 1999).

11. Schenk, F. and Parker, A. R. Iridescent colour: from nature to the painter's palette. *Leonardo* **44**, 109–115, 139 (2011).
12. Yablonovitch, E. Liquid versus photonic crystals. *Nature* **401**, 539–541 (1999).
13. Cohen, R.E., Zhai, L., Nolte, A. and Rubner, M.F. pH gated porosity transitions of polyelectrolyte multilayers in confined geometries and their applications as tunable Bragg reflectors. *Macromol.* **37**, 6113 (2004).
14. Ghiradella, H. Structure and development of iridescent butterfly scales: lattices and laminae. *J. Morph.* **202**, 69–88 (1989).
15. Berthier, S. Les coulers des papillons ou l'imperative beauté. *Proprietes optiques des ailes de papillons*. Springer, Paris. 142 pp. (2005).
16. DeSilva, L., Hodgkinson, I., Murray, P., Wu, Q., Arnold, M., Leader, J. and Mcnaughton, A. Natural and nanoengineered chiral reflectors: structural colour of manuka beetles and titania coatings. *Electromagnetics* **25**, 391–408 (2005).
17. Kinoshita, S., Yoshioka, S., Fujii, Y. and Okamoto, N. *Forma* **17**, 103 (2002).
18. Watanabe, K., Hoshino, T., Kanda, K., Haruyama, Y. and Matsui, S. Brilliant blue observation from a *Morpho*-butterfly-scale quasi-structure. *Jap. J. Appl. Phys.* **44**, L48–L50 (2005).
19. Zhang, W., Zhang, D., Fan, T., Ding, J., Gu, J., Guo, Q. and Ogawa, H. Biomimetic zinc oxide replica with structural color using butterfly (*Ideopsis similis*) wings as templates. *Bioinspir. Biomim.* Manuscript submitted.
20. Fuhrmann, T., Lanwehr, S., El Rharbi-Kucki, M. and Sumper, M. Diatoms as living photonic crystals. *Appl. Phys. B* **78**, 257–260 (2004).
21. Lehmann, V. On the origin of electrochemical oscillations at silicon electrodes. *J. Electrochem. Soc.* **143**, 1313 (1993).
22. Parker, A.R. Conservative photonic crystals imply indirect transcription from genotype to phenotype. *Rec. Res. Develop. Entomol.* **5**, 1–10 (2006).
23. Threadgold, L. T. *The ultrastructure of the animal cell*. Pergamon Press, Oxford, 313 pp. (1967).
24. Rosi, N.L., Thaxton, C.S. and Mirkin, C.A. Control of nanoparticle assembly by using DNA-modified diatom templates. *Agnew Chem. Int. Ed.* **43**, 5500–5503 (2004).
25. Cullis, A.G., Canham, L.T. and Calcott, P.D.J. The structural and luminescence properties of porous silicon. *J. Appl. Phys.* **82**, 909–965 (1997).
26. De Stefano, L., Rendina, I., De Stefano, M., Bismuto, A. and Maddalena, P. Marine diatoms as optical chemical sensors. *Appl. Phys. Let.* **87**, 233902 (2005).
27. Sandhage, K.H., Dickerson, M,B., Huseman, P.M., Caranna, M.A., Clifton, J.D., Bull, T.A., Heibel, T.J., Overton, W.R. and Schoenwaelder, M.E.A. Novel, bioclastic route to self-assembled, 3D, chemically tailored meso/nanostructures: Shape-preserving reactive conversion of biosilica (diatom) microshells. *Adv. Mater.* **14**, 429–433 (2002).
28. Rorrer, G.L., Chang, C.H., Liu, S.H., Jeffryes, C., Jiao, J. and Hedberg, J.A. Biosynthesis of silicon-germanium oxide nanocomposites by the marine diatom Nitzschia frustulum. *J. Nanosci. Nanotechnol.* **5**, 41–49 (2004).
29. Hildebrand, M. and Palenik, B. Grant report '*Investigation into the optical properties of nanostructured silica from diatoms*' (2003).
30. Radloff, C., Vaia, R.A., Brunton, J., Bouwer, G.T. and Ward, V.K. Metal nanoshell assembly on a virus bioscaffold. *Nano Letters* **5**, 1187–1191 (2005).

31. Schmid, A.M.M. Aspects of morphogenesis and function of diatom cell walls with implications for taxonomy. *Protoplasma* **181**, 43–60 (1994).
32. Smith, M.A. The effect of heavy metals on the cytoplasmic fine structure of Skeletonema costatum (Bacillariophyta). *Protoplasma* **116**, 14–23 (1983).
33. Saboski, E. Effects of mercury and tin on frustular ultrastructure of the marine diatom *Nitzschia liebethrutti. Water, Air Soil Pollut.* **8**, 461–466 (1977).
34. Corstjens, P.L.A.M. and Gonzales, E.L. Effects of nitrogen and phosphorus availability on the expression of the coccolith-vesicle v-ATPase (subunit C) of Pleurochrysis (Haptophyta). *J. Phycol.* **40**, 82–87 (2004).
35. Klaveness, D. and Paasche, E. Physiology of coccolithophorids. In: *Biochemistry and physiology of protozoa*, ed. 2, vol. 1. Academic Press, New York (1979).
36. Klaveness, D. and Guillard, R.R.L. The requirement for silicon in *Synura petersenii* (Chrysophyceae). *J. Phycol.* **11**, 349–355 (1975).
37. Overton, J. Microtubules and microfibrils in morphogenesis of the scale cells of *Ephestia kuhniella. J. Cell Biol.* **29**, 293–305 (1966).
38. Vukusic, P. Structural colour in Lepidoptera. Current Biology 16, R621–R623 (2006).
39. Ingram, A. and Parker, A.R. A review of the diversity and evolution of photonic structures in butterflies. *Phil. Trans. R. Soc. Lond. B* **363**, 2465–2480 (2008).

2
Optical applications of biomolecules

M. J. RANAGHAN, N. L. WAGNER, M. N. SANDBERG and R. R. BIRGE, University of Connecticut, USA

Abstract: From kinematics to molecular machines, biologically inspired technologies harness and enhance the intrinsic properties of naturally occurring materials and systems for applied technologies. Bacteriorhodopsin (BR) represents the most studied protein for photonic applications, and has found use in artificial retinas, associative and volumetric memories, optical limiters, photovoltaic cells and other devices. The native BR protein is rarely optimal for device applications, and genetic engineering plays an important role in the optimization process. In addition, new retinal proteins such as proteorhodopsin and channelrhodopsin-2 have been discovered which provide new options and opportunities.

Key words: bacteriorhodopsin, cyclicity, biophotonic devices, retinal analogs, genetic engineering.

2.1 Introduction: biomimetics and biotechnology

The natural world has long inspired the work of artists and inventors. From kinematics to molecular machines, biologically inspired technologies harness and enhance the intrinsic properties of naturally occurring materials and systems for applied technologies (Tian *et al.*, 2006; Large *et al.*, 2007; Di Natale *et al.*, 2008; Gorb, 2008; Cutkosky and Kim, 2009; Tamerler and Sarikaya, 2009). Natural selection, in particular, has optimized each of these molecules for a specific function in a highly competitive environment. Hence, the evolution of these systems has provided an excellent model for numerous technologies, most of which have yet to be realized on a commercial level.

Molecular electronics is a field that encompasses the storage and retrieval of information at the molecular level. Because these systems can be created from the 'bottom-up,' these technologies are believed to be the next phase of data storage and manipulation once lithographic technologies become size limited. This changeover is predicted to occur within the next two decades based on extrapolation of Moore's Law (Birge, 1994b).

Designs for both organic and inorganic biomimetic devices exist, but bacteriorhodopsin (BR) is often the exemplary biophotonic material for a vast array of applications (see section 2.3 for a comprehensive list of BR-based devices proposed to date). Soviet scientists first explored the unique photophysical properties of BR as a real-time holographic medium, which was termed *Biochrome* (Hong, 1994; Birge *et al.*, 1999). Many of the details related to the materials produced during this time are vague, because they are classified by the Russian military, but a historical account of this period is available (Vsevolodov, 1998). This reference is also the most comprehensive description of the history, application, and optimization of BR and is an excellent introduction into the field of biomolecular electronic devices.

2.2 Retinylidene proteins for optical devices

Retinylidene proteins are a family of photoactive, integral transmembrane proteins that are found in all three domains of life (Béjà *et al.*, 2000; Spudich *et al.*, 2000; Béjà *et al.*, 2001; Terakita, 2005). These proteins encompass a vast array of functions that include vision (Goldsmith, 1994; Yoshizawa, 1994; Shichida and Imai, 1998; Nathans, 1999), light-transduction (Birge, 1981; Keszthelyi, 1988; Sineshchekov and Spudich, 2004), ion translocation (Lanyi, 1986; Nagel *et al.*, 2003), and phototaxis (Spudich *et al.*, 1989; Takahashi *et al.*, 1990; Spudich *et al.*, 1997; Ren *et al.*, 2001) among others (Max *et al.*, 1995; Max *et al.*, 1998; Provencio *et al.*, 1998; Terakita, 2005). Despite their function, all retinylidene proteins contain an organic chromophore (i.e. vitamin A aldehyde or retinal) that is covalently bound to a conserved lysine residue in helix 7 (Luecke *et al.*, 1999). Upon the absorption of a photon, the retinal chromophore isomerizes and induces a change in the protein structure. This excited state then undergoes a series of thermally driven conformational changes as the protein relaxes back to the resting state.

This family of proteins are divided into two main classes based on their specific role: (1) microbial ion-channel pumps and sensors, and (2) G-protein coupled receptors (Terakita, 2005; Sharma *et al.*, 2007). The type I proteins, which are often represented by BR, have the ability to photocycle and are capable of many such events without regeneration of the photoactive element (Oesterhelt *et al.*, 1991; Hirayama *et al.*, 1992; Sasaki and Spudich, 1999). Most of the type II proteins undergo a photobleaching process that results in an inactive form of the protein. These proteins are often represented by bovine rhodopsin and are not considered useful for devices. Hence, it is the photocyclic mechanism of the type I proteins, specifically BR, that makes the protein a useful biomaterial. Application of this photocyclic mechanism is discussed further in section 2.3.

2.2.1 The bacteriorhodopsin photocycle

The photocycle of BR represents a series of transient photochemical states experienced by the protein after the absorption of a photon. Figure 2.1 shows the currently accepted progression of photostates through the BR photocycle (Lanyi, 2000; Gillespie *et al.*, 2002). Each state is represented by a spectrally discrete conformation of the protein as a proton is transported through the molecule and across the membrane. This process is well characterized, after almost four decades of research, and several models have been devised to explain the complex nature of the BR photocycle. Such models include, but are not limited to, the Isomerization-Switch-Transfer (IST) model (Lanyi, 1998b; Haupts *et al.*, 1997; Tittor *et al.*, 1997),

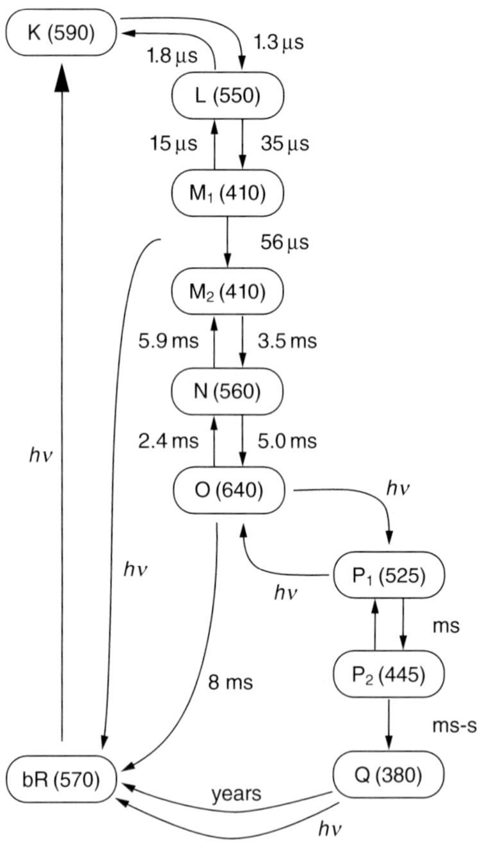

2.1 The main and branched photocycle of native bacteriorhodopsin with approximate lifetimes at ambient temperature. Photointermediates are shown with the respective absorption maximum in parenthesis (in nanometers).

the Local-Access model (Lanyi, 1998a; Brown *et al.*, 1998a; Brown *et al.*, 1998b) and the Hydration model (Kandori, 2004).

Regardless of the model, the **M**, **O** and **Q** photostates are demonstrated to be useful targets for numerous biophotonic and bioelectronic applications. Specific discussions of these devices, however, are reserved for section 2.3 of this chapter. The next sections will discuss the main and branched photocycles that result from absorption of a single green photon or sequential green and red photons, respectively.

The main photocycle

The residues within the active site of BR require an exact geometry and set of protonation states for proper functionality (Brown *et al.*, 1994; Tallent *et al.*, 1998; Kusnetzow *et al.*, 1999; Luecke *et al.*, 1999). Figure 2.2 illustrates the quadrupole geometry of the BR active site. This arrangement is well resolved to contain the positively charged Schiff base nitrogen of the chromophore-lysine linkage, two negative aspartic acids and one positive arginine that are all coordinated around three water molecules. Experimental evidence of this arrangement is upheld by a combination of theoretical

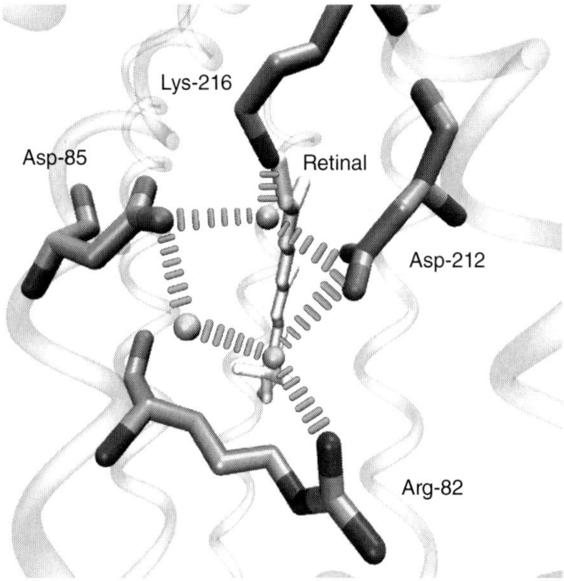

2.2 The quadrupole geometry of the bacteriorhodopsin active site comprises four residues and three water molecules. The model was created with the Visual Molecular Dynamics software package (http://www.ks.uiuc.edu/Research/vmd/) and using the 1C3W crystal structure (Luecke *et al.*, 1999).

(Birge and Zhang, 1990; Tallent *et al.*, 1998; Kusnetzow *et al.*, 1999), mutagenic (Duñach *et al.*, 1990; Heberle *et al.*, 1993; Balashov *et al.*, 1995a; Hutson *et al.*, 2000; Shibata *et al.*, 2003) and crystallographic data (see Table 2.1). Perturbation of this system, via chemical or mutagenic manipulation, greatly reduces the ability of the protein to function properly (Kobayashi *et al.*, 1983; Mogi *et al.*, 1988; Heberle *et al.*, 1993; Moltke and Heyn, 1995).

The main photocycle, comprising of the **bR**, **K**, **L**, **M**, **N** and **O** photostates, is a fast event (~10–15 ms) that translocates a proton across the membrane (Birge, 1994a). Under continuous illumination, this process is capable of producing and maintaining a proton motive force sufficient for the anaerobic synthesis of adenosine triphosphate (ATP) (Oesterhelt and Stoeckenius, 1973). The high quantum efficiency ($\phi \sim 0.65$) of the primary photochemical event makes this process very efficient and a detailed discussion of the photophysics is presented by Stuart and Birge (1996). This absorption event produces an excited state, often recognized as the **K** intermediate, and involves the isomerization of the all-*trans* retinal to a 13-*cis* configuration after the absorption of a photon. Additionally, it should be noted that other

Table 2.1 X-ray crystallographic determinations of the bacteriorhodopsin photointermediates

Photostate	PDB[a] code	Resolution (Å)	Reference
bR	1C3W	1.55	(Luecke *et al.*, 1999)
	1IW6	2.30	(Matsui *et al.*, 2002)
	1KGB	1.60	(Facciotti *et al.*, 2001)
K	1IXF	2.60	(Matsui *et al.*, 2002)
	1M0K	1.43	(Schobert *et al.*, 2002)
L[b]	2NTU	1.53	(Lanyi and Schobert, 2007)
	1VJM	2.30	(Edman *et al.*, 2004)
	1UCQ	2.40	(Kouyama *et al.*, 2004)
	1E0P	2.10	(Royant *et al.*, 2000)
M	2ZZL	2.03	(Yamamoto *et al.*, 2009)
	1IW9	2.50	(Takeda *et al.*, 2004)
	1P8H[c]	1.52	(Schobert *et al.*, 2003)
	1M0M	1.43	(Lanyi and Schobert, 2002)
	1DZE	2.50	(Takeda *et al.*, 2004)
N′	1P8U[d]	1.62	(Schobert *et al.*, 2003)
O-like	1X0I[e]	2.30	(Okumura *et al.*, 2005)
	1JV7[f]	2.25	(Rouhani *et al.*, 2001)

[a] Protein Data Bank (http://www.pdb.org)
[b] The differences observed between the L-state structures are compared to spectroscopic data and discussed in Lanyi (2004a)
[c] M_1 photostate produced at room temperature
[d] V49A mutant
[e] Acid blue form of native bacteriorhodopsin with a resting state at 603 nm
[f] D85S mutant with a resting state at 603 nm

photointermediates (i.e. **I** and **J**) are reported to precede **K** with submicrosecond lifetimes (Kobayashi *et al.*, 1983; Doig *et al.*, 1991; Shim *et al.*, 2008). Attempts to thermally trap these states, however, have been unsuccessful at temperatures as low as four Kelvin (Birge *et al.*, 1999). Many of the other photostates have been trapped and studied spectrally (Smith *et al.*, 1983; Bressler *et al.*, 1999; Kobayashi *et al.*, 2001; Shim *et al.*, 2008) or via X-ray crystallography (Table 2.1). Hence, these untrappable states can be explained as a vibrationally hot **K** state that contains a mixture of ground and excited states (Birge *et al.*, 1987).

Proton transport through BR is conventionally divided into five steps (Fig. 2.3). The following discussion highlights the key transfer events that follow photoexcitation of the protein and the reader is advised to investigate the select reviews by Balashov (2000), Lanyi (2000, 2004) and Luecke *et al.* (2000) for a more detailed discussion of the photocycle. The first proton transfer event occurs during the **L→M** conversion where a proton is transferred from the Schiff base nitrogen to Asp-85 (Metz *et al.*, 1992; Zimányi *et al.*, 1992; Dickopf and Heyn, 1997). Retinal is in a distorted 13-*cis*, 15-*anti* configuration during the **M** state (Schobert *et al.*, 2003). The

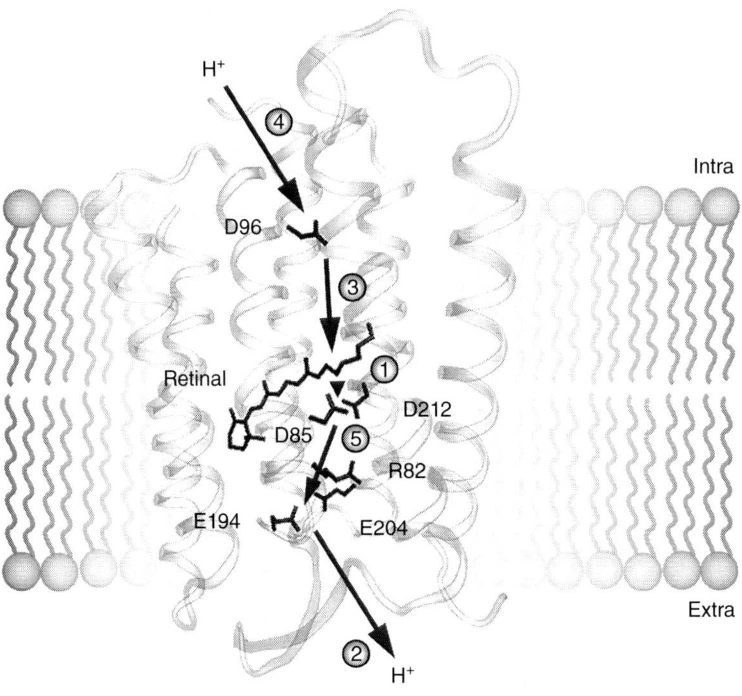

2.3 Proton transport through bacteriorhodopsin. The details of this mechanism are described in section 2.2.1 under 'The main photocycle'.

protein then releases a proton from the proton release complex as a dark reaction. This dark reaction is not represented by an observed photointermediate and occurs as the protein undergoes a conformational change from the **M→N** state (Balashov *et al.*, 1993; Heberle *et al.*, 1993). Debate over the residues involved in the proton release complex, or even its existence, is present in the literature, but mutational studies of the glutamic acids at the extracellular surface demonstrate that Glu-194 and Glu-204 are vital to the structure and function of BR (Richter *et al.*, 1996a; Richter *et al.*, 1996b; Balashov *et al.*, 1997; Dioumaev *et al.*, 1998; Sanz *et al.*, 2001). Water, as a single molecule or complex of molecules, is also essential for proton release (Luecke *et al.*, 1998; Rammelsberg *et al.*, 1998). The third step involves reprotonation of Schiff base nitrogen via Asp-96 (Cao *et al.*, 1991; Cao *et al.*, 1993). Translocation of this proton must span a distance of seven to nine angstroms and it is generally accepted that a hydrogen-bonded network of water molecules is involved (Cao *et al.*, 1991; le Coutre *et al.*, 1995; Sass *et al.*, 2000). The fourth step involves reprotonation of Asp-96 from the cellular cytoplasm and isomerization of retinal from a 13-*cis* to all-*trans* configuration (Holz *et al.*, 1989; Otto *et al.*, 1989; Brown *et al.*, 1999). This step coincides with the **O** state. The final proton transfer event is the reprotonation of the proton release complex via Asp-85 (Richter *et al.*, 1996c). This step reforms the **bR** resting state, where the protein will remain until another photon is absorbed by the chromophore.

The branched photocycle

Access to the branched photocycle is possible by several routes (Popp *et al.*, 1993; Tallent *et al.*, 1998; Birge *et al.*, 1999; Masthay *et al.*, 2002), as illustrated in Fig. 2.4, but the sequential absorbance of green and red photons is the best understood mechanism. When a red photon is absorbed during the **O** state, the protonated all-*trans* retinal photoisomerizes to a protonated 9-*cis* configuration that is unstable in the binding pocket (Popp *et al.*, 1993). The bound 9-*cis* chromophore is represented by the **P** state. This state was originally defined as having a λ_{max} ~ 490 nm (Chang *et al.*, 1987), but was later shown to comprise the $\mathbf{P_1}$ (λ_{max} = 525 nm) and $\mathbf{P_2}$ (λ_{max} = 445 nm) states that undergo a dynamic equilibrium after photon absorption (Gillespie *et al.*, 2002). Hydrolysis of the Schiff base produces an inactive form of the protein, termed the **Q** state (λ_{max} ~ 390 nm), that results from the unfavorable 9-*cis* geometry of the chromophore within the BR binding pocket (Popp *et al.*, 1993). The free retinal is trapped within the protein binding site and, with an activation barrier of ~ 190 kJ mol^{-1}, the **Q** state is stable for 7–12 years at ambient temperature (Birge *et al.*, 1999). Reversion of **Q** to **bR** is possible by the absorption of a blue photon by the 9-*cis* chromophore (Dancshazy and Tokaji, 2000).

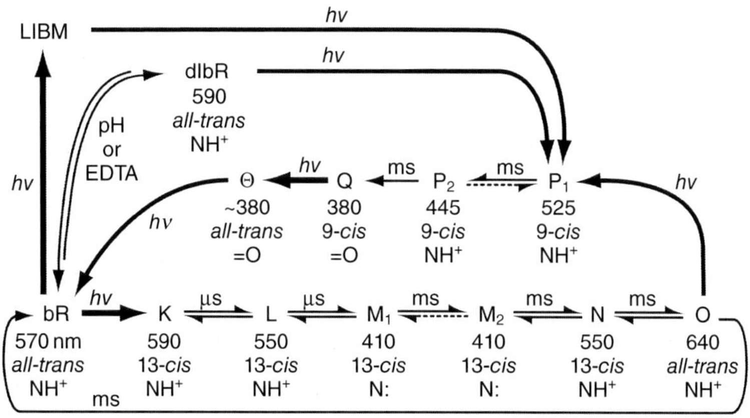

2.4 Scheme of the various methods for accessing the branched photocycle of bacteriorhodopsin. The Laser Induced Blue Membrane (LIBM) and Deionized Blue membrane (dIBR) are discussed in section 2.4.1 under 'Experimental pH'. The absorption maximum (in nanometers), chromophore configuration and state of the Schiff base linkage between retinal and Lysine 216 are noted beneath each photostate.

2.2.2 Other retinylidene proteins with the potential for application

Hundreds of type I retinylidene proteins have been reported in recent years, many of which are putative opsins, from all domains of life (Spudich *et al.*, 2000; Adamian *et al.*, 2006; Rusch *et al.*, 2007). We note that type II retinylidene proteins of higher eukaryotes have been investigated for their potential as a biomaterial (see Vsevolodov (1998) and references therein), but their application is limited because many such proteins photobleach upon activation (Shichida, 1986). The mechanism by which these molecules function is also more complex than type I retinylidene proteins, which are often simple ion pumps or channels. As G-protein coupled receptors, many type II retinylidene proteins interact with soluble proteins, by means of their intracellular loop regions, and initiate a cascade mechanism that is yet to be useful in applications, biomimetic or otherwise. The type I retinylidene proteins that demonstrate potential usefulness are described in the following sections.

Eubacterial proteorhodopsins

Proteorhodopsin (PR), a eubacterial photoactive membrane protein, was discovered in 2000 and is identified as a proton pump with a *rhodopsin-like* topology (Béjà *et al.*, 2000; Rangarajan *et al.*, 2007; Gourdon *et al.*, 2008). Found throughout the oceans of the Earth (Béjà *et al.*, 2000; Béjà *et al.*,

2001; Man *et al.*, 2003; Sabehi *et al.*, 2004; Rusch *et al.*, 2007), all of the identified variants are classified as either blue-absorbing (BPR; λ_{max} ~ 494 nm) or green-absorbing (GPR; λ_{max} ~ 520 nm) based on the relative absorption maximum to one another (Béjà *et al.*, 2001; Wang *et al.*, 2003; Rusch *et al.*, 2007). Variants of both PR sub-families share highly similar primary sequences (~80% identity) and exhibit photocycle speeds that differ by an order of magnitude (Béjà *et al.*, 2001; Man *et al.*, 2003); however, only the GPR photointermediates have been well resolved to date (Lakatos *et al.*, 2003; Váró *et al.*, 2003; Huber *et al.*, 2005; Xiao *et al.*, 2005).

Despite literature bias of GPR characterization, the analogous photocycles of BR and PR has drawn significant attention for the striking similarities between these proteins (Váró and Lanyi, 1991; Gillespie *et al.*, 2002; Váró *et al.*, 2003). Specifically, the potential applications of PR in protein-based devices are currently boundless because these proteins are both stable and functional when non-natively expressed and detergent solubilized. Bacteriorhodopsin, which exhibits extraordinary photochemical and thermal stabilities, becomes significantly fragile when removed from the native lipids of *Halobacterium salinarum* (Brouillette *et al.*, 1989; Hendler and Dracheva, 2001; Heyes and El-Sayed, 2002). Thus, the ability of PR to remain functional and highly stable in a non-native lipid environment is a significant advantage over BR.

From a materials standpoint, PR exhibits several strategic advantages over BR. First, GPR exhibits a fast photocycle (~15–50 ms), with photointermediates that are comparable to those of BR, when suspended in a phospholipid environment (Béjà *et al.*, 2001; Xiao *et al.*, 2005; Bergo *et al.*, 2009). Like BR, however, the photocycle kinetics of GPR are affected by the experimental conditions, as evidenced by the lifetimes shown in Fig. 2.5. No detailed study of the BPR photocycle has been conducted to date. Second, PR exhibits *BR-like* photophysical properties when suspended in various lipid environments either from the host of non-native expression methods (Béjà *et al.*, 2001; Lörinczi *et al.*, 2009) or detergent solubilization (Kim *et al.*, 2008; Xi *et al.*, 2008). The ability to generate stable detergent-PR suspensions marks a significant advantage over BR, which is structurally and functionally altered without the native host lipids (Dencher *et al.*, 1983; Brouillette *et al.*, 1989; Hendler and Dracheva, 2001; Heyes and El-Sayed, 2002b). We do not suggest that the current photophysical properties of PR are equivalent to those exhibited in the native lipid environment, which is presently unknown, but rather that the detergent solubilized form is comparable to native BR and is currently useful in device applications.

Many of the proposed PR-based technologies are logical extensions of existing BR-based devices. For example, the holographic (Xi *et al.*, 2008) and photovoltaic (Dioumaev *et al.*, 2003; Tamogami *et al.*, 2009) properties of PR have been reported with similar efficiencies to BR. There are also

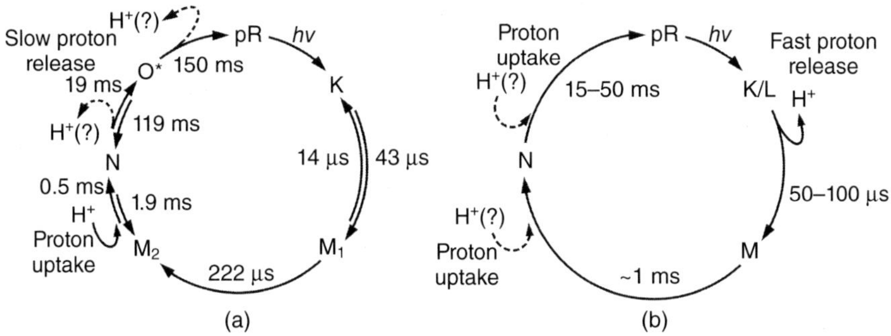

2.5 The photocycle of green proteorhodopsin in a (a) non-ionic detergent (Dioumaev *et al.*, 2002; Varo *et al.*, 2003), or (b) phospholipid environment (Krebs *et al.*, 2002; Xiao *et al.*, 2005; Bergo *et al.*, 2009) at ambient temperature. Lifetimes are shown for select photostates that have been resolved in the cited references. The **O*** state of the detergent photocycle represents the **pR′(O)** state, which is structurally identical to the pR resting state, and is only observed by kinetic analysis of UV-vis spectra (Varo *et al.*, 2003) or FT-IR spectroscopy (Dioumaev *et al.*, 2002). Similarly, the $\mathbf{M_2}$ state of the detergent photocycle was kinetically resolved from the absorption data at 420 nm (Varo *et al.*, 2003). The *dashed trajectories* of proton uptake or release represent an approximate assignment of the event.

patents that employ PR in security inks (Jensen *et al.*, 2008), binary optical memories (Stuart, 2008) and solar cells (Delong, 2007). Although the numerous similarities between BR and PR will expedite the development of PR-based devices, much remains to be explored before such devices can become a reality. For example, few studies address the structural nature (i.e. quaternary structure) of PRs. The existing PR structural literature, although sparse, agrees on a *rhodopsin-like* structure (Béjà *et al.*, 2000; Rangarajan *et al.*, 2007; Gourdon *et al.*, 2008) that can assemble into stacked lamellar sheets of two-dimensional (2D) oligomeric protein arrays when suspended in a cationic detergent (Liang *et al.*, 2007).

Halorhodopsin

Halorhodopsin (HR), discovered in 1980 as a light-driven ion-pump belonging to the protein family of archaeal rhodopsins (Matsuno-Yagi and Mukohata, 1980), has recently emerged as a potentially useful photosensitive protein for device applications. In 2000, Kolbe *et al.* determined the X-ray structure of HR at 1.8 Å resolution, revealing a 31% sequence identity with BR. Despite the high degree of homology, HR functions as a chloride ion pump in which Cl^- ions move from the extracellular milieu to the cell cytoplasm. The direction of ion movement, which is opposite of BR, produces an electrochemical ion gradient that enables pH control in

H. salinarum (Spudich, 2000; Essen, 2002). Subsequent studies on HRs of *H. salinarum* and *Natronomonas pharaonis* determined that HR pumps Cl^- ions into the cell cytoplasm after illumination with yellow (~580 nm) light. The majority of device applications that employ HR (discussed below) use *N. pharaonis* HR (NpHR) due to its enhanced stability and chloride affinity over *Halobacterium* species (see Zhang *et al.* (2007) and references within). However, one of the first applications of HR, the development of an anion-sensitive biosensor by Seki *et al.* (1994), used halophilic bacteria containing HR.

When NpHR is exogenously expressed in neurons, the photoactive protein can be used to silence neuronal electrical activity by hyperpolarizing the neurons rapidly and reversibly upon illumination with intense yellow light (Lynagh and Lynch, 2010). Arrenberg *et al.* (2009) generated a transgenic zebrafish expressing enhanced NpHR in order to investigate the silencing of neurons. Concurrently, Tønnesen *et al.* (2009) studied the use of NpHR in organotypic hippocampal cultures to inhibit excessive hyperexcitability and epileptiform activity.

Channelrhodopsin-2

Channelrhodopsin-2 (ChR2) is a retinylidene protein that functions as a light-gated non-specific cation channel, which conducts H^+, Na^+, K^+ and Ca^{+2} ions into the cell cytoplasm after exposure to blue light (~460 nm). This microbial-type rhodopsin is found in the green algae *Chlamydomonas reinhardtii* and is responsible for phototaxis and photophobic responses (Nagel *et al.*, 2003). Although there is low sequence homology to BR, several amino acids that define the retinal binding site and the proton transport network are conserved (Nagel *et al.*, 2005).

In the mid 2000s, ChR2 emerged as one of several light sensitive molecules that could potentially photostimulate excitable cells such as neurons, endocrine, cardiac or skeletal cells (Zhang *et al.*, 2006). Investigation into the engineering aspects of ChR2 revealed that the fast kinetics, high conductance and good temporal resolution of the photoresponse of ChR2 made the photoactive protein a promising candidate for future neuron circuit models (Nikolic *et al.*, 2006). Initial studies by Boyden *et al.* (2005) demonstrated that illumination of ChR2-expressing neurons with blue light generated action potentials with millisecond precision. Subsequent studies on neurons expressing ChR2 proved that ChR2 could be used rapidly and non-invasively to control neural circuit activity with optical stimulation (see Campagnola *et al.* (2008) and references within).

The combination of NpHR and ChR2 as a neuroengineering tool was first proposed by Zhang *et al.* (2007) in order to investigate neural circuit function. The goal was to develop precise neuromodulation technologies for the treatment of neuropsychiatric disorders, which may result from circuit-level effects of malfunctioning neurons. Because the wavelengths of activation for each protein are separated by nearly 100 nm, the neurons, expressing either NpHR or ChR2, could be controlled independently to either drive action potential firing (ChR2-expressing neurons) or suppress neural activity (NpHR-expressing neurons) (Zhang *et al.*, 2007). Recently, Bernstein *et al.* (2008) genetically expressed ChR2 and NpHR into neuronal cells in the cortex of a non-human primate, a key step for translating previous advances in optogenetic neuromodulation vertebrate and invertebrate therapies to human therapies.

Several applications have emerged for light-based cell stimulation including optogenetic neuromodulation therapy (Zhang and Deisseroth, 2009), a photo-ionic pacemaker (Sergot *et al.*, 2010), an artificial retina (Balya *et al.*, 2008) and restoration of ON/OFF responses in retinas with photoreceptor degeneration (Zhang *et al.*, 2009). The use of both NpHR- and ChR2-expressing excitable cells allows for the facilitation or inhibition of the flow of cations or anions through cellular membranes and results in the hyperpolarization or depolarization of the photosensitized cells. There are several advantages of optical stimulation over electrical stimulation. First, by using light as the stimulus there are no unwanted side effects from extraneous stimulation of adjacent non-targeted neurons. Second, electrodes have limited, sometimes inadequate, mechanical stability as well as issues with migrating away from targeted areas. Third, encapsulation by glial cells over time increases the resistance of the electrode and consequently the voltage needed to reach the targeted cells. For reviews on optogenetic technologies and photochemical tools for controlling neuronal activity see Kramer *et al.* (2009) and also Baler and Scott (2009).

2.3 Applications of bacteriorhodopsin

An ideal photochromic material should exhibit excellent resistance to chemical, photonic and thermal stress. Bacteriorhodopsin represents such a material because it retains both structural and functional integrity at high temperatures (Jackson and Sturtevant, 1978; Wang and El-Sayed, 1999; Muller *et al.*, 2000; Wang and El-Sayed, 2000) and over a broad range of pH (Brouillette *et al.*, 1987; Kono *et al.*, 1993; Rammelsberg *et al.*, 1998; Balashov, 2000). Furthermore, the protein exhibits increased stability, up to 140°C, when prepared as a dry film (Shen *et al.*, 1993). Such stability is uncommon in most proteins, which denature or become inactive outside of their physiological conditions, and results from several structural features of the protein that include: the semi-crystalline organization of the protein within the membrane bilayer (Henderson *et al.*, 1990; Grigorieff *et al.*, 1996; Heyes and El-Sayed, 2002), cationic cofactors (e.g., Ca^{2+}, Mg^{2+}) (El-Sayed *et al.*, 1995; Heyes and El-Sayed, 2001), an α_{II} type helical structure (Krimm and Dwivedi, 1982; Wang and El-Sayed, 1999; Wang and El-Sayed, 2000) and specific interactions with the native *H. salinarum* lipids (Heyes and El-Sayed, 2002a; Kresheck *et al.*, 1990). (See Heyes and El-Sayed (2003) for a review of how these factors stabilize BR.) These factors have indirectly optimized BR for application as a robust biomaterial by exposing the protein to a brine environment. Nonetheless, the extraordinary stability of BR makes it an excellent biomaterial for which it has found application in a vast array of optical devices (Table 2.2).

Table 2.2 Technologies derived from the bacteriorhodopsin photocycle

Application	Type	Reference
Artificial retinas	Bioelectronic	(Miyasaka *et al.*, 1992; Takei *et al.*, 1992; Chen and Birge, 1993a)
Associative memories	Biophotonic	(Birge, 1990; Hillebrecht *et al.*, 2005)
Dynamic time-average interferometers	Biophotonic	(Renner and Hampp, 1992)
Holographic correlators	Biophotonic	(Thoma and Hampp, 1992; Zhang *et al.*, 1995)
Multi-level logic gates	Biophotonic	(Gu *et al.*, 1996)
Neural-type logic gates	Bioelectronic	(Mobarry and Lewis, 1986)
Non-linear optical filters	Biophotonic	(Thoma *et al.*, 1991)
Optical computing	Biophotonic	(Lewis *et al.*, 1997)
Optical limiters	Biophotonic	(Huang *et al.*, 2004; Song *et al.*, 1993)
Pattern recognition systems	Biophotonic	(Hampp *et al.*, 1994)
Photon counters and photovoltaic converters	Bioelectronic	(Marwan *et al.*, 1988; Sasabe *et al.*, 1989; Hong, 1994)
Photovoltaics	Bioelectronic	(Hong, 1997b)
Picosecond photodetectors	Bioelectronic	(Rayfield, 1989; Rayfield, 1994)
Random access thin film memories	Biophotonic	(Lawrence and Birge, 1984; Schick *et al.*, 1988; Birge *et al.*, 1989)
Reversible holographic media	Biophotonic	(Vsevolodov *et al.*, 1989; Hampp *et al.*, 1990)
Spatial light modulators	Biophotonic	(Birge *et al.*, 1990; Song *et al.*, 1993)
Two-photon volumetric memories	Biophotonic	(Birge *et al.*, 1994; Vought and Birge, 1999; Stuart *et al.*, 1996; Hillebrecht *et al.*, 2005; Birge *et al.*, 1990; Birge, 1992)

The proposed BR-based devices are defined as either biophotonic or bioelectronic. Biophotonic devices utilize the photophysical properties of the transient intermediates produced during the BR photocycle (see section 2.2.1). Recall that the protein translocates a proton after activation and, under prolonged illumination, is capable of creating a proton motive force that sustains anaerobic respiration (Oesterhelt and Stoeckenius, 1973). This force results in a physical charge displacement that translates into an electrical signal that is useful in bioelectronic devices (Hong, 1997). Select biophotonic and bioelectronic devices are discussed in the subsequent sections and the reader is encouraged to explore the cited references for specific details pertaining to each application.

2.3.1 Associative memories

The encoding, manipulation and retrieval of data from biomacromolecules defines the field of biomolecular electronics. Biological molecules and proteins offer inherent advantages over conventional mechanical and computer engineering. These advantages stem from natural selection. Evolutionary processes select for organisms that are efficient and robust machines, able to carry out specific functions in nature. Bacteriorhodopsin, a protein with a high photochemical efficiency, thermal stability and cyclicity, serves as an excellent example of a protein that is useful for application in biomolecular devices (Stuart *et al.*, 2003). Unfortunately, nature has little stake in optimizing proteins for non-native functions and environments. Chemical modification or mutagenesis is necessary for these biomolecules to function in applied technologies such as holographic associative memories (Wise *et al.*, 2002). The most successful of these applications is the Fringemaker holographic interferogram developed by Juchem and Hampp (Juchem and Hampp, 2000; Hampp and Juchem, 2000).

Associative memories function differently from serial memories that dominate current computer designs. These memories take an input 'image', and independently 'scan' the entire memory for a page of data that matches the input. In most implementations, the memory finds the closest match when a perfect match is not available. Finally, the memory will return the memory page that fulfills the matching criteria. The most sophisticated associative memories allow for variable-sized blocks, which make the association process adaptable to the amount of information returned. Because the human brain operates in a neural, associative mode, the implementation of large capacity associative memories will likely be a necessary component of any computer architecture that achieves artificial intelligence (Birge, 1990; Birge *et al.*, 1999).

Optical associative memories implement memory recall by using Fourier holographic association (Birge *et al.*, 1999). An example of the optical configuration of a protein-based optical associative memory is presented in Birge *et al.* (1999). The use of PRs (Xi *et al.*, 2008) and BR (Hampp *et al.*, 1994) in holographic applications is best explored in terms of the refractive index changes that can be generated by converting the protein from the resting state to a blue-shifted intermediate (e.g. the **M** or **Q** photostate) (Fig. 2.6). One of the quintessential devices for holography, developed by Nikolai Vsevolodov, the principle scientist of Starzent, uses BR in a prototype holographic memory (Vsevolodov, 1998).

Historically, the **M** state has gained the most attention for associative memories, and some of the earliest applications of BR in devices utilized the ability of the protein photochemically to convert from the resting state to the **M** state (Birge *et al.*, 1992). This transition results in a stable, but short-lived binary system that allows for real-time holography. Additionally, the **M** state has a λ_{max} of 410 nm, which is important because it induces a large change in the refractive index of the medium following photoconversion. The **bR/M** holographic efficiency is calculated using the Kramer's Kronig transformation (Birge *et al.*, 1991; Gross *et al.*, 1992). Because the **M** state is not permanent, holograms made with the **bR/M** transition will only last as long as the **M** state lifetime and the application of this state is limited to real-time dynamic holography.

The **Q** state has also been considered for holographic memories because, when formed, it is stable for years, and it has a blue-shifted λ_{max} (~380 nm) in comparison to the **M** state, as illustrated in Fig. 2.6 (Birge *et al.*, 1999). Furthermore, devices based on the **Q** state remove the time-sensitive nature of devices based on the **M** state because the **M** state is only stable for minutes (Stuart *et al.*, 2007). Formation of the **Q** state occurs via a sequential two-photon event that is described in section 2.2.1 under 'The branched photocycle' and methods for enhancing **Q** state formation are discussed below in section 2.4.3.

Manipulation of the **M** and **Q** photostates allows for the generation of more efficient memory systems. Alterations in pH, temperature, solvent environment, chromophore configuration and the addition of chemicals are some of the ways in which the protein is modified (Schmidt *et al.*, 1998; Stuart *et al.*, 2002). Mutagenesis of BR to enhance **M** state formation usually targets residues involved in the proton transfer from the intracellular surface to the Schiff base (Wise and Birge, 2004). The **M** → **N** transition is associated with reprotonation of the Schiff base, and therefore disruptions, such has a decrease in the pH of the system, will stabilize the **M** state. Site-directed mutagenesis by Hampp *et al.* (Hampp *et al.*, 1990; Hampp, 2000b) and Holz *et al.* (1989) reveals that the mutant D96N yields a two-fold increase in the diffraction efficiency of the protein as well as an increase in the photosensitivity of thin films prepared with the mutant in comparison to wild type. Additionally, the lifetime of the **M** state increases

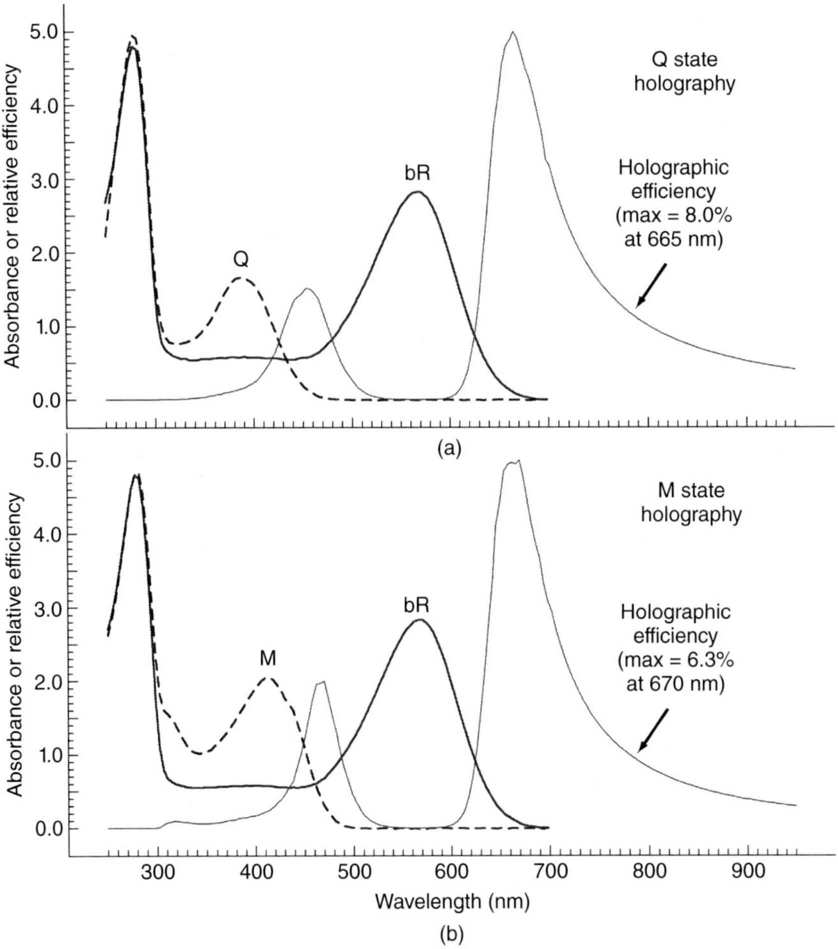

2.6 The relative holographic efficiency (*thin line*) is plotted for 50% conversion of the **bR** resting state (*thick line*) to either the (a) **Q** state, or (b) **M** state. *Dashed lines* represent both the **M** and **Q** states in the respective figures. The **bR→Q** state conversion yields a maximum holographic efficiency of 8.0% at 665 nm, whereas the **bR→M** state conversion yields a maximum holographic efficiency of 6.3% at 670 nm. In general, real-time holographic associative processors or pattern recognition systems use the **M** state while long-term holographic associative memories use the **Q** state. All spectra represent data collected at ambient temperature.

from approximately nine milliseconds in wild type BR to nearly 750 ms in the D96N mutant (Pandey *et al.*, 1999).

2.3.2 Volumetric memories

Three-dimensional (3D) memories store information in a volumetric memory medium and offer as much as a 1000-fold improvement in data storage capacity for a given enclosure size. However, optical and reliability considerations of these devices tend to reduce the comparative advantage factor to values closer to a 300-fold improvement. The branched-photocycle architecture discussed herein is made possible by an unusual photochemical characteristic of BR that provides the ability quickly and reliably to store and retrieve information. A prototype device for this memory is shown holding a 'data cuvette' in Fig. 2.7.

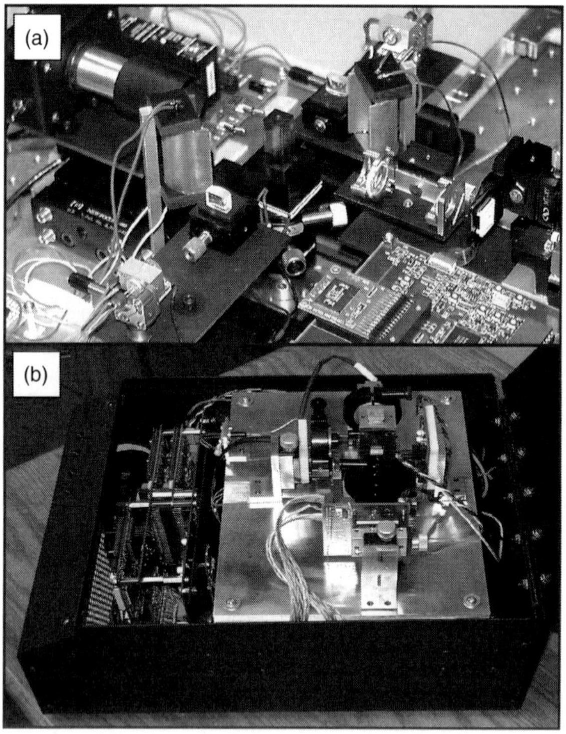

2.7 Two versions of the sequential two-photon volumetric memory (a) in the form of the original prototype, and (b) as built for the United States Air Force. The protein is immobilized in a polyacrylamide matrix within a sealed three-milliliter cuvette. Details of the theory and methods for preparing these memories are discussed by Birge *et al.* (1999) and Stuart *et al.* (2002).

Exploitation of the branching reaction, which was introduced in section 2.2.1 under 'The branched photocycle', is achieved by assigning the resting state (**bR**) to bit 0 and both **P** and **Q** photostates to bit 1 (Fig. 2.8). This model indicates the wavelength maximum (in nanometers), the configuration and the protonation state of the chromophore directly underneath the intermediate label. The 3D aspect of this branching reaction derives from the fact that the green (λ_{max} ~ 570 nm) paging beam is orthogonal to the red (λ_{max} > 600 nm) write beam (Stuart *et al.*, 1996; Stuart *et al.*, 2000; Stuart *et al.*, 2003). To be specific, the paging beam selects a thin page inside the memory medium, and the write beam arrives a few milliseconds later with the data imposed upon it via a spatial light modulator. The red beam interacts with the **O** state and converts the irradiated voxels, and only the irradiated voxels, from **O→P**. The **P** state then spontaneously converts to the **Q** state, and it is **Q** that stores the data in a format that is long-term and shifted in wavelength space so that they are not erased during other read-write cycles. Those voxels within the page that were not irradiated with the write beam continue the photocycle and return to the **bR** resting state. Thus, a page of data can be written without disturbing data written elsewhere in the volumetric memory medium. A write process proceeds in a comparable fashion, but the spatial light modulator turns on all of the pixels but at a very low level sufficient to image the page onto a detector. The process is described in detail in Birge *et al.* (1999).

The key requirement of a protein to be used in a sequential two-photon **Q**-based memory is efficient **O→P** photochemistry. The native protein is quite inefficient at this process, and memories based on the native protein require 100–200 mW lasers to carry out the conversion process. By using directed evolution, the efficiency of the **Q** state photochemistry has been

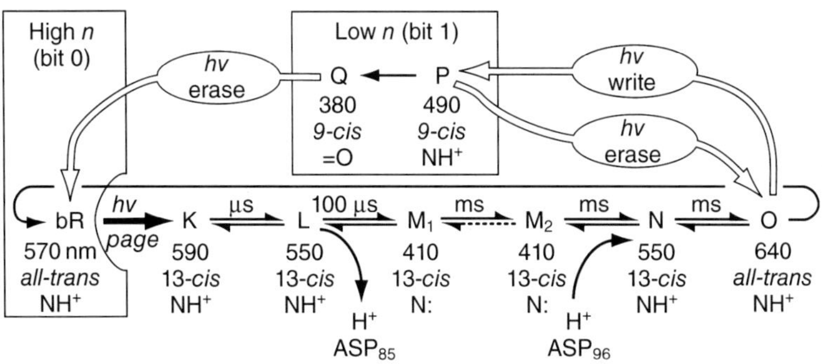

2.8 Branching reaction of bacteriorhodopsin. This model indicates the wavelength maximum (in nanometers), the conformation and the protonation state of the chromophore directly underneath the intermediate label.

extensively enhanced to where it is useful in volumetric memories. Optimization of the **Q** state by this method is described in section 2.4.3.

2.3.3 Photovoltaics

The native function of BR is the light-activated construction and maintenance of a proton gradient across the membrane that drives ATP-synthase under anaerobic conditions (Oesterhelt and Stoeckenius, 1973; Oesterhelt, 1998). This translocation event results in a surplus of positive charge, up to 4 pH units on the extracellular membrane surface (Calimet and Ulmann, 2004), that can be measured as an electrical response to a light stimulus (Hong, 1994). Magnification of the electrical signal occurs when the protein is condensed into a film on an electrically conductive surface (Váró, 1981; Váró and Keszthelyi, 1983). Proper orientation of the protein within a film, however, enhances both the signal and reproducibility of the experiment. Methods for orienting the protein films include, but are not limited to: sol-gel encapsulation (Chen *et al.*, 1991), antibody-mediated orientation (Koyama *et al.*, 1994), electric field sedimentation (Váró and Keszthelyi, 1983), Langmuir–Blodgett deposition (Miyasaka and Koyama, 1992) and electrostatic layer-by-layer adsorption (He *et al.*, 1998). The duration and magnitude of the photovoltaic signal are sensitive experimental conditions (e.g. light intensity, humidity). A comprehensive review by Hong outlines the conditions, via electrical circuit models, that control the photoelectric signal of BR (Hong, 1999).

Illumination of BR films results in one of two photovoltaic responses: a fast photoelectric pulse or a slow differential response. The fast event is observed at high temporal resolution following flash photolysis (Fig. 2.9). This signal is conventionally divided into the B1, B2 and B3 components.

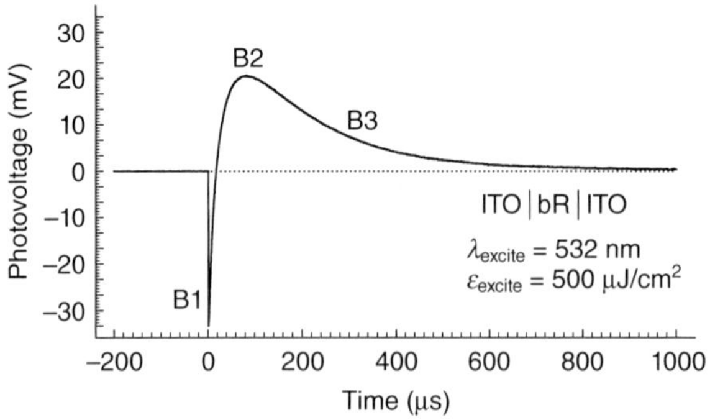

2.9 Fast electrical response that is produced by flash photolysis of a bacteriorhodopsin film that is sedimented onto glass coated with indium-tin oxide.

The B1 signal is a sharp negative voltage spike with a rise time of ~6 ps and is aligned with the photoisomerization of the retinal chromophore from 13-*s-trans* to 13-*s-cis* (Xu *et al.*, 2003). The relative magnitude of the negative signal is controlled by the molecular motion of Arg-82 and can theoretically be mediated via substitution with other amino acid residues (Xu *et al.*, 2003). Conversion of this negative signal to the positive B2 phase is associated with the **K→L→M** photointermediates of the BR photocycle (Fig. 2.1) (Keszthelyi and Ormos, 1980; Xu *et al.*, 2003). Enhancing the B2 signal is accomplished by manipulation of the **M** state lifetime. The B3 signal, which is responsible for the long tail in the signal, is associated with the **M→N→O→bR** portion of the photocycle (Keszthelyi and Ormos, 1980).

The slow differential responsivity of BR thin films results from long periods of illumination (Fig. 2.10). Comprising transient positive and negative signals, each of which correlate with the onset and termination of the light stimulus, this amalgamated photoresponse presents as either a square wave or a differential response (Hong, 1997). Realize that the shape represents the creation and maintenance of a displaced charge that is directly

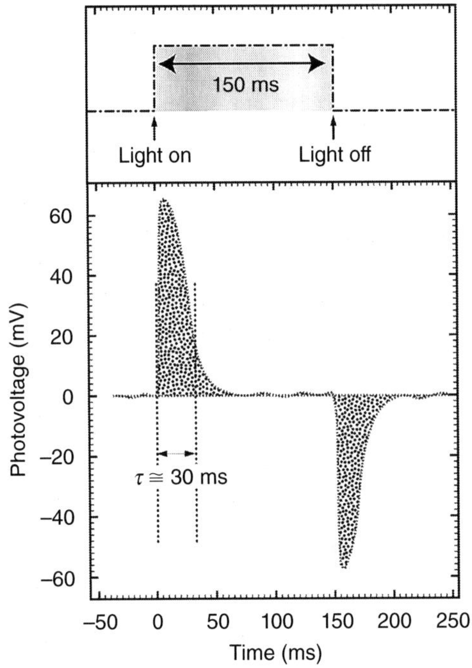

2.10 Differential photovoltage produced by a bacteriorhodopsin film that was exposed to 150 ms of continuous illumination of filtered (550–600 nm) light.

reflected in the signal. Symmetrical positive and negative transitions are produced when the protein is uniformly aligned and is capable of maintaining, or storing, the charge at one surface of the film. This phenomenon has been occasionally termed chemical capacitance and has been likened to an electrical circuit (Hong, 1997).

2.3.4 Photosensors and artificial retinas

The photoelectric properties of BR, which are derived from the ability of the protein to generate a pH gradient upon light absorption, make the protein a viable candidate for photoelectric sensors (Birge, 1990; Haronian and Lewis, 1991; Takei *et al.*, 1992; Boyer *et al.*, 1995; Knopf *et al.*, 2009), motion detectors (Miyasaka and Koyama, 1993; Fukuzawa, 1994; Yao *et al.*, 1997) and complex artificial retina devices (Miyasaka and Koyama, 1992; Chen and Birge, 1993; Hong, 1997). The proton pumping capability of BR transports a proton roughly five nanometers from the cytoplasmic to the extracellular milieu to generate a pH gradient across the membrane (Oesterhelt and Hess, 1973; Birge, 1990). This photoelectrical signal can be amplified when the protein is condensed into a thin film (Váró, 1981; Váró and Keszthelyi, 1983). The resulting photovoltage is complex in nature and can be supported by an embedded electrolyte-containing layer within a photoelectric cell (see detailed discussion in section 2.3.3) (Hampp, 2000a). The intrinsic properties of BR that make it uniquely qualified for use in photoelectric devices include: (1) long-term thermal and photochemical stability, (2) high forward and reverse quantum yields, which allow for activation at low light levels, (3) picosecond photochemical response times, (4) wavelength-independent quantum yields, (5) generation of photoelectric signal that has an opposite polarity for the forward and reverse photoreactions, (6) differential responsivity that mimics *in vivo* photoreceptors under certain conditions and (7) the ability to reproducibly form oriented thin films or volumetric cubes of the protein (Birge, 1990; Chen and Birge, 1993; Tukiainen *et al.*, 2007).

Photosensors

One of the earliest applications of the photoelectric properties of BR was implemented by Trissl *et al.* in the development of a super-fast photodetection device (Trissl, 1987; Trissl *et al.*, 1989). Shortly after, Haronian and Lewis (1991) demonstrated that the electrical effects of BR are analogous to conventional neural network architectures through the development of a reprogrammable BR-based neural network. In conventional systems, the energy (i.e. voltage) is supplied by the neuron, whereas in the device proposed by Haronian and Lewis (1991) the BR molecule, upon photoexcitation, supplies the energy. This neural network device was one of the first examples of electrooptical erasable memory and also encouraged the development of BR-based image sensing (Haronian and Lewis, 1991).

While investigating the strange behavior of BR at the electrode-electrolyte interface, Boyer *et al.* (1995) discovered the possibility of color discrimination by BR-based photosensors. Similar to the differential photosensitivity of BR (see section 2.3.3), forward and reverse photocurrents, having opposite signs, were measured on the green absorbing ground state (bR570) and a structurally and spectrally modified bR480. The altered pigment (bR480) was generated by the addition of halogenated anesthetics and maintained full functionality. Results of these experiments indicated that BR readily self-orients itself at the electrode interface (Boyer *et al.*, 1995).

Motion detection and edge detection

Neural networks are fundamentally characterized by their ability to elicit excitatory and inhibitory responses. In retinal ganglion receptive fields, the ability to detect a sharp intensity variation in image space, i.e. edge detection, as well as the ability to detect motion of objects is feasible due to the different response times for the excitatory and inhibitory regions of the receptive fields. Significant efforts have been made to develop a biomaterial that can mimic the inherent ability of the retinal neural network. In the early 1990s, several reports of optoelectric devices using BR demonstrated that receptive fields of BR are capable of excitatory and inhibitory responses as well as edge detection at zero crossings (Takei *et al.*, 1992; Martin *et al.*, 1997; Yao *et al.*, 1997; Yang and Wang, 1998). Additionally, a wet-type 256-pixel artificial photoreceptor created from immobilized thin films of BR at the solid–liquid interface of an electrode surface was used to demonstrate the ability of BR to detect motion and edge information of images in real time (Miyasaka and Koyama, 1993). This unique ability of BR to detect motion is a result of a differential responsivity to light intensity (see section 2.3.3) (Hong, 1997). Although the aforementioned BR-based motion sensors were successfully shown to sense the motion and direction of movement from an object, the sensors were unable to detect the displacement of a moving object without the use of an external memory device. The image-recording capability of BR was first demonstrated by treating BR thin films with a high pH buffer. A key advantage of the high pH BR sensor was the ability of the sensor to detect an object's current and previous position without the use of an external memory device. The authors noted this unique advantage of a BR-based motion sensitive position sensor could find useful application in robotic vision (Fukuzawa, 1994).

Artificial retinas

Protein-based artificial retinas have been proposed based on the use of BR, which converts light into an electrical signal, to activate the retinal ganglion or bipolar cells. The complex photovoltaic signal produced by BR during its photocycle has been reported to be as large as 5 VDC (Crittenden *et al.*, 2003; Zhang *et al.*, 2003), while experimental data show that a signal of only 1.6 VDC can stimulate nerve impulse (unpublished data). Through genetic engineering it is believed that the photovoltaic signal of BR can be enhanced three-to-five fold. The proposed motion-sensitive artificial retinas of Miyasaka (Miyasaka and Koyama, 1992) and Birge (Chen and Birge, 1993) use the differential light responsivity of BR to sense an object's motion as well as the direction of movement.

Color vision is a result of the combination of three types of cone pigments in humans and animals. Once the visual pigments absorb photons, color may be assigned by comparing the signals from the three cone pigments. Specific details about color vision and BR-based color sensitive biosensors are discussed by Hong (1997, 1999), Lensu (Lensu *et al.*, 2007; Tukiainen *et al.*, 2007), Ogawa (1990) and Martin *et al.* (1997). Recently, a flexible BR-based photodetector and imaging array, which provides a substantial advantage over previously developed photosensors, was developed by Knopf *et al.* (2009). Additionally, intelligent artificial retinas with color sensitivity have been proposed by using a combination of native BR with BR analogs that contain modified chromophores (see section 2.4.2) (Frydrych *et al.*, 2000). For a detailed review on color vision and BR-based color sensitive biosensors see Ogawa (1990), Lensu *et al.* (2007) and references within.

2.3.5 Chemical sensors

Research groups throughout the world are working to develop biosensors with an increased sensitivity and affinity for the detection of biomolecules, including olfactory proteins, DNA and toxic agents involved in biological weapons (Delahanty and Ligler, 2002; Fivash *et al.*, 1998; O'Shannessy *et al.*, 1992). Bacteriorhodopsin is currently being examined as a biomaterial for the aforementioned application (Bryl and Yoshihara, 2001). The protein is shown to have sensitivity to a variety of hydrophobic, polar and charged compounds (Nakagawa *et al.*, 1994; Taneva *et al.*, 1995; Boucher *et al.*, 1996; Lanyi and Luecke, 2001; Heyes *et al.*, 2002). Through molecular modeling of BR with a variety of ligands, specific residues can be targeted for mutagenesis. Mutants can then be created and experimentally tested to generate a library of biomolecules that are capable of chemical detection.

Time-resolved UV-visible spectroscopy is used to monitor BR in response to various concentrations of the target ligand (Fig. 2.11b). The **M** and **O** states are used to screen BR for chemical responsivity because these photochemical states are most affected by various chemical additives (Lin *et al.*, 1992; Nakagawa *et al.*, 1994; Lanyi and Luecke, 2001; Heyes *et al.*, 2002). Mutants that show the significant sensitivity are then selected for application in biological sensor devices.

2.4 Enhancing bacteriorhodopsin for device applications

Bacteriorhodopsin is prepared as large membrane fragments, of gigadalton size, and is commonly referred to as the purple membrane. Comprising only BR and lipids with a 10:1 lipid-to-protein ratio (Dracheva *et al.*, 1996; Renner *et al.*, 2005), this structure is approximately 0.5–1.0 μm in diameter and five nanometers thick (Henderson, 1975). This feature makes the protein easy to purify in high yield (~20–30 mg L^{-1}) without the use of expensive chemicals or instrumentation (Oesterhelt and Stoeckenius, 1974; Becher and Cassim, 1975; Lorber and DeLucas, 1990). A general scheme of the protein preparation method is shown in Fig. 2.11. Modulating the photophysical properties of BR is possible by several methods. The most common methods for modulating these properties are through chemical manipulation of the bulk structure, substitution of the chromophore with synthetic chromophore analogs and genetic engineering of the native protein.

2.4.1 Chemical manipulation

Chemical modification of BR is accomplished by any number of methods that expose the protein to a non-native solute or solvent, solubilize the protein in a detergent, reconstitute BR in an alternate lipid environment or modify the organic chromophore. These conditions will be discussed in the subsequent section with exception to the studies that integrate synthetic retinal analogs within the active site of BR. This last condition represents a large subfield of BR optimization and will be reviewed in section 2.4.2.

Lipid environment

Bacteriorhodopsin is natively expressed by *Halobacterium salinarum*, a halophilic archaea, when oxygen levels become too low to sustain respiration (Oesterhelt and Stoeckenius, 1973). The methods for cultivating the protein in this way recreate such an environment in the laboratory (Oesterhelt and Stoeckenius, 1974; Becher and Cassim, 1975; Lorber and DeLucas, 1990). Although *H. salinarum* is slow growing, with a doubling

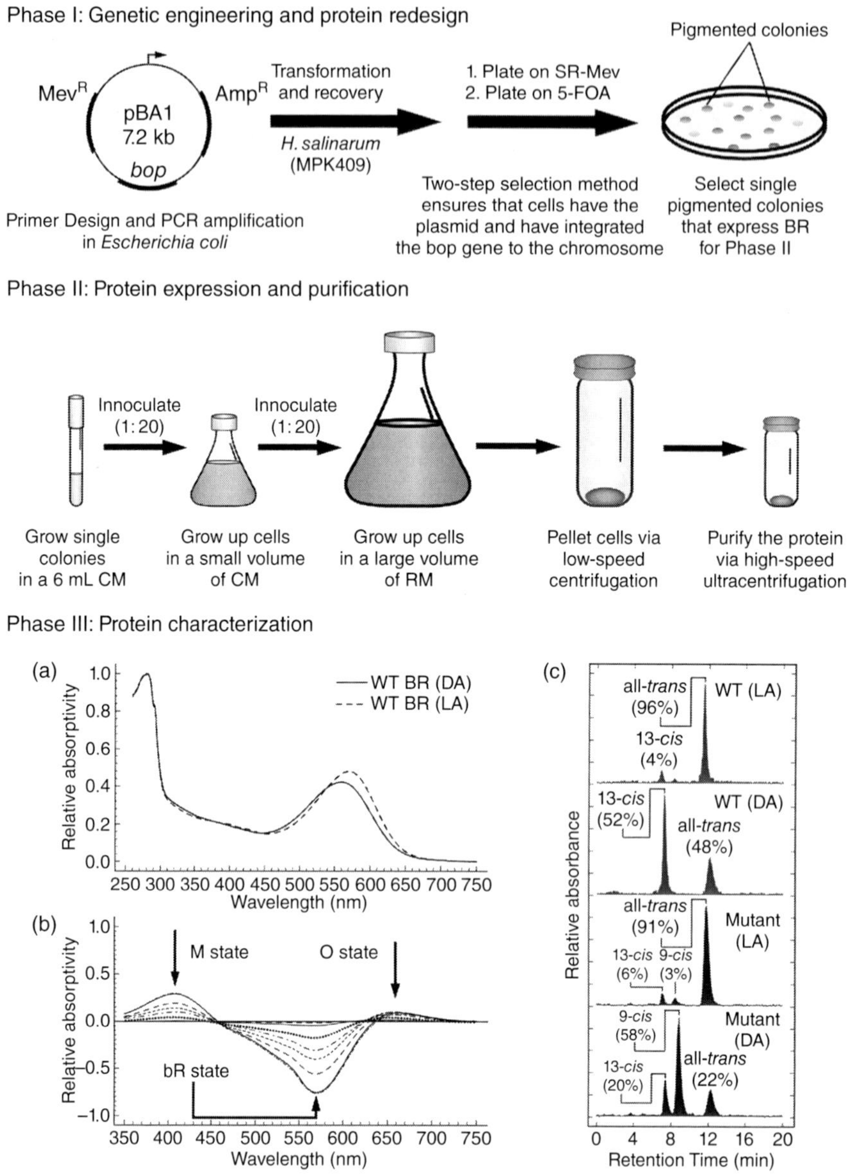
Phase I: Genetic engineering and protein redesign
Pigmented colonies
MevR
pBA1
7.2 kb
bop
AmpR
Transformation
and recovery
H. salinarum
(MPK409)
1. Plate on SR-Mev
2. Plate on 5-FOA
Primer Design and PCR amplification in Escherichia coli
Two-step selection method ensures that cells have the plasmid and have integrated the bop gene to the chromosome
Select single pigmented colonies that express BR for Phase II
Phase II: Protein expression and purification
Innoculate (1:20)
Innoculate (1:20)
Grow single colonies in a 6 mL CM
Grow up cells in a small volume of CM
Grow up cells in a large volume of RM
Pellet cells via low-speed centrifugation
Purify the protein via high-speed ultracentrifugation
Phase III: Protein characterization
(a)
WT BR (DA)
WT BR (LA)
Relative absorptivity
Wavelength (nm)
(b)
M state
O state
bR state
Relative absorptivity
Wavelength (nm)
(c)
all-trans (96%)
13-cis (4%)
WT (LA)
13-cis (52%)
WT (DA)
all-trans (48%)
all-trans (91%)
Mutant (LA)
13-cis (6%)
9-cis (3%)
9-cis (58%)
Mutant (DA)
13-cis (20%)
all-trans (22%)
Relative absorbance
Retention Time (min)

2.11 The three stages involved in the preparation of bacteriorhodopsin for device applications. Phase I involves the genetic engineering of the bacterioopsin (*bop*) gene. The *bop* gene is engineered into the pBA1 plasmid, which contains both ampicillin and mevinolin resistance. Transformation of the plasmid into *Escherichia coli*, a shuttle vector, generates multiple copies of the construct. *Escherichia coli* is chosen for this step because it reproduces exponentially faster (18 minutes at 37°C) than the native organism, *Halobacterium salinarum* (18 hours at 40°C). Incorporation of the bop gene into *H. salinarum* is critical for proper expression of bacterioopsin and is accomplished by transforming the resulting purified DNA into the MPK409 strain of *H. salinarum*. Next, cells are plated on 5-FOA, which selects for recombinants because the MPK 409 cell line lacks the ura3 cassette. Non-recombinants will die on 5-FOA because they will metabolize the 5-FOA into a toxic byproduct. The details of the mutagenesis process are described in Ni *et al.* (1990), Wise *et al.* (2002) and Hillebrecht *et al.* (2004). Phase II involves the expression of the protein in *H. salinarum*, which is vital to the structure and function of bacteriorhodopsin (Heyes and El-Sayed, 2002). Recipes for the media and conditions for optimal expression and purification of bacteriorhodopsin are described by Hillebrecht *et al.* (2004). Phase III involves the biochemical and biophysical characterization of bacteriorhodopsin. Examples of such characterization are: (a) light/dark adaptation (Becher and Cassim, 1976; Dencher *et al.*, 1983; Duñach *et al.*, 1990); (b) temporal measurements of the transient photostates that result from photoactivation (Zimányi *et al.*, 1989; Váró and Lanyi, 1991); and (c) analysis of the retinal configurations within the protein sample (Scherrer *et al.*, 1989; Gillespie *et al.*, 2005).

time of 18 hours at 40°C, expression of the protein within the native membrane is absolutely necessary for proper production of this biomaterial. Non-native expression, via *Escherichia coli* (Braiman *et al.*, 1987; Dunn *et al.*, 1987; Hackett *et al.*, 1987; Karnik *et al.*, 1987; Mogi *et al.*, 1987; Nassal *et al.*, 1987), and reconstitution of BR in non-native lipid structures e.g. liposomes (Sugihara *et al.*, 1982; Torres *et al.*, 1995; Kalaidzidis *et al.*, 1999), unilamellar vesicles (Piknova *et al.*, 1993), mixed micelles (Allen *et al.*, 2001; Brouillette *et al.*, 1989; Booth *et al.*, 1997), and detergent suspensions (Dencher and Heyn, 1978; London and Khorana, 1982; Muccio and DeLucas, 1985; del Rio *et al.*, 1991; Massotte and Aghion, 1991) has been attempted. However, the function and structure of BR are significantly altered without the native archaeal lipids (Dencher *et al.*, 1983; Brouillette *et al.*, 1989; Milder *et al.*, 1991; Dracheva *et al.*, 1996; Hendler and Dracheva, 2001; Heyes and El-Sayed, 2002b).

Nanodiscs are a recently developed technology that allows the isolation of membrane proteins within a soluble discoidal structure (see Nath *et al.* (2007) and references therein). These self-assembling structures offer a novel tool for stabilizing membrane proteins in a phospholipid environment and are quickly finding application with many membrane proteins that include bacteriorhodopsin (Bayburt *et al.*, 2006), cytochrome P450 (Denisov *et al.*, 2007), G-protein coupled receptors (Bayburt *et al.*, 2007) and others (Alami *et al.*, 2007; Bayburt and Sligar, 2009; Borch and Hamann, 2009). Furthermore, isolation of membrane proteins within a nanodisc has minimal effects on the physical properties (e.g. fluidity) of lipids (Shaw *et al.*, 2004; Denisov *et al.*, 2005).

The implementation of these structures in biotechnology is an obvious and simple transition because these structures readily adsorb to surfaces such as glass or mica (Carlson *et al.*, 2000; Goluch *et al.*, 2008; Vinchurkar *et al.*, 2008) and are easily modified with affinity tags (Marin *et al.*, 2007; Shaw *et al.*, 2007). Utilization of such features also offers a means of controlling the protein structure in a realistic and minimally invasive manner. Specifically, BR has been removed from the native purple membrane, an oligomeric structure of gigadalton size, and incorporated into nanodiscs as a monomer and a trimer (Bayburt and Sligar, 2003; Bayburt *et al.*, 2006). Recall that disruption of the native purple membrane lipids, by detergent solubilization or non-native protein expression, often results in undesirable alteration of the photophysical properties of BR (Dencher and Heyn, 1978; Dencher *et al.*, 1983; Massotte and Aghion, 1991). Functional BR was confirmed to exist within nanodisc assemblies, however, by the presence of active trimers as exemplified by the characteristic exciton splitting in visible circular dichroism spectra (Cassim, 1992).

Experimental pH

Altering the environmental pH, which is the simplest form of chemical manipulation, results in three predominant forms of the photocycle (Fig. 2.12). The main photocycle exists between pH 6 and 9.5 and proton transport under these conditions is discussed in section 2.2.1. Application of the photophyscial properties around neutral pH is discussed in section 2.3. Acidification of the protein produces a blue bacteriorhodopsin (bBR) membrane that does not translocate any protons and alters the photophysical properties of BR (Mowery *et al.*, 1979; Maeda *et al.*, 1981; Liu and Ebrey, 1987; Váró and Lanyi, 1989; Moltke and Heyn, 1995). This pigment results from a red shift of the absorption maximum, from 568 nm to 603 nm, that is controlled by protonation of Aspartic acid 85 (Balashov *et al.*, 1995a; Balashov *et al.*, 1996). An acid purple form of the protein is formed when the pH is below 2 and Cl^- ions are present (Váró and Lanyi, 1989; Logunov *et al.*, 1996; Tokaji *et al.*, 1997). Sulfuric acid is preferentially used over hydrochloric acid for accurate investigations of the acidified bBR

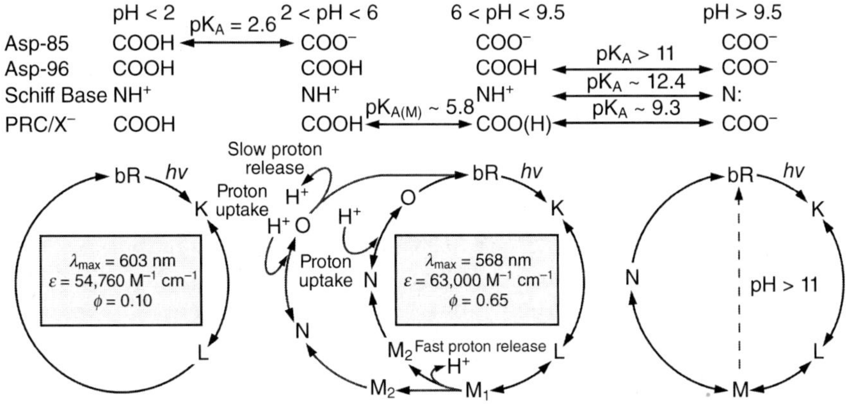

2.12 Modulation of the bacteriorhodopsin photocycle via alteration of the experimental pH. The protonation state of Aspartic acid 85 (Asp-85), Aspartic acid 96 (Asp-96), the Schiff base nitrogen of the chromophore-protein linkage and proton release complex (PRC/X^-) are defined for each experimental condition. The pK_A of each residue or complex is described in the resting (**bR**) state. Below pH 6, which is close to the pK_A of the PRC during the **M** state (Lanyi and Varo, 1995), proton release is delayed until the **O→bR** transition (Zimányi *et al.*, 1992). A summary of the photophysical properties of bacteriorhodopsin is listed at the center of each photocycle. The values for BR at alkaline pH are nearly identical to those at neutral pH. Image was adapted from (Balashov, 2000).

membranes (Moltke and Heyn, 1995; Okumura *et al.*, 2005). When illuminated under a specific set of conditions, the acid bBR membranes will produce a 9-*cis* photoproduct (Maeda *et al.*, 1980; Fischer *et al.*, 1981; Maeda *et al.*, 1981). One should note, however, that this form of BR is less thermally stable and precipitates (Kresheck *et al.*, 1990). The acidified protein is often encased in a matrix (e.g. gelatin, agarose) to prevent precipitation. Deionization or mutagenesis of native BR will produce soluble protein that offers analogous photophysical properties to the acid bBR membranes (Hampp *et al.*, 1992; Tallent *et al.*, 1996; Tallent *et al.*, 1998; Millerd *et al.*, 1999). Deionized bBR membranes will be discussed in the following section.

For comparison, an alkaline form of the photocycle occurs above pH 9.5 where no fast proton release is observed (Kono *et al.*, 1993). Application of this photocycle in devices has yet to be realized because of the truncated photocycle. Upon illumination, the fast formation of the **M** state is observed (~1 μs) with minimal, if any, formation of an **O** state. This truncation of the photocycle is reported to result from a slow reprotonation of Asp-96 and deprotonation of the proton release complex (often referred to as PRC or X′) (Váró and Lanyi, 1990; Chizov *et al.*, 1992; Govindjee *et al.*, 1996; Richter *et al.*, 1996b). The residues involved in the proton release complex are yet to be absolutely resolved, but significant evidence indicates that several amino acids are involved in proton release during the BR photocycle (Balashov *et al.*, 1995b; Balashov *et al.*, 1996; Bressler *et al.*, 1999; Li *et al.*, 2000).

Deionized membranes

The biological cations that associate with the purple membrane, the oligomeric structure of BR, are vital for the structural (Cladera *et al.*, 1988; Kresheck *et al.*, 1990) and functional (Chronister and El-Sayed, 1987; Tallent *et al.*, 1998) properties of BR. The surface charge effects and functional changes of cation binding, both monovalent and divalent, to BR have been well characterized experimentally (Chang *et al.*, 1986; Duñach *et al.*, 1988; Alexiev *et al.*, 1994; Birge *et al.*, 1996; Váró *et al.*, 1999; Sanz *et al.*, 2001) and theoretically (Szundi and Stoeckenius, 1987; Szundi and Stoeckenius, 1989). Native BR binds approximately 4 mol Ca^{2+} and Mg^{2+} per mol BR at pH 6 (Chang *et al.*, 1985; Chang *et al.*, 1986; Chang *et al.*, 1987) and two high affinity sites (K_D = 0.25 and 35 μM) have been identified from isothermal titration calorimetry experiments with Ca^{2+} (Váró *et al.*, 1999). Displacement of these cations produces a deionized form of the protein (dIBR) that exhibits many similar photophysical properties to the acid bBR membranes (Smith and Mathies, 1985; Chronister and El-Sayed, 1987; Váró and Lanyi, 1989). The properties of the acidified protein are discussed in the previous section 'Experimental pH'. Deionization of BR is accomplished via acidification (Váró and Lanyi, 1989), laser induction (Masthay *et al.*, 2002) or treatment with ion-exchange resin (Váró *et al.*, 1999), sugar (Rhinow and Hammp, 2008) or chelating agent (Chang *et al.*, 1987; Tallent *et al.*, 1998). An extensive review of the effects of pH and cation binding to the surface charge of the purple membrane is found in Jonas *et al.* (1990).

Photoexcitation of dIBR with red light (>640 nm) produces a pink membrane (λ_{max} ~ 490 nm). This photoproduct contains 80% 9-*cis* retinal (Chang *et al.*, 1987) that is produced in a yield comparable to the 9-*cis* of the acid blue membranes (Maeda *et al.*, 1980; Fischer *et al.*, 1981) and is similar to the **P** state of the branched BR photocycle (see 'The branched photocycle' in section 2.2.1). Formation of the 9-*cis* photoproduct is ideal for application in optical memories (see sections 2.3.1 and 2.3.2). Fabrication of dried thin films, comprising dIBR and polyvinyl alcohol, enhances the photochromic properties of the biomaterial (Tallent *et al.*, 1996). Encapsulation of dIBR in a polymer matrix also precludes aggregation of the protein if the pH becomes acidic. No formation of a hydrolyzed 9-*cis* chromophore is reported in these films, however, and is presumably due to the dehydration of the material. Hydrolysis of the 9-*cis* chromophore is, by nature, dependent upon hydration of the active site (Popp *et al.*, 1993).

2.4.2 Synthetic retinal analogs

Significant efforts have been made chemically to modify the photoactive chromophore, all-*trans* retinal, to optimize the properties of BR for device applications (Singh and Hota, 2007). By altering the chromophore structure, the spectral and photochemical characteristics (e.g. absorption maximum, photocycle speed, proton transportation) of native and mutant BRs can be fine-tuned for photochromic, electrochromic and holographic technologies. A summary of the process for developing and fine-tuning BR analogs with modified chromophores is shown in Fig. 2.13 (Khodonov *et al.*, 2000). To date there are over 30 different retinal chromophores available in the literature with modifications to the retinal moiety including: removal of the β-ionone ring, demethylation, lengthening of the polyene chain at the β-ionone ring and locking the retinal into specific configurations (Fig. 2.14) (Vsevolodov, 1998; Singh and Hota, 2007). While most modifications to the retinal have provided valuable insight into the protein/chromophore interaction (Balogh-Nair *et al.*, 1980; Nakanishi *et al.*, 1980; Kakitani *et al.*, 1983; Muradin-Szweykowska *et al.*, 1984; Lugtenburg *et al.*, 1986; Spudich *et al.*, 1986; van der Steen *et al.*, 1986; Zingoni *et al.*, 1986; Drachev *et al.*, 1989; Asato *et al.*, 1990; Liu *et al.*, 1991; Liu *et al.*, 1993; Nakanishi *et al.*, 1995; Yan *et al.*, 1995; Druzhko *et al.*, 1998; Weetall *et al.*, 2000; Haacke *et al.*, 2002; Singh and Manjula, 2003; Bismuth *et al.*, 2007) or the structure/function of the light-absorbing chromophore within the binding site of BR (Marcus *et al.*, 1977; Cookingham *et al.*, 1978; Marcus and Lewis, 1978; Bayley *et al.*, 1981; Huang *et al.*, 1982; Crouch *et al.*, 1984; Crouch *et al.*, 1985; Schiffmiller *et al.*, 1985; Boehm *et al.*, 1990; Büldt *et al.*, 1991; Bhattacharya *et al.*, 1992; Delaney *et al.*, 1995; Druzhko *et al.*, 1996; Tuzi *et al.*, 1996; Rousso *et al.*, 1997; Weetall *et al.*, 1997; Sakamoto *et al.*, 1998; Weetall *et al.*, 2000; Aharoni *et al.*, 2003; Laptev *et al.*, 2008; Gross *et al.*, 2009; Khitrina *et al.*, 2009; Das *et al.*, 1999), few synthetic analogs have been used in device applications. The most promising BRs containing retinal analogs for optoelectronic devices are discussed below.

Bacteriorhodopsin containing a 4-keto retinal or the more commonly labeled 4-keto BR has found great application in the area of photochromic and electrochromic materials (Druzhko *et al.*, 1995; Druzhko and Chamorovsky, 1995; Khodonov *et al.*, 2000). The 4-keto retinal analog contains an elongated polyene chain extending from the β-ionone ring (see analog **2** of Fig. 2.14). The resting state of 4-keto BR displays a single absorption maximum (~500 nm) that progresses into three blue-shifted absorption maxima, at 390, 420 and 440 nm, after photoexcitation. These peaks have been described as three distinct ***M**-like* states and are suggested to be intermediates in the three photocycles of 4-keto BR (Barmenkov *et al.*, 2000). In addition to containing multiple photocycles, 4-keto BR displays

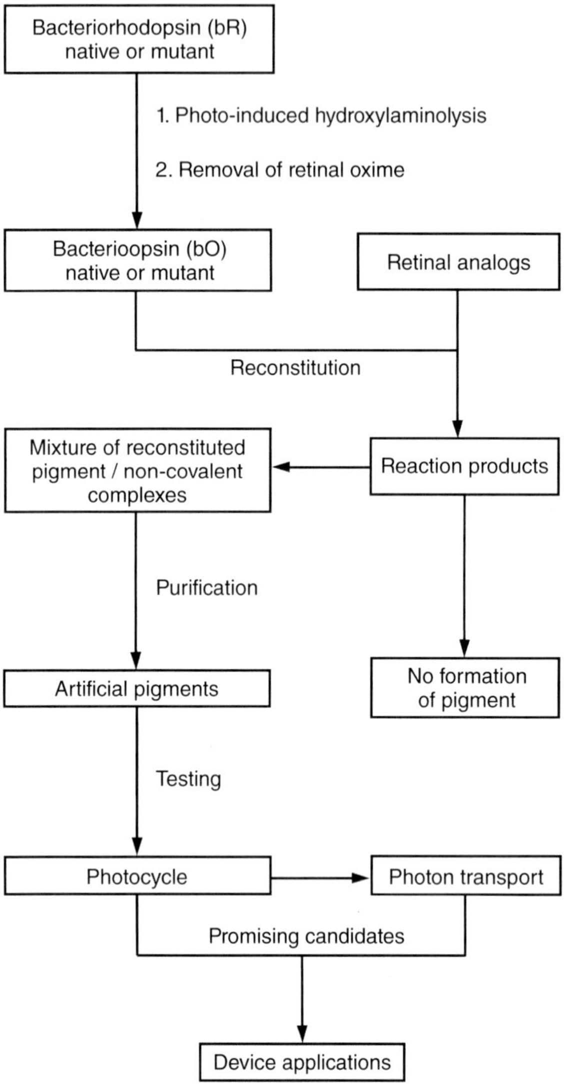

2.13 Procedure for the preparation of bacteriorhodopsin analogs (adapted from Khodonov *et al.* (2000)).

photocycle kinetics roughly three times slower than that of native BR. These properties of 4-keto BR may allow for longer storage time of optical information and gives the BR analog a distinct advantage over its native protein counterpart (Druzhko *et al.*, 1995).

The combination of 4-keto BR ($\lambda_{max} = 505$ nm), native BR ($\lambda_{max} = 568$ nm) and A2 BR (BR containing a 3,4-didehydro retinal, $\lambda_{max} = 590$ nm) has

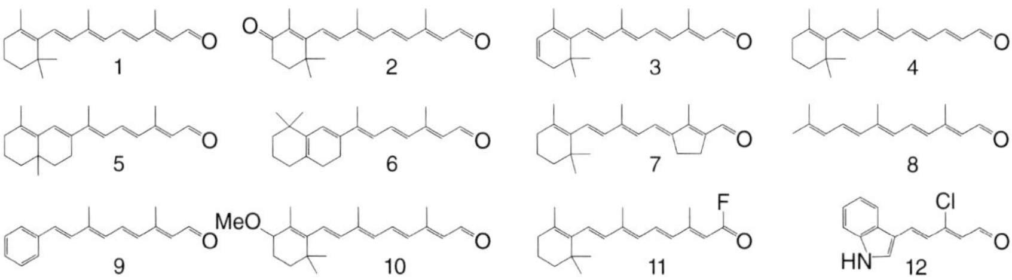

2.14 Structures of retinal and synthetic analogs. (1) all-*trans* retinal, (2) 4-keto retinal, (3) A2-retinal, (4) 13-desmethyl retinal, (5) locked 6-s-*trans* retinal, (6) locked 6-s-*cis* retinal, (7) 12,14-ethanoretinal, (8) 3,7,11-trimethyl-2,4,6,8-dodecatetraenal, (9) phenylretinal, (10) 4-methoxyretinal, (11) all-*trans* 14-fluororetinal, (12) (2*Z*,4*E*)-3-Chloro-5-(1H-indol-3-yl)-2,4-pentadienal.

shown application in color detection (Lensu *et al.*, 2004; Frydrych *et al.*, 2005) and in a color-sensitive retina (Frydrych *et al.*, 2000). The A2-retinal contains an elongated polyene chain within the β-ionone ring and the A2-BR analog retains optoelectronic activity, similar to 4-keto BR. By altering the absorption maximum of the photocycle states, color sensitivity can be achieved. The three-protein component system is analogous to the color system within the retina as well as the RGB color display system commonly employed in electronic displays (Lensu *et al.*, 2004).

Red-shifted BR pigments have become of particular interest for device applications because they would allow for the use of inexpensive semiconductor lasers as part of the optoelectronic device. Several chromophore analogs (e.g. analog **3** of Fig. 2.14) have been prepared that red-shift the absorption maximum while maintaining the photochromic activity of the BR analog (Muthyala *et al.*, 2001; Druzhko *et al.*, 1998). Additionally, BR analogs containing aromatic rings and heterocyclic rings (e.g. analogs **5**, **6**, **7** and **12** of Fig. 2.14) have been introduced for potential enhancement of BR-based electronic applications (see Singh *et al.* (1996), Singh and Manjula (2003), Hota and Singh (2007) and references within).

2.4.3 Genetic engineering

Protein engineering is a necessary step for the successful implementation of BR in applied technologies. Although nature has provided a robust photoactive protein, optimization of native BR is required for application as the primary photochromic material in non-biological environments. In most cases, the focus of mutagenesis is the optimization of the photointermediates, **M** and **O**, and the branched photoproduct **Q**, a characteristic rarely found in nature. However, mutagenesis of BR has also been employed to introduce gold-binding capabilities via cysteine mutations (Brizzolara *et al.*, 1997; Schranz *et al.*, 2007), enhance the innate dipole moment (unpublished data), and overall photocycle speed of the protein (unpublished data). A general timeline for producing BR mutants is shown in Fig. 2.15.

Directed evolution (DE) provides a cost-effective and time-efficient method for genetically manipulating biological macromolecules, specifically BR, to serve in non-native environments (Wise *et al.*, 2002; Hillebrecht *et al.*, 2004; Hillebrecht *et al.*, 2005). This process enhances a molecule toward a specific characteristic via repeated iterations of genetic mutation, screening and differential selection. Historically, DE has been used to modify properties of enzymes and biopolymers for industrial and pharmaceutical applications. In each case, a diversified genetic library was generated and appropriate screening methods were used to identify mutants with enhanced phenotypes that include substrate specificity, and chemical and thermal stability (Arnold, 1998; Kettling *et al.*, 1999; Lin and Cornish, 2002;

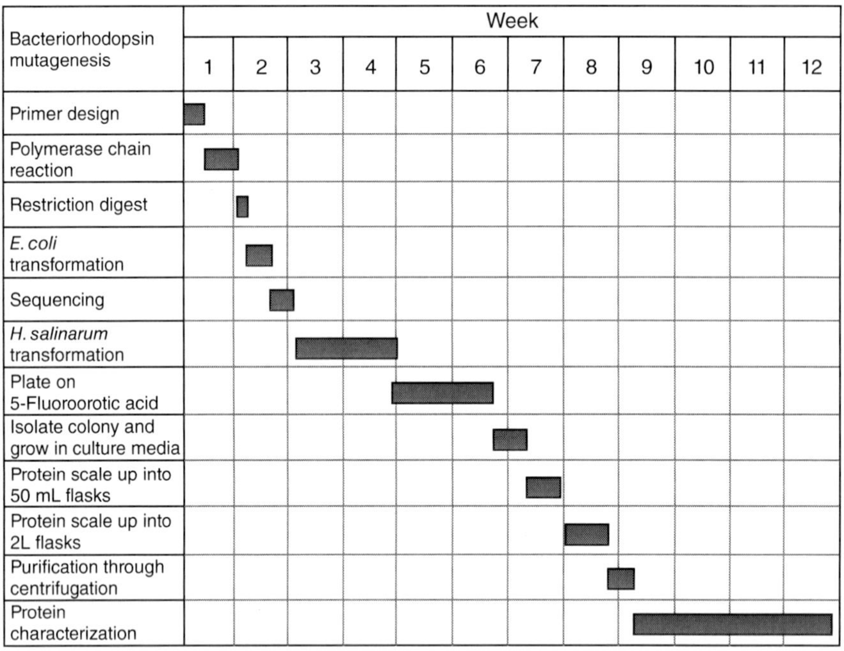

2.15 Timeline for the genetic engineering of bacteriorhodopsin through site-directed, site-specific saturation, semi-random and random mutagenesis. The chart represents a single round or stage of mutagenesis, which is accomplished in approximately ten weeks. After ten weeks, preliminary characterization of mutants is carried out, and the best mutants for a specific trait are selected to serve as the template for subsequent rounds of mutagenesis in a process referred to as type I directed evolution.

Jäckel *et al.*, 2008). This approach allows for the deliberate evolution of individual populations of molecules toward a specific trait.

Protein redesign of BR by DE is accomplished through a combination of site-directed, site-specific saturation, semi-random and random mutagenesis. Site-directed mutagenesis (SDM) exchanges a single amino acid, typically a key residue, to alter the structure or function of the molecule. Semi-random mutagenesis (SRM) and random mutagenesis (RM) produce a large number of indiscriminate mutants through the use of doped primers or oligonucleotides. Selection of prospective SRM mutants is feasible because *Halobacterium* colonies are pigmented when the protein is functionally expressed. Site-specific saturation mutagenesis (SSSM), a combination of SDM and SRM, allows for the exploration of multiple amino acid substitutions via the saturation of a key residue at a specific locus on the bacterio-opsin (*bop*) gene.

In the first round of DE, region-specific SRM is used to generate a large number of mutants that are then screened with respect to a specific property. For BR, mutants are typically screened for altered photophysical properties, such as **M** state lifetime or **Q** state formation, using 96 well plates. Because the deliberate engineering of BR for device applications necessitates several rounds of mutagenesis to identify a mutant that outperforms the native protein, an *in-house* automated screening system was developed based on microgram characterization (Wise *et al.*, 2002). Proteins exhibiting the desired trait are then grown up in sufficient quantities for further testing and are selected for subsequent rounds of mutagenesis via SDM, SSSM, SRM or RM. Through six-stages of DE, over 10 000 BR mutants have been screened and investigated for use in biomimetic devices. This process is illustrated for enhancing the **Q** state of BR in Fig. 2.16.

2.5 Conclusions and future trends

Biotechnology is now at a point where the macro- and the microstructures can be efficiently combined and improved upon with various techniques. The stable architecture of retinylidene proteins offers a template that provides significant biomimetic potential, not only because many of these proteins are very stable but also because of their unique photophysical properties. Bacteriorhodopsin has been one of the most successful targets because of the stability imparted by the semi-crystalline structural lattice. The recent applications of other retinylidene proteins, described in section 2.2.2, offer new and exciting advances in the field. Perhaps the greatest potential lies in the ability to manipulate retinylidene proteins via genetic engineering and protein redesign. Directed evolution provides the most significant potential when such can be implemented.

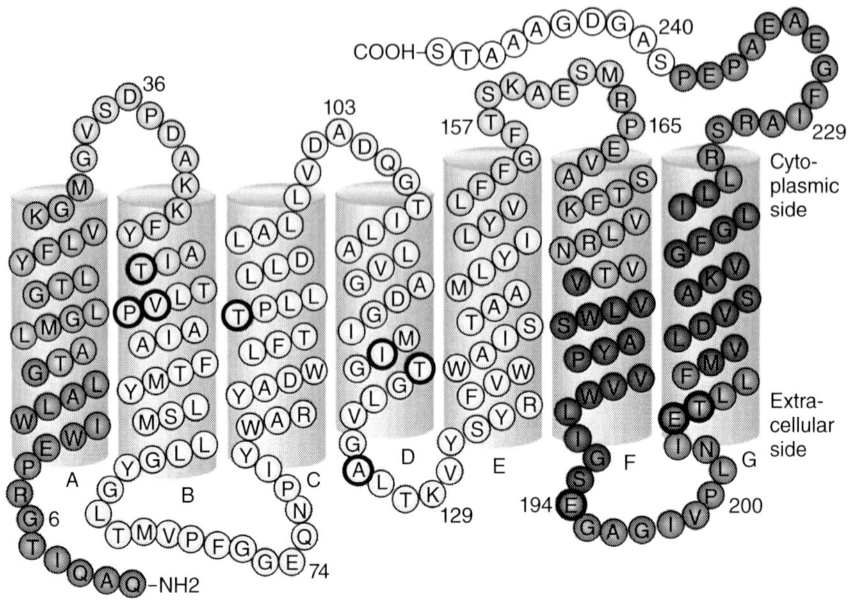
COOH-
240
36
103
157
165
229
Cyto-
plasmic
side
Extra-
cellular
side
A
B
C
D
E
F
G
6
74
129
194
200
-NH2

(a)

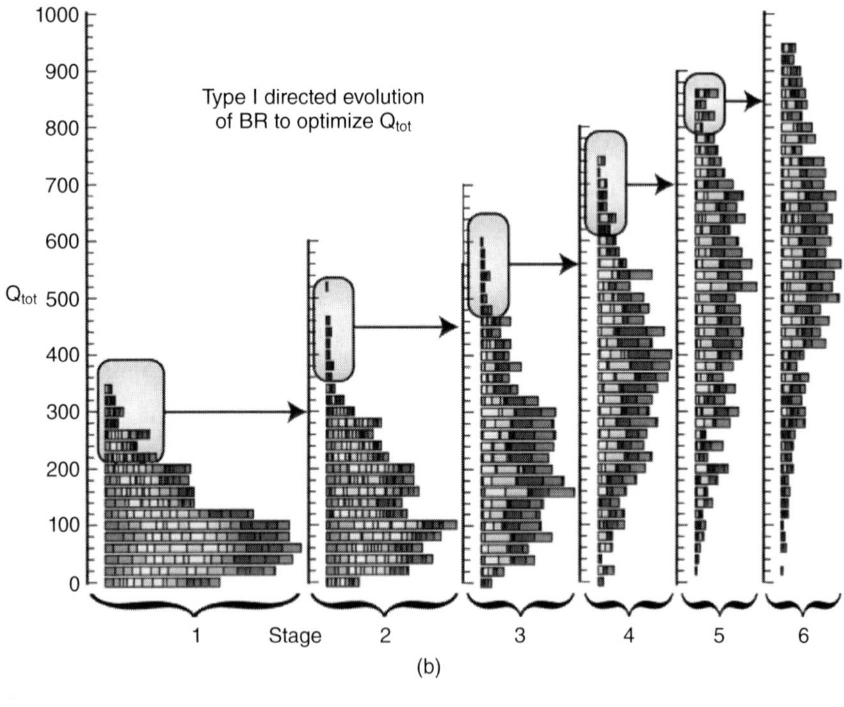
Type I directed evolution
of BR to optimize Q_{tot}
Q_{tot}
1000
900
800
700
600
500
400
300
200
100
0
1
Stage
2
3
4
5
6

(b)

2.16 Six stages of directed evolution of bacteriorhodopsin to enhance **Q** state formation. (a) A 2D structure of bacteriorhodopsin that was generated using the program Protein 3.6 (Professor Robert R. Birge, Department of Chemistry and Molecular and Cell Biology, University of Connecticut, Storrs, CT, USA). Bold residues represent the key sites responsible for the formation of the **Q** state. (b) Amino acids substitutions, which generated high amounts of **Q**, were used as the parents for the next generation of genetic progeny. After six rounds of mutagenesis, 1604 novel mutants were generated. The best mutants are currently being tested for application in 3D volumetric memories and holographic associative recall.

2.6 Acknowledgements

This work was supported in part by grants from the Defense Advanced Research Projects Agency (HR0011-05-1-0027), National Science Foundation (NSF-0829916), the National Institute of Health (GM-34548) and the Harold S. Schwenk Sr. Distinguished Chair in Chemistry.

2.7 References

Adamian, L., Ouyang, Z., Tseng, Y. Y. and Liang, J. (2006) Evolutionary patterns of retinal binding pockets of type I rhodopsins and their functions. *Photochemistry and Photobiology*, 82, pp. 1426–1435.

Aharoni, A., Khatchatouriants, A., Manevitch, A., Lewis, A. and Sheves, M. (2003) Protein-β-ionone ring interactions enhance the light-induced dipole of the chromophore in bacteriorhodopsin. *The Journal of Physical Chemistry B*, 107, pp. 6221–6225.

Alami, M., Dalal, K., Lelj-Garolla, B., Sligar, S. G. and Duong, F. (2007) Nanodiscs unravel the interaction between the SecYEG channel and its cytosolic partner SecA. *EMBO Journal*, 26, pp. 1995–2004.

Alexiev, U., Marti, T., Heyn, M., Khorana, H. and Scherrer, P. (1994) Surface charge of bacteriorhodopsin detected with covalently bound pH indicators at selected extracellular and cytoplasmic sites. *Biochemistry*, 33, pp. 298–306.

Allen, S. J., Kim, J. M., Khorana, H. G., Lu, H. and Booth, P. J. (2001) Structure and function in bacteriorhodopsin: The effect of the interhelical loops on the protein folding kinetics. *Journal of Molecular Biology*, 308, pp. 423–435.

Arnold, F. (1998) Design by directed evolution. *Accounts of Chemical Research*, 31, pp. 125–131.

Arrenberg, A. B., Del Bene, F. and Baler, H. (2009) Optical control of zebrafish behavior with halorhodopsin. *Proceedings of the National Academy of Sciences, USA*, 106, pp. 17968–17973.

Asato, A. E., Li, X.-Y., Mead, D., Patterson, G. M. L. and Liu, R. S. H. (1990) Azulenic retinoids and the corresponding bacteriorhodopsin analogues. Unusually red-shifted pigments. *Journal of the American Chemical Society*, 112, pp. 7398–7399.

Balashov, S. (2000) Protonation reactions and their coupling in bacteriorhodopsin. *Biochimica et Biophysica Acta*, 1460, pp. 75–94.

Balashov, S., Govindjee, R., Imasheva, E., Misra, S., Ebrey, T., Feng, Y., Crouch, R. and Menick, D. (1995a) The two pKa's of aspartate-85 and control of thermal isomerization and proton release in the arginine-82 to lysine mutant of bacteriorhodopsin. *Biochemistry*, 34, pp. 8820–8834.

Balashov, S., Govindjee, R., Kono, M., Imasheva, E., Lukashev, E., Ebrey, T., Crouch, R., Menick, D. and Feng, Y. (1993) Effect of the arginine-82 to alanine mutation in bacteriorhodopsin on dark adaptation, proton release, and the photochemical cycle. *Biochemistry*, 32, pp. 10331–10343.

Balashov, S., Imasheva, E., Ebrey, T., Chen, N., Menick, D. and Crouch, R. (1997) Glutamate 194 to cysteine mutation inhibits fast light-induced proton release in bacteriorhodopsin. *Biochemistry*, 36, pp. 8671–8676.

Balashov, S. P., Govindjee, R., Imasheva, E. S., Misra, S., Ebrey, T. G., Feng, Y., Crouch, R. K. and Menick, D. R. (1995b) Complex titration of Asp-85 in the unphotolyzed protein reflects dependence of another group. *Biochemistry*, 35, pp. 4054–4062.

Balashov, S. P., Imasheva, E. S., Govindjee, R. and Ebrey, T. G. (1996) Titration of aspartate-85 in bacteriorhodopsin: what it says about chromophore isomerization and proton release. *Biophysical Journal*, 70, pp. 473–481.

Baler, H. and Scott, E. K. (2009) Genetic and optical targeting of neural circuits and behavior – zebrafish in the spotlight. *Current Opinions in Neurobiology*, 19, pp. 553–560.

Balogh-Nair, V., Carriker, J. D., Honig, B., Kamat, V., Motto, M. G., Nakanishi, K., Sen, R., Sheves, M., Tanis, M. A. and Tsujimoto, K. (1980) The 'Opsin Shift' in bacteriorhodopsin: Studies with artificial bacteriorhodopsins. *Photochemistry and Photobiology*, 33, pp. 483–488.

Balya, D. C., Busskamp, V. C., Lagali, P. C., and Roska, B. C. (2008) *Use of light sensitive genes*. US Patent application 2009127705.

Barmenkov, Y. O., Kir'yanov, A. V., Starodumov, A. N., Leppanen, V.-P., Vanhanen, J., Jaaskelainen, T. and Kozhevnikov, N. M. (2000) Advanced model for a 4-keto bacteriorhodopsin photocycle based on phase grating recording analysis. *Laser Physics*, 10, pp. 60–63.

Bayburt, T. H., Grinkova, Y. V. and Sligar, S. G. (2006) Assembly of single bacteriorhodopsin trimers in bilayer nanodiscs. *Archives of Biochemistry and Biophysics*, 450, pp. 215–222.

Bayburt, T. H., Leitz, A. J., Xie, G., Oprian, D. D. and Dsligar, S. G. (2007) Transducin activation by nanoscale lipid bilayers containing one and two rhodopsins. *The Journal of Biological Chemistry*, 282, pp. 14875–14881.

Bayburt, T. H. and Sligar, S. G. (2003) Self-assembly of single integral membrane proteins into soluble nanoscale phospholipid bilayers. *Protein Science*, 12, pp. 2476–2481.

Bayburt, T. H. and Sligar, S. G. (2009) Membrane protein assembly into Nanodiscs. *FEBS Letters*, 584, pp. 1721–1727.

Bayley, H., Radhakrishnan, R., Huang, K.-S. and Khorana, H. G. (1981) Light-driven proton translocation by bacteriorhodopsin reconstituted with the phenyl analog of retinal. *The Journal of Biological Chemistry*, 256, pp. 3797–3801.

Becher, B. M. and Cassim, J. Y. (1975) Improved isolation procedures for the purple membrane of *Halobacterium halobium*. *Preparative Biochemistry and Biotechnology*, 5, pp. 161–178.

Becher, B. M. and Cassim, J. Y. (1976) Effects of light adaptation on the purple membrane structure of *Halobacterium halobium*. *Biophysical Journal*, 16, pp. 1183–1200.

Béjà, O., Aravind, L., Koonin, E. V., Suzuki, M. T., Hadd, A., Nguyen, L. P., Jovanovich, S. B., Gates, C. M., Feldman, R. A., Spudich, J. L., Spudich, E. N. and Delong, E. F. (2000) Bacterial rhodopsin: Evidence for a new type of phototrophy in the sea. *Science*, 289, pp. 1902–1906.

Béjà, O., Spudich, E. N., Spudich, J. L., Leclerc, M. and Delong, E. F. (2001) Proteorhodopsin phototrophy in the ocean. *Nature*, 411, pp. 786–789.

Bergo, V. B., Sineshchekov, O. A., Kralj, J. M., Partha, R., Spudich, E. N., Rothschild, K. J. and Spudich, J. L. (2009) His-75 in proteorhodopsin, a novel component in light-driven proton translocation by primary pumps. *The Journal of Biological Chemistry*, 284, pp. 2836–2843.

Bernstein, J. G., Han, X., Henninger, M. A., Ko, E. Y., Qian, X., Franzesi, G. T., McConnell, J. P., Stern, P., Desimone, R. and Boyden, E. S. (2008) Prosthetic systems for therapeutic optical activation and silencing of genetically-targeted neurons. *Proceedings of the Society of Photo-Optical Instrumentation Engineers*, 6854, p. 68540H.

Bhattacharya, S., Marti, T., Otto, H., Heyn, M. P. and Khorana, H. G. (1992) A bacteriorhodopsin analog reconstituted with a non-isomerizable 13-*trans* retinal derivative displays light insensitivity. *The Journal of Biological Chemistry*, 267, pp. 6757–6762.

Birge, R. R. (1981) Photophysics of light transduction in rhodopsin and bacteriorhodopsin. *Annual Review of Biophysics and Bioengineering*, 10, pp. 315–354.

Birge, R. R. (1990) Photophysics and molecular electronic applications of the rhodopsins. *Annual Review of Physical Chemistry*, 41, pp. 683–733.

Birge, R. R. (1994a) Bacteriorhodopsin, a non-linear proton pump. *Nature*, 371, pp. 659–660.

Birge, R. R. (1994b) 'Introduction to Molecular and Biomolecular Electronics.' *Advances in Chemistry*, 240, pp. 1–13.

Birge, R. R., Findsen, L. A. and Pierce, B. M. (1987) Molecular dynamics of the primary photochemical event in bacteriorhodopsin. Theoretical evidence for an excited singlet state assignment for the J intermediate. *Journal of the American Chemical Society*, 109, pp. 5041–5043.

Birge, R. R., Gillespie, N. B., Izaguirre, E. W., Kusnetzow, A., Lawrence, A. F., Singh, D., Song, Q. W., Schmidt, E., Stuart, J. A., Seetharaman, S. and Wise, K. J. (1999) Biomolecular Electronics: Protein-based associative processors and volumetric memories. *The Journal of Physical Chemistry B*, 103, pp. 10746–10766.

Birge, R. R., Govender, D. S. K., Izgi, K. C. and Tan, E. H. L. (1996) Role of calcium in the proton pump of bacteriorhodopsin. Microwave evidence for a cation-gated mechanism. *The Journal of Physical Chemistry*, 100, pp. 9990–10004.

Birge, R. R., Gross, R. B., Masthay, M. B., Stuart, J. A., Tallent, J. R. and Zhang, C. (1992) Non-linear optical properties of bacteriorhodopsin and protein-based two-photon three-dimensional memories. *Molecular Crystals and Liquid Crystals Science and Technology Section B Non-linear Optics 3,* pp. 133–147. Malaysia, Gordon and Breach Science Publishers S. A.

Birge, R. R., Izgi, K. C., Stuart, J. A. and Tallent, J. R. (1991) Wavelength dependence of the photorefractive and photodiffractive properties of holographic thin films based on bacteriorhodopsin. *Proceedings of the Materials Research Society*, 218, pp. 131–140.

Birge, R. R. and Zhang, C. F. (1990) Two-photon double resonance spectroscopy of bacteriorhodopsin. Assignment of the electronic and dipolar properties of the low-lying $^1A_g^{*-}$like and $^1B_u^{*+}$-like π, π* states. *The Journal of Chemical Physics*, 92, pp. 7178–7195.

Bismuth, O., Friedman, N., Sheves, M. and Ruhman, S. (2007) Photochemistry of a retinal protonated schiff-base analogue mimicking the opsin shift of bacteriorhodopsin. *The Journal of Physical Chemistry B*, 111, pp. 2327–2334.

Boehm, M. F., Gawinowicz, M. A., Foucault, A., Derguini, F. and Nakanishi, K. (1990) Photoaffinity labeling studies of bacteriorhodopsin with [15-^{3}H]-3-diazo-4-keto-all-*trans*-retinal. *Journal of the American Chemical Society*, 112, pp. 7779–7782.

Booth, P. J., Riley, M. L., Flitsch, S. L., Templer, R. H., Farooq, A., Curran, A. R., Chadborn, N. and Wright, P. (1997) Evidence that bilayer bending rigidity affects membrane protein folding. *Biochemistry*, 36, pp. 197–203.

Borch, J. and Hamann, T. (2009) The nanodisc: A novel tool for membrane protein studies. *Biological Chemistry*, 390, pp. 805–814.

Boucher, F., Taneva, S., Elouatik, S., Déry, M., Messaoudi, S., Harvey-Girard, E. and Beaudoin, N. (1996) Reversible inhibition of proton release activity and the anesthetic-induced acid-base equilibrium between the 480 and 570 nm forms of bacteriorhodopsin. *Biophysical Journal*, 70, pp. 948–961.

Boyden, E. S., Zhang, F., Bamberg, E., Nagel, G. and Deisseroth, K. (2005) Millisecond-timescale, genetically targeted optical control of neural activity. *Nature Neuroscience*, 8, pp. 1263–1268.

Boyer, A., Dery, M., Selles, P., Arbour, C. and Boucher, F. (1995) Colour discrimination by forward and reverse photocurrents in bacteriorhodopsin-based photosensor. *Biosensors and Bioelectronics*, 10, pp. 415–422.

Braiman, M. S., Stern, L. J., Chao, B. H. and Khorana, H. G. (1987) Structure-function studies on bacteriorhodopsin: IV. Purification and renaturation of bacterio-opsin polyepetide expressed in *Escherichia coli*. *The Journal of Biological Chemistry*, 262, pp. 9271–9276.

Bressler, S., Friedman, N., Li, Q., Ottolenghi, M., Saha, C. and Sheves, M. (1999) Generation of the O_{630} photointermediate of bacteriorhodopsin is controlled by the state of protonation of several protein residues. *Biochemistry*, 38, pp. 2018–2025.

Brizzolara, R. A., Boyd, J. L. and Tate, A. E. (1997) Evidence for covalent attachment of purple membrane to a gold surface via genetic modification of bacteriorhodopsin. *Journal of Vacuum Science and Technology*, 15, pp. 773–778.

Brouillette, C., Muccio, D. and Finney, T. (1987) pH dependence of bacteriorhodopsin thermal unfolding. *Biochemistry*, 26, pp. 7431–7438.

Brouillette, C. G., McMichens, R. B., Stern, L. J. and Khorana, H. G. (1989) Structure and thermal stability of monomeric bacteriorhodopsin in mixed pospholipid/detergent micelles. *Proteins: Structure, Function, and Genetics*, 5, pp. 38–46.

Brown, L., Needleman, R. and Lanyi, J. (1999) Functional roles of aspartic acid residues at the cytoplasmic surface of bacteriorhodopsin. *Biochemistry*, 38, pp. 6855–6861.

Brown, L. S., Dioumaev, A. K., Needleman, R. and Lanyi, J. K. (1998a) Connectivity of the retinal Schiff base to Asp85 and Asp96 during the bacteriorhodopsin photocycle: the local-access model. *Biophysical Journal*, 75, pp. 1455–1465.

Brown, L. S., Dioumaev, A. K., Needleman, R. and Lanyi, J. K. (1998b) Local-access model for proton transfer in bacteriorhodopsin. *Biochemistry*, 37, pp. 3982–3993.

Brown, L. S., Gat, Y., Sheves, M., Yamazaki, Y., Maeda, A., Needleman, N. and Lanyi, J. K. (1994) The retinal Schiff base-counterion complex of bacteriorhodopsin: Changed geometry during the photocycle is a cause of proton transfer to aspartate 85. *Biochemistry*, 33, pp. 12001–12011.

Bryl, K. and Yoshihara, K. (2001) Can purple membrane play the role of biosensor under extreme conditions? *NATO Science Series Sub Series I Life and Behavioural Sciences.*

Büldt, G., Konno, K., Nakanishi, K., Plohn, H.-J., Rao, B. N. and Dencher, N. A. (1991) Heavy-atom labelled retinal analogues located in bacteriorhodopsin by X-ray diffraction. *Photochemistry and Photobiology*, 54, pp. 873–879.

Calimet, N. and Ulmann, G. M. (2004) The influence of a transmembrane pH gradient on protonation probabilities of bacteriorhodopsin: the structural basis of the back-pressure effect. *Journal of Molecular Biology*, 339, pp. 571–589.

Campagnola, L., Wang, H. and Zylka, M. J. (2008) Fiber-coupled light-emitting diode for localized photostimulation of neurons expressing channelrhodopsin-2. *Journal of Neuroscience Methods*, 169, pp. 27–33.

Cao, Y., Váró, G., Chang, M., NI, B. F., Needleman, R. and Lanyi, J. K. (1991) Water is required for proton transfer from aspartate-96 to the bacteriorhodopsin Schiff base. *Biochemistry*, 30, pp. 10972–10979.

Cao, Y., Váró, G., Klinger, A. L., Czajkowsky, D. M., Braiman, M. S., Needleman, R. and Lanyi, J. K. (1993) Proton transfer from asp-96 to the bacteriorhodospin Schiff base is caused by a decrease of the pKa of asp-96 which follows a protein backbone conformational change. *Biochemistry*, 32, pp. 1981–1990.

Carlson, J. W., Bayburt, T. H. and Sligar, S. G. (2000) Nanopatterning phospholipid bilayers. *Langmuir*, 16, pp. 3927–3931.

Cassim, J. Y. (1992) Unique biphasic band shape of the visible circular dichroism of bacteriorhodopsin in purple membrane: Excitons, multiple transitions or protein heterogeneity? *Biophysical Journal*, 63, pp. 1432–1442.

Chang, C.-H., Jonas, R., Melchiore, S., Govindhee, R. and Ebrey, T. G. (1986) Mechanism and role of divalent cation binding of bacteriorhodopsin. *Biophysical Journal*, 49, pp. 731–739.

Chang, C. H., Chen, J. G., Govindjee, R. and Ebrey, T. (1985) Cation binding by bacteriorhodopsin. *Proceedings of the National Academy of Sciences, USA*, 82, pp. 396–400.

Chang, C. H., Liu, S. Y., Jonas, R. and Govindjee, R. (1987) The pink membrane: The stable photoproduct of deionized purple membrane. *Biophysical Journal*, 52, pp. 617–623.

Chen, Z. and Birge, R. R. (1993) Protein-based artificial retinas. *Trends in Biotechnology*, 11, pp. 292–300.

Chen, Z., Lewis, A., Takei, H. and Nabenzahl, I. (1991) Bacteriorhodopsin oriented in polyvinyl alcohol films as an erasable optical storage medium. *Applied Optics*, 30, pp. 5188–5196.

Chizov, I., Englehard, M., Chernavskii, D. S., Zubov, B. and Hess, B. (1992) Temperature and pH sensitivity of the O_{640} intermediate of the bacteriorhodopsin photocycle. *Biophysical Journal*, 61, pp. 1001–1006.

Chronister, E. L. and El-Sayed, M. M. (1987) Time-resolved resonance raman spectra of the photocycle intermediates of acid and deionized bacteriorhodopsin. *Photochemistry and Photobiology*, 45, pp. 507–513.
Cladera, J., Galisteo, M., Dunach, M., Mateo, P. and Padros, E. (1988) Thermal denaturation of deionized and native purple membranes. *Biochimica et Biophysica Acta*, 943, pp. 148–156.
Cookingham, R. E., Lewis, A. and Lemley, A. T. (1978) A vibrational analysis of rhodopsin and bacteriorhodopsin chromophore analogues: Resonance raman and infrared spectroscopy of chemically modified retinals and Schiff bases. *Biochemistry*, 17, pp. 4699–4711.
Crittenden, S., Howell, S., Reifenberger, R., Hillebrecht, J. and Birge, R. (2003) Humidity-dependent open circuit photovoltage from a bacteriorhodopsin/indium tin oxide bioelectronic heterostructure. *Nanotechnology*, 14, pp. 562–565.
Crouch, R. K., Or, Y. S., Ghent, S., Chang, C.-H., Govindjee, R., Ebrey, T. G., Callender, R. H. and Pande, A. (1984) Neither the retinal ring nor the ring double bond is required for proton pumping in bacteriorhodopsin: Acyclic retinal bacterioopsin analogues. *Journal of the American Chemical Society*, 106, pp. 8325–8327.
Crouch, R. K., Scott, R., Ghent, S., Govindjee, R., Chang, C.-H. and Ebrey, T. G. (1985) Properties of synthetic bacteriorhodopsin pigments. Further probes of the chromophore binding site. *Photochemistry and Photobiology*, 43, pp. 297–303.
Cutkosky, M. R. and Kim, S. (2009) Design and fabrication of multi-material structures for bioinspired robots. *Philosophical Transactions of the Royal Society A*, 367, pp. 1799–1813.
Dancshazy, Z. and Tokaji, Z. (2000) Blue light regeneration of bacteriorhodopsin bleached by continuous light. *FEBS Letters*, 476, pp. 171–173.
Das, J., Crouch, R. K., Govindjee, R., Balashov, S. and Ebrey, T. G. (1999) Studies on pyrylretinal analogues of bacteriorhodopsin. *Photochemistry and Photobiology*, 70, pp. 949–956.
Del Rio, E., Gonzalez-Manas, J. M., Gurtubay, J. I. and Goni, F. M. (1991) On the mechanism of bacteriorhodopsin solubilization by surfactants. *Archives in Biochemistry and Biophysics*, 291, pp. 300–306.
Delahanty, J. and Ligler, F. (2002) A microarray immunoassay for simultaneous detection of proteins and bacteria. *Analytical Chemistry*, 74, pp. 5681–5687.
Delaney, J. K., Brack, T. L., Atkinson, G. H., Ottolenghi, M., Steinberg, G. and Sheves, M. (1995) Primary picosecond molecular events in the photoreaction of the BR5.12 artificial bacteriorhodopsin pigment. *Proceedings of the National Academy of Sciences, USA*, 92, pp. 2101–2105.
Delong, E. F. M. and Béjà, O, (2007) *Light-driven energy generation using proteorhodopsin*. US Patent application 7253268.
Dencher, N. A. and Heyn, M. P. (1978) Formation and properties of bacteriorhodopsin monomers in the non-ionic detergents octyl-β-D-glucoside and triton X-100. *FEBS Letters*, 96, pp. 322–326.
Dencher, N. A., Kohl, K. D. and Heyn, M. P. (1983) Photochemical cycle and light-dark adaptation of monomeric and aggregated bacteriorhodopsin in various lipid environments. *Biochemistry*, 22, pp. 1323–1334.
Denisov, I. G., Baas, B. J., Grinkova, Y. V. and Sligar, S. G. (2007) Cooperativity in Cytochrome P450 3A4: Linkages in substrate binding, spin state, uncoupling,

and production formation. *The Journal of Biological Chemistry*, 282, pp. 7066–7076.

Denisov, I. G., McLean, M. A., Shaw, A. W., Grinkova, Y. V. and Sligar, S. G. (2005) Thermotropic phase transition in soluble nanoscale lipid bilayers. *The Journal of Physical Chemistry B*, 109, pp. 15580–15588.

Di Natale, C., Martinelli, E., Paolesse, R., D'Amico, A., Filippini, D. and Lundstrom, I. (2008) An experimental biomimetic platform for artificial olfaction. *PLoS ONE*, 3, p. e3139.

Dickopf, S. and Heyn, M. (1997) Evidence for the first phase of the reprotonation switch of bacteriorhodopsin from time-resolved photovoltage and flash photolysis experiments on the photoreversal of the M-intermediate. *Biophysical Journal*, 73, pp. 3171–3181.

Dioumaev, A., Brown, L., Shih, J., Spudich, E., Spudich, J. and Lanyi, J. (2002) Proton transfers in the photochemical reaction cycle of proteorhodopsin. *Biochemistry*, 41, pp. 5348–5358.

Dioumaev, A. K., Richter, H.-T., Brown, L. S., Michikazu, T., Tuzi, S., Saito, H., Kimura, Y., Needleman, R. and Lanyi, J. K. (1998) Existence of a proton transfer chain in bacteriorhodopsin: Participation of Glu-194 in the release of protons to the extracellular surface. *Biochemistry*, 37, pp. 2496–2506.

Dioumaev, A. K., Wang, J. M., Balint, Z., Váró, G. and Lanyi, J. K. (2003) Proton transport by proteorhodopsin requires that the retinal Schiff base counterion Asp-97 be anionic. *Biochemistry*, 42, pp. 6582–6587.

Doig, S. J., Reid, P. J. and Mathies, R. A. (1991) Picosecond time-resolved resonance Raman spectroscopy of bacteriorhodopsin's J, K, and KL intermediates. *The Journal of Physical Chemistry*, 95, pp. 6372–6379.

Drachev, L. A., Drachev, A. L., Chekulaeva, L. N., Evstigneeva, R. P., Kaulen, A. D., Khitrina, L. V., Khodonov, A. A., Lazarova, Z. R. and Mitsner, B. I. (1989) An investigation of the electrochemical cycle of bacteriorhodopsin analogs with the modified ring. *Archives of Biochemistry and Biophysics*, 270, pp. 184–187.

Dracheva, S., Bose, S. and Hendler, R. W. (1996) Chemical and functional studies on the importance of purple membrane lipids in bacteriorhodopsin photocycle behavior. *FEBS Letters*, 382, pp. 209–212.

Druzhko, A. B. and Chamorovsky, S. K. (1995) The cycle of photochromic reactions of a bacteriorhodopsin analog with 4-keto-retinal. *Biosystems*, 35, pp. 133–136.

Druzhko, A. B., Chamorovsky, S. K., Lukashev, E. P., Kononenko, A. A. and Vsevolodov, N. N. (1995) 4-keto-bacteriorhodopsin films as a promising photochromic and electrochromic biological material. *Biosystems*, 35, pp. 129–132.

Druzhko, A. B., Robertson, B., Alvarez, R., De Lera, A. R. and Weetall, H. H. (1998) Phototransformation and proton pumping activity of the 14-fluoro bacteriorhodopsin derivatives. *Biochemistry et Biophys Acta*, 1371, pp. 371–381.

Druzhko, A. B., Vanderah, D. J., Robertson, B. and Weetall, H. H. (1996) An azulenic bacteriorhodopsin analog has photoinduced activity. *Photochemistry and Photobiology*, 64, 867–869.

Duñach, M., Marti, T., Khorana, H. and Rothschild, K. (1990) UV-Visible spectroscopy of bacteriorhodopsin mutants: Substitution of Arg-82, Asp-85, Tyr-185, and Asp-212 results in abnormal light-dark adaptation. *Proceedings of the National Academy of Sciences, USA*, 87, pp. 9873–9877.

Duñach, M., Seigneuret, M., Riguad, J.-L. and Padrós, E. (1988) Influence of cations on the blue to purple transition of bacteriorhodopsin. *The Journal of Biological Chemistry*, 263, pp. 17378–17384.

Dunn, R., Hackett, N. R., McCoy, J., Chao, B., Kimura, K. and Khorana, G. (1987) Structure-function studies on bacteriorhodopsin I. Expression of the bacterio-opsin gene in *Escherischia coli. The Journal of Biological Chemistry*, 262, pp. 9246–9254.

Edman, K., Royant, A., Larsson, G., Jacobson, F., Taylor, T., van der Spoel, D., Landau, E. M., Pebay-Peyroula, E and Neutze, R. (2004) Deformation of helix C in the low temperature L-intermediate of bacteriorhodopsin. *Journal of Biological Chemistry*, 279, pp. 2147–2158.

El-Sayed, M. A., Yang, D., Yoo, S.-K. and Shang, N. (1995) The effect of different metal cation binding on the proton pumping in bacteriorhodopsin. *Israel Journal of Chemistry*, 35, pp. 465–474.

Essen, L. O. (2002) Halorhodopsin: light-driven ion pumping made simple? *Current Opinions in Structural Biology*, 12, pp. 516–522.

Facciotti, M. T., Rouhani, S., Burkard, F. T., Betancourt, F. M., Downing, K. H., Rose, R. B., Mcdermott, G. and Glaeser, R. M. (2001) Structure of an early intermediate in the M-state phase of the bacteriorhodopsin photocycle. *Biophysical Journal*, 81, pp. 3442–3455.

Fischer, U. C., Towner, P. and Oesterheldt, D. (1981) Light induced isomerization, at acidic pH, initiates hydrolysis of bacteriorhodopsin to bacterio-opsin and 9–cis-retinal. *Photochemistry and Photobiology*, 33, pp. 529–537.

Fivash, M., Towler, E. M. and Fisher, R. J. (1998) BIAcore for macromolecular interaction. *Current Opinions in Biotechnology*, 9, pp. 97–101.

Frydrych, M., Lensu, L., Parkkinen, S., Parkkinen, J. and Jaaskelainen, T. (2005) 'Photoelectric response of bacteriorhodopsin in thin PVA films and its model.' In De Gregorio, M., Di Maio, V., Frucci, M. and Musio, C. (Eds.) *Brain, Vision, and Artificial Intelligence: Lecture Notes in Computer Science.* Heidelberg, Springer.

Frydrych, M., Silfsten, P., Parkkinen, S., Parkkinen, J. and Jaaskelainen, T. (2000) Color sensitive retina based on bacteriorhodopsin. *Biosystems*, 54, pp. 131–140.

Fukuzawa, K. (1994) Motion-sensitive position sensor using bacteriorhodopsin. *Applied Optics*, 33, pp. 7489–7495.

Gillespie, N. B., Ren, L., Ramos, L., Daniell, H., Dews, D., Utzat, K. A., Stuart, J. A., Buck, C. H. and Birge, R. R. (2005) Characterization and photochemistry of 13-desmethyl bacteriorhodopsin. *The Journal of Physical Chemistry B*, 109, pp. 16142–16152.

Gillespie, N. B., Wise, K. J., Ren, L., Stuart, J. A., Marcy, D. L., Hillebrecht, J., Li, Q., Ramos, L., Jordan, K., Fyvie, S. and Birge, R. R. (2002) Characterization of the branched-photocycle intermediates P and Q of bacteriorhodopsin. *The Journal of Physical Chemistry B*, 106, pp. 13352–13361.

Goldsmith, T. H. (1994) Ultraviolet receptors and color vision: evolutionary implications and a dissonance of paradigms. *Vision Research*, 34, pp. 1479–1487.

Goluch, E. D., Shaw, A. W., Sligar, S. G. and Liu, C. (2008) Microfluidic patterning of nanodisc lipid bilayers and multiplexed analysis of protein interaction. *Lab Chip*, 8, pp. 1723–1728.

Gorb, S. N. (2008) Biological attachment devices: exploring nature's diversity for biomimetics. *Philosophical Transactions of the Royal Society A*, 366, pp. 1557–1574.

Gourdon, P., Alfredsson, A., Pedersen, A., Malmerberg, E., Nyblom, M., Widell, M., Berntsson, R., Pinhassi, J., Braiman, M., Hansson, O., Bonander, N., Karlsson, G. and Neutze, R. (2008) Optimized *in vitro* and *in vivo* expression of proteorhodopsin: A seven-transmembrane proton pump. *Protein Expression and Purification*, 58, pp. 103–113.

Govindjee, R., Misra, S., Balashov, S. P., Ebrey, T. G., Crouch, R. K. and Menick, D. R. (1996) Arginine-82 regulates the pK(a) of the group responsible for the light-driven proton release in bacteriorhodopsin. *Biophysical Journal*, 71, pp. 1011–1023.

Grigorieff, N., Ceska, T. A., Downing, K. H., Baldwin, J. M. and Henderson, R. (1996) Electron-crystallographic refinement of the structure of bacteriorhodopsin. *Journal of Molecular Biology*, 259, pp. 393–421.

Gross, R., Schumann, C., Wolf, M. M. N., Herbst, J., Diller, R., Friedman, N. and Sheves, M. (2009) Ultrafast protein conformational alterations in bacteriorhodopsin and its locked analogue BR5.12. *The Journal of Physical Chemistry B*, 113, pp. 7851–7860.

Gross, R. B., Izgi, K. C. and Birge, R. R. (1992) Holographic thin films, spatial light modulators and optical associative memories based on bacteriorhodopsin. *Proceedings of SPIE*, 1662, pp. 186–196.

Haacke, S., Schenkl, S., Vinzani, S. and Chergui, M. (2002) Femtosecond and picosecond fluorescence of native bacteriorhodopsin and a non-isomerizing analog. *Biopolymers*, 67, pp. 306–309.

Hackett, N. R., Stern, L. J., Chao, B. H., Kronis, K. A. and Khorana, H. G. (1987) Structure-function studies on bacteriorhodopsin: V. Effects of amino acid substitutions in the putative helix F. *The Journal of Biological Chemistry*, 269, pp. 9277–9284.

Hampp, N. (2000a) Bacteriorhodopsin as a photochromic retinal protein for optical memories. *Chemical Reviews*, 100, pp. 1755–1776.

Hampp, N., Bräuchle, C. and Oesterhelt, D. (1990) Bacteriorhodopsin wildtype and variant aspartate-96 to asparagine as reversible holographic media. *Biophysical Journal*, 58, pp. 83–93.

Hampp, N. and Juchem, T. (2000) 'Fringemaker – the first technical system based on bacteriorhodopsin.' In Dér, A. and Keszthelyi, L. (Eds.) *Bioelectronic Applications of Photochromic Pigments.* Szeged, Hungary, IOS Press.

Hampp, N., Popp, A., Braeuchle, C. and Oesterhelt, D. (1992) Diffraction efficiency of bacteriorhodopsin films for holography containing bacteriorhodopsin wildtype BR_{WT} and its variants BR_{D85E} and BR_{D96N}. *The Journal of Physical Chemistry*, 96, pp. 4679–4685.

Hampp, N., Thoma, R., Zeisel, D. and Bräuchle, C. (1994) Bacteriorhodopsin variants for holographic pattern recognition. *Advances in Chemistry*, 240, pp. 511–526.

Hampp, N. A. (2000b) Bacteriorhodopsin: mutating a biomaterial into an optoelectronic material. *Applied Biochemistry and Biotechnology*, 53, pp. 633–639.

Haronian, D. and Lewis, A. (1991) Elements of a unique bacteriorhodopsin neural network architecture. *Applied Optics*, 30, pp. 597–608.

Haupts, U., Tittor, J., Bamberg, E. and Oesterhelt, D. (1997) General concept for ion translocation by halobacterial retinal proteins: The isomerization/switch/transfer (IST) model. *Biochemistry*, 36, pp. 2–7.

He, J., Samuelson, L., Li, L., Kumar, J. and Tripathy, S. K. (1998) Photoelectric properties of oriented bacteriorhodopsin/polycation multilayers by electrostatic layer-by-layer assembly. *The Journal of Physical Chemistry B*, 102, pp. 7067–7072.

Heberle, J., Oesterhelt, D. and Dencher, N. A. (1993) Decoupling the photo- and proton cycle in the Asp85 > Glu mutant of bacteriorhodopsin. *EMBO Journal*, 12, pp. 3721–3727.

Henderson, R. (1975) The structure of the purple membrane from *Halobacterium halobium*. Analysis of the X-ray diffraction pattern. *Journal of Molecular Biology*, 93, pp. 123–138.

Henderson, R., Baldwin, J. M., Ceska, T. A., Zemlin, F., Beckmann, E. and Downing, K. H. (1990) Model for the structure of bacteriorhodopsin based on high-resolution electron cryo-microscopy. *Journal of Molecular Biology*, 213, pp. 899–929.

Hendler, R. W. and Dracheva, S. (2001) Importance of lipids for bacteriorhodopsin structure, photocycle, and function. *Biochemistry (Moscow)*, 66, pp. 1311–1314.

Heyes, C. and El-Sayed, M. (2001) Effect of temperature, pH, and metal ion binding on the secondary structure of bacteriorhodopsin: FT-IR study of the melting and premelting temperatures. *Biochemistry*, 40, pp. 11819–11827.

Heyes, C. D. and El-Sayed, M. A. (2002a) The role of native lipids and lattice structure in bacteriorhodopsin protein conformation and stability as studied by temperature-dependent Fourier transform-infrared spectroscopy. *The Journal of Biological Chemistry*, 277, pp. 29437–29443.

Heyes, C. D. and El-Sayed, M. (2003) Thermal properties of bacteriorhodopsin. *The Journal of Physical Chemistry B*, 107, pp. 12045–12053.

Heyes, C. D., Wang, J., Sanii, L. S. and El-Sayed, M. (2002) Fourier transform infrared study of the effect of different cations on bacteriorhodopsin protein thermal stability. *Biophysical Journal*, 82, pp. 1598–1606.

Hillebrecht, J., Wise, K. J., Koscielecki, J. F. and Birge, R. R. (2004) Directed evolution of bacteriorhodopsin for device applications. *Methods in Enzymology*, 388, pp. 333–347.

Hillebrecht, J. R., Koscielecki, J. F., Wise, K. J., Marcy, D. L., Tetley, W., Rangarajan, R., Sullivan, J., Brideau, M., Krebs, M. P., Stuart, J. A. and BIRGE, R. R. (2005) Optimization of protein-based volumetric optical memories and associative processors by using directed evolution. *NanoBiotechnology*, 1, pp. 141–152.

Hirayama, J., Imamoto, Y., Shichida, Y., Kamo, N., Tomioka, H. and Yoshizawa, T. (1992) Photocycle of phoborhodopsin from haloalkaliphilic bacterium (*Natronobacterium pharaonis*) studied by low-temperature spectrophotometry. *Biochemistry*, 31, pp. 2093–2098.

Holz, M., Drachev, L. A., Mogi, T., Otto, H., Kaulen, A. D., Heyn, M. P., Skulachev, V. P. and Khorana, H. G. (1989) Replacement of aspartic acid-96 by asparagine in bacteriorhodopsin slows both the decay of the M intermediate and the associated proton movement. *Proceedings of the National Academy of Sciences, USA*, 86, pp. 2167–2171.

Hong, F. (1997) Molecular sensors based on the photoelectric effect of bacteriorhodopsin: Origin of differential responsivity. *Materials Science and Engineering: C*, 4, pp. 267–285.

Hong, F. T. (1994) 'Retinal proteins in photovoltaic devices.' *Advances in Chemistry*, 240, pp. 527–560.

Hong, F. T. (1999) Interfacial photochemistry of retinal proteins. *Progress in Surface Science*, 62, pp. 1–237.

Hota, P. K. and Singh, A. K. (2007) Bacteriorhodopsin analogue from indolic chromophores. *Letters in Organic Chemistry*, 4, pp. 300–305.

Huang, K.-S., Radhakrishnan, R., Bayley, H. and Khorana, H. G. (1982) Orientation of retinal in bacteriorhodopsin as studied by cross-linking using a photosensitive analog of retinal. *The Journal of Biological Chemistry*, 257, pp. 13616–13623.

Huber, R., Kohler, T., Lenz, M. O., Bamberg, E., Kalmbach, R., Engelhard, M. and Wachtveitl, J. (2005) pH-dependent photoisomerization of retinal in proteorhodopsin. *Biochemistry*, 44, pp. 1800–1806.

Hutson, M. S., Alexiev, U., Shilov, S. V., Wise, K. J. and Braiman, M. S. (2000) Evidence for a perturbation of arginine-82 in the bacteriorhodopsin photocycle from time-resolved infrared spectra. *Biochemistry*, 39, pp. 13189–13200.

Jäckel, C., Kast, P. and Hilvert, D. (2008) Protein design by directed evolution. *Annual Reviews of Biophysics*, 37, pp. 153–173.

Jackson, M. and Sturtevant, J. (1978) Phase transitions of the purple membranes of *Halobacterium halobium*. *Biochemistry*, 17, pp. 911–915.

Jensen, R. B., Kelemen, B. R., McAuliffe, J. C. and Smith, W. C. (2008) *Optical information carrier comprising immobilized proteorhodopsin.* US Patent application 7378219.

Jonas, R., Koutalos, Y. and Ebrey, T. G. (1990) Purple membrane: Surface charge density and the multiple effect of pH and cations. *Photochemistry and Photobiology*, 52, pp. 1163–1177.

Juchem, T. and Hampp, N. (2000) Interferometric system for non-destructive testing based on large diameter bacteriorhodopsin films. *Optics and Lasers in Engineering*, 34, pp. 87–100.

Kakitani, H., Kakitani, T., Rodman, H. and Honig, B. (1983) Correlation of vibrational frequencies with absorption maxima in polyenes, rhodopsin, bacteriorhodopsin, and retinal analogues. *The Journal of Physical Chemistry*, 87, pp. 3620–3628.

Kalaidzidis, I. V., Belevich, I. N., Kalaidzidis, Y. L. and Kaulen, A. D. (1999) Membrane potential stabilizes the O intermediate in liposomes containing bacteriorhodopsin. *FEBS Letters*, 459, pp. 143–147.

Kandori, H. (2004) Hydration switch model for the protonation transfer in the Schiff base region of bacteriorhodopsin. *Biochimica et Biophysica Acta*, 1658, pp. 72–79.

Karnik, S. S., Nassal, M., Doi, T., Jay, E., Sgaramella, V. and Khorana, G. (1987) Structure-function studies on bacteriorhodopsin II. Improved expression of the bacterio-opsin gene in *Escherischia coli*. *The Journal of Biological Chemistry*, 262, pp. 9255–9263.

Keszthelyi, L. (1988) Primary charge motions and light-energy transduction in bacteriorhodopsin. *Biophysical Chemistry*, 29, pp. 127–136.

Keszthelyi, L. and Ormos, P. (1980) Electric signals associated with the photocycle of bacteriorhodopsin. *FEBS Letters*, 109, pp. 189–193.

Kettling, U., Koltermann, A. and Eigen, M. (1999) Evolutionary biotechnology – reflections and perspectives. *Current Topics in Microbiology and Immunology*, 243, pp. 173–174.

Khitrina, L. V., Mironova, E. V. and Khodonov, A. A. (2009) Bacteriorhodopsin containing modified chromophores: A study on the wild-type and D96N mutant of *Halobacterium salinarum*. *Biochemistry (Moscow)*, 3, pp. 128–133.

Khodonov, A. A., Shevyakov, S. V., Mironova, E. V., Shvets, V. I., Alexeeva, S. G., Demina, O. V., Khitrina, L. V. and Kaulen, A. D. (2000) Bacteriorhodopsin analogs, bearing modified chromophore as a basis for the photochromic materials. *Molecular Crystals and Liquid Crystals*, 345, pp. 317–322.

Kim, S. Y., Waschuk, S. A., Brown, L. S. and Jung, K. H. (2008) Screening and characterization of protorhodopsin color-tuning mutations in *Escherichia coli* with endogenous retinal synthesis. *Biochimica et Biophysica Acta*, 1777, pp. 504–513.

Knopf, G. K., Wang, W. W. and Bassi, A. S. (2009) *Flexible bioelectronic photodetector and imaging arrays based on bacteriorhodopsin (BR) thin films*. US Patent application US20090032683.

Kobayashi, T., Ohtani, H., Iwai, J., Ikegami, A. and Uchiki, H. (1983) Effect of pH on the photoreaction cycles of bacteriorhodopsin. *FEBS Letters*, 162, pp. 197–200.

Kobayashi, T., Salto, T. and Ohtani, H. (2001) Real-time spectroscopy of transition states in bacteriorhodopsin during retinal isomerization. *Nature*, 414, pp. 531–534.

Kolbe, M., Besir, H., Essen, L. O. and Oesterhelt, D. (2000) Structure of the light-driven chloride pump halorhodopsin at 1.8 Å resolution. *Science*, 288, pp. 1390–1396.

Kono, M., Misra, S. and Ebrey, T. G. (1993) pH dependence of light-induced proton release by bacteriorhodopsin. *FEBS Letters*, 331, pp. 31–34.

Kouyama, T., Nichikawa, T. and Okumura, H. (2004) Crystal structure of the L intermediate of bacteriorhodopsin: Evidence for vertical translocation of a water molecule during the proton pumping cycle. *Journal of Molecular Biology*, 335, pp. 531–546.

Koyama, K., Yamaguchi, N. and Miyasaka, T. (1994) Antibody-mediated bacteriorhodopsin orientation for molecular device architectures. *Science*, 265, pp. 762–765.

Kramer, R. H., Fortin, D. L. and Trauner, D. (2009) New photochemical tools for controlling neuronal activity. *Current Opinions in Neurobiology*, 19, pp. 544–552.

Krebs, R. A., Alexiev, U., Partha, R., Devita, A. M. and Braiman, M. S. (2002) Detection of fast light-activated H^+ release and M intermediate formation from proteorhodopsin. *BMC Physiology*, 2, pp. 5.

Kresheck, G., Lin, C., Williamson, L., Mason, W., Jang, D. and El-Sayed, M. (1990) The thermal stability of native, delipidated, deionized and regenerated bacteriorhodopsin. *Journal of Photochemistry and Photobiology B*, 7, pp. 289–302.

Krimm, S. and Dwivedi, A. M. (1982) Infrared spectrum of the purple membrane: Clue to a proton conduction mechanism? *Science*, 216, pp. 407–408.

Kusnetzow, A., Singh, D. L., Martin, C. H., Barani, I. and Birge, R. R. (1999) Nature of the chromophore binding site of bacteriorhodopsin. The potential role of Arg-82 as a principal counterion. *Biophysical Journal*, 76, pp. 2370–2389.

Lakatos, M., Lanyi, J. K., Szakacs, J. and Váró, G. (2003) The photochemical reaction cycle of proteorhodopsin at low pH. *Biophysical Journal*, 84, pp. 3252–3256.

Lanyi, J. and Luecke, H. (2001) Bacteriorhodopsin. *Current Opinions in Structural Biology*, 11, pp. 415–419.
Lanyi, J. K. (1986) Halorhodopsin: A light-driven chloride ion pump. *Annual Review of Biophysics and Biophysical Chemistry*, 15, pp. 11–28.
Lanyi, J. K. (1998a) The local-access mechanism of proton transport by bacteriorhodopsin. *Biochimica et Biophysica Acta*, 1365, pp. 17–22.
Lanyi, J. K. (1998b) Understanding structure and function in the light-driven proton pump bacteriorhodopsin. *Journal of Structural Biology*, 124, pp. 164–178.
Lanyi, J. K. (2000) Molecular mechanism of ion transport in bacteriorhodopsin: Insights from crystallographic, spectroscopic, kinetic, and mutational studies. *The Journal of Physical Chemistry B*, 104, pp. 11441–11448.
Lanyi, J. K. (2004) X-ray diffraction of bacteriorhodopsin photocycle intermediates. *Molecular Membrane Biology*, 21, pp. 143–150.
Lanyi, J. K. (2004) What is the real crystallographic structure of the L photo-intermediate of bacteriorhodopsin? *Biochimica et Biophysica Acta*, 1658, pp. 14–22.
Lanyi, J. K. and Schobert, B. (2002) Crystallographic structure of the retinal and the protein after deprotonation of the Schiff base: the switch in the bacteriorhodopsin photocycle. *Journal of Molecular Biology*, 321, pp. 727–737.
Lanyi, J. K. and Schobert, B. (2007) Structural changes in the L photointermediate of bacteriorhodopsin. *Journal of Molecular Biology*, 365, pp. 1379–1392.
Lanyi, J. and Váró, G. (1995) The photocycles of bacteriorhodopsin. *Israel Journal of Chemistry*, 35, pp. 365–385.
Laptev, A. V., Belikov, N. E., Lukin, A. Y., Strokach, Y. P., Barachevsky, V. A., Alfimov, M. V., Demina, O. V., Shvets, V. I., Skladnev, D. A. and Khodonov, A. A. (2008) Synthesis of spiropyran analogues of retinal and study of their interaction with bacterioopsin from *Halobacterium salinarum*. *Russian Journal of Bioorganic Chemistry*, 34, pp. 252–260.
Large, M., Wickham, S., Hayes, J. and Poladian, L. (2007) Insights from nature: Optical biomimetics. *Physica B*, 394, pp. 229–232.
Le Coutre, J., Tittor, J., Oesterhelt, D. and Gerwert, K. (1995) Experimental evidence for hydrogen-bonded network proton transfer in bacteriorhodopsin shown by Fourier-transform infrared spectroscopy using azide as catalyst. *Proceedings of the National Academy of Sciences, USA*, 92, pp. 4962–4966.
Lensu, L., Frydrych, M., Parkkinen, J., Parkkinen, S. and Jaaskelainen, T. (2004) Photoelectric properties of bacteriorhodopsin analogs for color-sensitive optoelectronic devices. *Optical Materials*, 27, pp. 57–62.
Lensu, L., Frydrych, M., Parkkinen, S., Jaaskelainen, T. and Parkkinen, J. (2007) Color-sensitive biosensors for imaging applications. In Knopf, G. K. and Bassi, A. S. (Eds.) *Smart Biosensor Technology.* Boca Raton, FL USA, CRC Press.
Li, Q., Bressler, S., Ovrutsky, D., Ottolenghi, M., Friedman, N. and Sheves, M. (2000) On the protein residues that control the yield and kinetics of O(630) in the photocycle of bacteriorhodopsin. *Biophysical Journal*, 78, pp. 354–362.
Liang, H., Whited, G., Nguyen, C. and Stucky, G. D. (2007) The directed cooperative assembly of proteorhodopsin into 2D and 3D polarized arrays. *Proceedings of the National Academy of Sciences, USA*, 104, pp. 8212–8217.
Lin, C. T., Mertz, C. J., Bitting, H. C. and El-Sayed, M. A. (1992) Fluorescence anisotropy studies of dibucaine-HCl in micelles and bacteriorhodopsin. *Journal of Photochemistry and Photobiology B: Biology*, 13, pp. 169–185.

Lin, H. and Cornish, V. W. (2002) Screening and selection methods for large-scale analysis of protein function. *Angewandte Chemie International Edition*, 41, pp. 4402–4425.

Liu, R. S. H., Krogh, E., Li, X.-Y., Mead, D., Colmenares, L. U., Thiel, J. R., Ellis, J., Wong, D. and Asato, A. E. (1993) Analyzing the red-shift characteristics of azulenic, naphthyl, other ring-fused and retinyl pigment analogs of bacteriorhodopsin. *Photochemistry and Photobiology*, 58, pp. 701–705.

Liu, R. S. H., Liu, C. W., Li, X.-Y. and Asato, A. E. (1991) Butyl conformational reorganization as a possible explanation for the longitudinal flexibility of the binding site of bacteriorhodopsin. The azulene and C-22 retinoid analogs. *Photochemistry and Photobiology*, 54, pp. 625–631.

Liu, S. Y. and Ebrey, T. G. (1987) The quantum efficiency for the interphotoconversion of the blue and pink forms of purple membrane. *Photochemistry and Photobiology*, 46, pp. 263–267.

Logunov, S. L., El-Sayed, M. A. and Lanyi, J. K. (1996) Catalysis of the retinal subpicosecond photoisomerization process in acid purple bacteriorhodopsin and some bacteriorhodopsin mutants by chloride ions. *Biophysical Journal*, 71, pp. 1545–1553.

London, E. and Khorana, H. G. (1982) Denaturation and renaturation of bacteriorhodopsin in detergents and lipid-detergent mixtures. *The Journal of Biological Chemistry*, 257, pp. 7003–7011.

Lorber, B. and Delucas, L. J. (1990) Large scale preparation of homogeneous bacteriorhodopsin. *FEBS Letters*, 261, pp. 14–18.

Lörinczi, E., Verhoefen, M. K., Wachtveitl, J., Woerner, A. C., Glaubitz, C., Engelhard, M., Bamberg, E. and Friedrich, T. (2009) Voltage- and pH-dependent changes in vectoriality of photocurrents mediated by wild-type and mutant proteorhodopsins upon expression in *Xenopus* oocytes. *Journal of Molecular Biology*, 393, pp. 320–341.

Luecke, H., Richter, H. T. and Lanyi, J. K. (1998) Proton transfer pathways in bacteriorhodopsin at 2.3 Angstrom resolution. *Science*, 280, pp. 1934–1937.

Luecke, H., Schobert, B., Richter, H. T., Cartailler, J. P. and Lanyi, J. K. (1999) Structure of bacteriorhodopsin at 1.55 Å resolution. *Journal of Molecular Biology*, 291, pp. 899–911.

Luecke, H., Schobert, B., Cartailler, J.-P., Richter, H.-T., Rosengarth, A., Needleman, R. and Lanyi, J. K. (2000) Coupling photoisomerization of retinal to directional transport in bacteriorhodopsin. *Journal of Molecular Biology*, 300, pp. 1237–1255.

Lugtenburg, J., Muradin-Szweykowska, M., Heeremans, C., Pardoen, J. A., Harbison, G. S., Herzfeld, J., Griffin, R. G., Smith, S. O. and Mathies, R. A. (1986) Mechanism for the opsin shift of retinal's absorption in bacteriorhodopsin. *Journal of the American Chemical Society*, 108, pp. 3104–3105.

Lynagh, T. and Lynch, J. W. (2010) An improved Ivermectin-activated chloride channel receptor for inhibiting electrical activity in defined neuronal populations. *The Journal of Biological Chemistry*, 285, pp. 14890–14897.

Maeda, A., Iwasa, T. and Yoshizawa, T. (1980) Formation of 9-*cis*- and 11-*cis*-retinal pigments from bacteriorhodopsin by irradiating purple membrane in acid. *Biochemistry*, 19, pp. 3825–3831.

Maeda, A., Tatsuo, I. and Yoshizawa, T. (1981) Photoreaction of the acidified form of bacteriorhodopsin and its 9-*cis* derivative in purple membrane at low temperatures. *Photochemistry and Photobiology*, 33, pp. 559–565.

Man, D., Wang, W., Sabehi, G., Aravind, L., Post, A. F., Massana, R., Spudich, E. N., Spudich, J. L. and Béjà, O. (2003) Diversification and spectral tuning in marine proteorhodopsins. *EMBO Journal*, 22, pp. 1725–1731.

Marcus, M. A. and Lewis, A. (1978) Resonance raman spectroscopy of the retinylidene chromophore in bacteriorhodopsin (bR570), bR560, M412, and other intermediates: Structural conclusions based on kinetics, analogues, models, and isotopically labeled membranes. *Biochemistry*, 17, pp. 4722–4735.

Marcus, M. A., Lewis, A., Racker, E. and Crespi, H. (1977) Physiological and structural investigations of bacteriorhodopsin analogs. *Biochemical and Biophysical Research Communications*, 78, pp. 669–675.

Marin, V. L., Bayburt, T. H., Sligar, S. G. and Mrksich, M. (2007) Functional assays of membrane-bound proteins with SAMDI-TOF mass spectroscopy. *Angewandte Chemie*, 119, pp. 8952–8954.

Martin, C. H., Chen, Z. P. and Birge, R. R. (Eds.) (1997) *Towards a Bacteriorhodopsin-Silicon Neuromorphic Photosensor.* Maui, World Scientific, pp. 268–279.

Massotte, D. and Aghion, J. (1991) Bacteriorhodopsin 'detergent-monomers,' blue shift and velocity of light-dark adaptation. *Biochemistry and Biophysical Research Communications*, 181, pp. 1301–1305.

Masthay, M. B., Sammeth, D. M., Helvenston, M. C., Buckman, C. B., Li, W., CDE-Baca, M. J. and Kofron, J. T. (2002) The laser-induced blue state of bacteriorhodopsin: Mechanistic and color regulatory roles of protein-protein interactions, protein-lipid interactions, and metal ions. *Journal of the American Chemical Society*, 124, pp. 3418–3430.

Matsui, Y., Sakai, K., Shiro, Y., Adachi, S., Okumura, H. and Kouyama, T. (2002) Specific damage induced by X-ray radiation and structural changes in the primary photoreaction of bacteriorhodopsin. *Journal of Molecular Biology*, 324, pp. 469–481.

Matsuno-Yagi, A. and Mukohata, Y. (1980) ATP-synthesis linked to light-dependent proton-uptake in a red mutant strain of *Halobacterium* lacking bacteriorhodopsin. *Archives of Biochemistry and Biophysics*, 199, pp. 297–303.

Max, M., McKinnon, P. J., Seidenman, K. J., Barrett, R. K., Applebury, M. L., Takahashi, J. S. and Margolskee, R. F. (1995) Pineal opsin: A non-visual opsin expressed in chick pineal. *Science*, 267, pp. 1502–1506.

Max, M., Surya, A., Takahashi, J., Margolskee, R. and Knox, B. (1998) Light-dependent activation of rod transducin by pineal opsin. *The Journal of Biological Chemistry*, 273, pp. 26820–26826.

Metz, G., Siebert, F. and Engelhard, M. (1992) Asp85 is the only internal aspartic acid that gets protonated in the M intermediate and the purple-to-blue transition of bacteriorhodopsin. A solid-state ^{13}C CP-MAS NMR investigation. *FEBS Letters*, 303, pp. 237–241.

Milder, S. J., Thorgeirsson, T. E., Miercke, L. J. W., Stroud, R. M. and Kilger, D. S. (1991) Effects of detergent environments on the photocycle of purified monomeric bacteriorhodopsin. *Biochemistry*, 30, pp. 1751–1761.

Millerd, J. E., Rohrbacher, A., Brock, N. J., Chau, C.K., Smith, P. and Needleman, R. (1999) Improved sensitivity in blue-membrane bacteriorhodopsin films. *Optics Letters*, 24, pp. 1355–1357.

Miyasaka, T. and Koyama, K. (1992) Rectified photocurrents from purple membrane Lanmuir-Blodgett films at the electrode-electrolyte interface. *Thin Solid Films*, 210–211, pp. 146–149.

Miyasaka, T. and Koyama, K. (1993) Image sensing and processing by a bacteriorhodopsin-based artificial photoreceptor. *Applied Optics*, 32, pp. 6371–6379.

Mogi, T., Stern, L. J., Hackett, N. R. and Khorana, H. G. (1987) Structure-function studies on bacteriorhodopsin. VI. Bacteriorhodopsin mutants containing single tyrosine to phenylalanine substitutions are all active in proton translocation. *Proceedings of the National Academy of Sciences, USA*, 84, pp. 5595–5599.

Mogi, T., Stern, L. J., Marti, T., Chao, B. H. and Khorana, H. G. (1988) Aspartic acid substitutions affect proton translocation by bacteriorhodopsin. *Proceedings of the National Academy of Sciences, USA*, 85, pp. 4148–4152.

Moltke, S. and Heyn, M. P. (1995) Photovoltage kinetics of the acid-blue and acid-purple forms of bacteriorhodopsin: Evidence for no net charge transfer. *Biophysical Journal*, 69, pp. 2066–2073.

Mowery, P. C., Lozier, R. H., Chae, Q., Tseng, Y. W., Taylor, M. and Stoeckenius, W. (1979) Effect of acid pH on the absorption spectra and photoreactions of bacteriorhodopsin. *Biochemistry*, 18, pp. 4100–4107.

Muccio, D. D. and Delucas, L. J. (1985) Isolation of detergent-solubilized monomers of bacteriorhodopsin by size-exclusion high-performance liquid chromatography. *Journal of Chromatography A*, 326, pp. 243–250.

Muller, J., Munster, C. and Salditt, T. (2000) Thermal denaturing of bacteriorhodopsin by X-ray scattering from oriented purple membranes. *Biophysical Journal*, 78, pp. 3208–3217.

Muradin-Szweykowska, M., Pardoen, J. A., Dobbelstein, D., Van Amsterdam, L. J. P. and Lugtenburg, J. (1984) Bacteriorhodopsins with chromophores modified at the β-ionone site. *European Journal of Biochemistry*, 140, pp. 173–176.

Muthyala, R., Watanabe, D., Asato, A. E. and Liu, R. S. H. (2001) The nature of the delocalized cations in azulenic bacteriorhodopsin analogs. *Photochemistry and Photobiology*, 74, pp. 837–845.

Nagel, G., Szellas, T., Huhn, W., Kateriya, S., Adeishvili, N., Berthold, P., Ollig, D., Hegemann, P. and Bamberg, E. (2003) Channelrhodopsin-2, a directly light-gated cation-selective membrane channel. *Proceedings of the National Academy of Sciences, USA*, 100, pp. 13940–13945.

Nagel, G., Szellas, T., Kateriya, S., Adeishvili, N., Hegemann, P. and Bamberg, E. (2005) Channelrhodopsins: directly light-gated cation channels. *Biochemical Society Transactions*, 33, pp. 863–866.

Nakagawa, T., Hamanaka, T., Nishimura, S., Uruga, T. and Kito, Y. (1994) The specific binding site of the volatile anesthetic diiodomethane to purple membrane by X-ray diffraction. *Journal of Molecular Biology*, 238, pp. 297–301.

Nakanishi, K., Balogh-Nair, V., Arnaboldi, M., Tsujimoto, K. and Honig, B. (1980) An external point-charge model for bacteriorhodopsin to account for its purple color. *Journal of the American Chemical Society*, 102, pp. 7945–7947.

Nakanishi, K., Zhang, H., Lerro, K. A., Takekuma, S., Yamamoto, T., Lien, T. H., Sastry, L., Baek, D.-J., Moquin-Pattey, C., Boehm, M. F., Derguini, F. and Gawinowicz, M. A. (1995) Photoaffinity labeling of rhodopsin and bacteriorhodopsin. *Biophysical Chemistry*, 56, pp. 13–22.

Nassal, M., Mogi, T., Karnik, S. S. and Khorana, G. (1987) Structure-function studies on bacteriorhodopsin III. Total synthesis of a gene for bacterio-opsin and its expression in *Escherischia coli. The Journal of Biological Chemistry*, 262, pp. 9264–9270.

Nath, A., Atkins, W. M. and Sligar, S. G. (2007) Applications of phospholipid bilayer nanodiscs in the study of membranes and membrane proteins. *Biochemistry*, 46, pp. 2059–2069.

Nathans, J. (1999) The evolution and physiology of human color vision: Insights from molecular genetic studies of visual pigments. *Neuron*, 24, pp. 299–312.

Ni, B., Chang, M., Duschl, A., Lanyi, J. and Needleman, R. (1990) An efficient system for the synthesis of bacteriorhodopsin in *Halobacterium halobium*. *Gene*, 90, pp. 169–172.

Nikolic, K., Degenaar, P. and Toumazou, C. (2006) Modeling and engineering aspects of channelrhodopsin-2 system for neural photostimulation. *Proceedings of the 28th IEEE Engineering in Medicine and Biology Society,* pp. 1626–1629.

O'Shannessy, D. J., Brigham-Burke, M. and Peck, K. (1992) Immobilization chemistries suitable for use in the BIAcore surface plasmon resonance detector. *Analytical Biochemistry*, 205, pp. 132–136.

Oesterhelt, D. (1998) The structure and mechanism of the family of retinal proteins from halophilic archaea. *Current Opinions in Structural Biology*, 8, pp. 489–500.

Oesterhelt, D., Bräuchle, C. and Hampp, N. (1991) Bacteriorhodopsin: A biological material for information processing. *Quarterly Review of Biophysics*, 24, pp. 425–478.

Oesterhelt, D. and Hess, B. (1973) Reversible photolysis of the purple complex in the purple membrane of *Halobacterium halobium*. *European Journal of Biochemistry*, 37, pp. 316–326.

Oesterhelt, D. and Stoeckenius, W. (1973) Functions of a new photoreceptor membrane. *Proceedings of the National Academy of Sciences, USA*, 70, pp. 2853–2857.

Oesterhelt, D. and Stoeckenius, W. (1974) Isolation of the cell membrane of *Halobacterium halobium* and its fractionation into red and purple membrane. *Methods in Enzymology*, 31, pp. 667–678.

Ogawa, K. (1990) *Color image sensor obtained from visual photosensitive materials derived from biological substances.* US Patent 4896049.

Okumura, H., Murakami, M. and Kouyama, T. (2005) Crystal structures of acid blue and alkaline purple forms of bacteriorhodopsin. *Journal of Molecular Biology*, 351, pp. 481–495.

Otto, H., Marti, T., Holz, M., Lindau, M., Khorana, H. G. and Heyn, M. P. (1989) Aspartic acid-96 is the internal proton donor in the repronation of the Schiff base of bacteriorhodopsin. *Proceedings of the National Academy of Sciences, USA*, 86, pp. 9228–9232.

Pandey, P. C., Upadhyay, B. C., Pandey, C. M. D. and Pathak, H. C. (1999) Electrochemical studies on D96N bacteriorhodopsin and its application in the development of photosensors. *Sensors and Actuators B,* B56, pp. 112–120.

Peck, R. F., Dassarma, S. and Krebs, M. P. (2000) Homologous gene knockout in the archaeon *Halobacterium salinarum* with *ura3* as a counterselectable marker. *Molecular Microbiology*, 35, pp. 667–676.

Piknova, B., Perochon, E. and Tocanne, J. F. (1993) Hydrophobic mismatch and long-range protein/lipid interactions in bacteriorhodopsin/phosphatidylcholine vesicles. *European Journal of Biochemistry*, 218, pp. 385–396.

Popp, A., Wolperdinger, M., Hampp, N., Bräuchle, C. and Oesterhelt, D. (1993) Photochemical conversion of the O-intermediate to 9-*cis*-retinal-containing products in bacteriorhodopsin films. *Biophysical Journal*, 65, pp. 1449–1459.

Provencio, I., Jiang, G., De Grip, W. J., Hayes, W. P. and Rollag, M. D. (1998) Melanopsin: An opsin in melanophores, brain, and eye. *Proceedings of the National Academy of Sciences, USA*, 95, pp. 340–345.

Rammelsberg, R., Huhn, G., Lubben, M. and Gerwert, K. (1998) Bacteriorhodopsin's intramolecular proton-release pathway consists of a hydrogen-bonded network. *Biochemistry*, 37, pp. 5001–5009.

Rangarajan, R., Galan, J. F., Whited, G. and Birge, R. R. (2007) Mechanism of spectral tuning in green-absorbing proteorhodopsin. *Biochemistry*, 46, pp. 12679–12686.

Ren, L., Martin, C. H., Wise, K. J., Gillespie, N. B., Luecke, H., Lanyi, J. K., Spudich, J. L. and Birge, R. R. (2001) Molecular mechanism of spectral tuning in sensory rhodopsin II. *Biochemistry*, 40, pp. 13906–13914.

Renner, C., Kessler, B. and Oesterhelt, D. (2005) Lipid composition of integral purple membrane by ^{1}H and ^{31}P NMR. *Journal of Lipid Research*, 46, pp. 1755–1764.

Rhinow, D. and Hammp, N. (2008) Sugar-induced blue membrane: release of divalent cations during phase transition of purple membranes observed in sugar-derived glasses. *The Journal of Physical Chemistry B*, 112, pp. 4613–4619.

Richter, H., Needleman, R. and Lanyi, J. (1996a) Perturbed interaction between residues 85 and 204 in Tyr185Phe and Asp85Glu bacteriorhodopsins. *Biophysical Journal*, 71, pp. 3392–3398.

Richter, H.-T., Brown, L. S., Needleman, R. and Lanyi, J. K. (1996b) A linkage of the pKa's of asp-85 and glu-204 forms part of the reprotonation switch of bacteriorhodopsin. *Biochemistry*, 35, pp. 4054–4062.

Richter, H.-T., Needleman, R., Kandori, H., Maeda, A. and Lanyi, J. K. (1996c) Relationship of retinal configuration and internal proton transfer at the end of the bacteriorhodopsin photocycle. *Biochemistry*, 35, pp. 15461–15466.

Rouhani, S., Cartailler, J., Facciotti, M., Walian, P., Needleman, R., Lanyi, J., Glaeser, R. and Luecke, H. (2001) Crystal structure of the D85S mutant of bacteriorhodopsin: Model of an O-like photocycle intermediate. *Journal of Molecular Biology*, 313, pp. 615–628.

Rousso, I., Khachatryan, E., Gat, Y., Brodsky, I., Ottolenghi, M., Sheves, M. and Lewis, A. (1997) Microsecond atomic force sensing of protein conformational dynamics: Implications for the primary light-induced events in bacteriorhodopsin. *Proceedings of the National Academy of Sciences, USA*, 94, pp. 7937–7941.

Royant, A., Edman, K., Ursby, T., Pebay-Peyroula, E., Landau, E. M. and Neutze, R. (2000) Helix deformation is coupled to vectorial proton transport in the photocycle of bacteriorhodopsin. *Nature*, 406, pp. 645–648.

Rusch, D. B., Halpern, A. L., Sutton, G., Heidelberg, K. B., Williamson, S., Yooseph, S., Wu, D., Eisen, J. A., Hoffman, J. M., Remington, K., Beeson, K., Tran, B., Smith, H., Baden-Tillson, H., Stewart, C., Thorpe, J., Freeman, J., Andrews-Pfannkoch,

C., Venter, J. E., Li, K., Kravitz, S., Heidelberg, J. F., Utterback, T., Rogers, Y.-H., Falcon, L. I., Souza, V., Bonilla-Rosso, G., Eguiarte, L. E., Karl, D. M., Sathyendranath, S., Platt, T., Bermingham, E., Gallardo, V., Tamayo-Castillo, G., Ferrari, M. R., Strausberg, R. L., Nealson, K., Friedman, R., Frazier, M. and Venter, J. C. (2007) The Sorcerer II global ocean sampling expedition: Northwest Atlantic through Eastern Tropical Pacific. *PLoS Biology*, 5, pp. 398–431.

Sabehi, G., Béjà, O., Suzuki, M., Preston, C. and Delong, E. F. (2004) Different SAR86 subgroups harbour divergent proteorhodopsins. *Environmental Microbiology*, 6, pp. 903–910.

Sakamoto, M., Wada, A., Akai, A., Ito, M., Goshima, T. and Takahashi, T. (1998) Evidence for the archaebacterial-type conformation about the bond between the β-ionone ring and the polyene chain of the chromophore retinal in chlamyrhodopsin. *FEBS Letters*, 434, pp. 335–338.

Sanz, C., Marquez, M., Peralvarez, A., Elouatik, S., Sepulcre, F., Querol, E., Lazarova, T. and Padros, E. (2001) Contribution of extracellular Glu residues to the structure and function of bacteriorhodopsin. Presence of specific cation-binding sites. *The Journal of Biological Chemistry*, 276, pp. 40788–40794.

Sasaki, J. and Spudich, J. L. (1999) Proton circulation during the photocycle of sensory rhodopsin II. *Biophysical Journal*, 77, pp. 2145–2152.

Sass, H. J., Buldt, G., Gessenich, R., Hehn, D., Neff, D., Schlesinger, R., Berendzen, J. and Ormos, P. (2000) Structural alterations for proton translocation in the M state of wild-type bacteriorhodopsin. *Nature*, 406, pp. 649–653.

Scherrer, P., Mathew, M. K., Sperling, W. and Stoeckenius, W. (1989) Retinal isomer ratio in dark-adapted purple membrane and bacteriorhodopsin monomers. *Biochemistry*, 28, pp. 829–834.

Schiffmiller, R., Callender, R. H., Waddell, W. H., Govindjee, R., Ebrey, T. G., Kakitani, H., Honig, B. and Nakanishi, K. (1985) Resonance raman studies of bacteriorhodopsin analogues. *Photochemistry and Photobiology*, 41, pp. 563–567.

Schmidt, E. J., Stuart, J. A., Singh, D. and Birge, R. R. (1998) Bacteriorhodopsin-based volumetric optical memory. *International NonVolatile Memory Technology Conference.*

Schobert, B., Brown, L. S. and Lanyi, J. K. (2003) Crystallographic structures of the M and N intermediates of bacteriorhodopsin: assembly of a hydrogen-bonded chain of water molecules between Asp-96 and the retinal Schiff base. *Journal of Molecular Biology*, 330, pp. 553–570.

Schobert, B., Cupp-Vickery, J., Hornak, V., Smith, S. O. and Lanyi, J. K. (2002) Crystallographic structure of the K intermediate of bacteriorhodopsin: conservation of free energy after photoisomerization of the retinal. *Journal of Molecular Biology*, 321, pp. 715–726.

Schranz, M., Noll, F. and Hampp, N. (2007) Oriented purple membrane monolayers covalently attached to gold by multiple thiole linkages analyzed by single molecule force spectroscopy. *Langmuir*, 23, pp. 11134–11138.

Seki, A., Kubo, I., Sasabe, H. and Tomioka, H. (1994) A new anion-sensitive biosensor using an ion-sensitive field effect transistor and a light-driven chloride pump, halorhodopsin. *Applied Biochemistry and Biotechnology*, 48, pp. 205–211.

Sergot, L., Oliver, N., Degenaar, P. and Grossman, N. (2010) *Photo-ionic pacemakers*. GB Patent GB2009/002166.

Sharma, A., Walsh, D., Bapteste, E., Rodriguez-Valera, F., Ford Doolittle, W. and Papke, R. T. (2007) Evolution of rhodopsin ion pumps in haloarchaea. *BMC Evolutionary Biology*, 7, p. 79.

Shaw, A. W., McLean, M. A. and Sligar, S. G. (2004) Phospholipid phase transitions in homogenous nanometer scale bilayer discs. *FEBS Letters*, 556, pp. 260–264.

Shaw, A. W., Pureza, V. S., Sligar, S. G. and Morrissey, J. H. (2007) A local phospholipid environment modulates the activation of blood clotting. *The Journal of Biological Chemistry*, 282, pp. 6556–6563.

Shen, Y., Safinya, C. R., Liang, K. S., Ruppert, A. F. and Rothschild, K. J. (1993) Stabilization of the membrane protein bacteriorhodopsin to 140°C in two-dimensional films. *Nature*, 366, pp. 48–50.

Shibata, M., Tanimoto, T. and Kandori, H. (2003) Water molecules in the Schiff base region of bacteriorhodopsin. *The Journal of the American Chemical Society*, 125, pp. 13312–13313.

Shichida, Y. (1986) Primary intermediates of photobleaching of rhodopsin. *Photobiochemistry and Photobiophysics*, 13, pp. 287–307.

Shichida, Y. and Imai, H. (1998) Visual pigment: G-protein-coupled receptor for light signals. *Cellular and Molecular Life Sciences*, 54, pp. 1299–1315.

Shim, S., Dasgupta, J. and Mathies, R. A. (2008) Femtosecond time-resolved stimulated raman reveals the birth of bacteriorhodopsin's J and K intermediates. *Journal of the American Chemical Society*, 131, pp. 7592–7597.

Sineshchekov, O. A. and Spudich, J. L. (2004) Light-induced intramolecular charge movements in microbial rhodopsins in intact *E. coli* cells. *Photochemical and Photobiological Sciences*, 3, pp. 548–554.

Singh, A. K., Das, J. and Majumdar, N. (1996) Novel bacteriorhodopsin analogues based on azo chromophores. *Journal of the American Chemical Society*, 118, pp. 6185–6191.

Singh, A. K. and Hota, P. K. (2007) Development of bacteriorhodopsin analogues and studies of charge separated excited states in the photoprocesses of linear polyenes. *Photochemistry and Photobiology*, 83, pp. 50–62.

Singh, A. K. and Manjula, D. (2003) Bacteriorhodopsin analogs from diphenylpolyene chromophores. *Photochemistry and Photobiology*, 78, pp. 503–510.

Smith, S. O. and Mathies, R. A. (1985) Resonance raman spectra of the acidified and deionized forms of bacteriorhodopsin. *Biophysical Journal*, 47, pp. 251–254.

Smith, S. O., Pardoen, J. A., Mulder, P., Curry, B., Lugtenburg, J. and Mathies, R. A. (1983) Chromophore structure in bacteriorhodopsin's O640 photointermediate. *Biochemistry*, 22, pp. 6141–6148.

Spudich, E. N., Takahashi, T. and Spudich, J. L. (1989) Sensory rhodopsins I and II modulate a methylation/demethylation system in *Halobacterium halobium* phototaxis. *Proceedings of the National Academy of Sciences, USA*, 86, pp. 7746–7750.

Spudich, E. N., Zhang, W., Alam, M. and Spudich, J. L. (1997) Constitutive signaling by the phototaxis receptor sensory rhodopsin II from disruption of its protonated Schiff base-Asp-73 interhelical salt bridge. *Proceedings of the National Academy of Sciences, USA*, 94, pp. 4960–4965.

Spudich, J. L. (2000) A chloride pump at atomic resolution. *Science*, 288, pp. 1358–1359.

Spudich, J. L., McCain, D. A., Nakanishi, K., Okabe, M., Shimizu, N., Rodman, H., Honig, B. and Bogomolni, R. A. (1986) Chromophore/protein interaction in bacterial sensory rhodopsin and bacteriorhodopsin. *Biophysical Journal*, 49, pp. 479–483.

Spudich, J. L., Yang, C.-S., Jung, K.-H. and Spudich, E. N. (2000) Retinylidene proteins: Structures and Functions from Archaea to Humans. *Annual Review of Cell and Developmental Biology*, 16, pp. 365–392.

Stuart, J. A. and Birge, R. R. (1996) 'Characterization of the primary photochemical events in bacteriorhodopsin and rhodopsin.' In Lee, A. G. (Ed.) *Biomembranes* 2A, pp. 33–139. London, JAI Press.

Stuart, J. A., Birge, R. R., Krebs, M. P., Xi, B., Tetley, W., Marcy, D. L., Koscielecki, J. F. and Hillebrecht, J. R. (2007) Protein-based optical memories. *Nano and Molecular Electronics Handbook*. London, JAI Press, 16, pp. 16.1–16.23.

Stuart, J. A., Marcy, D. L. and Birge, R. R. (2000) 'Bacteriorhodopsin-based three-dimensional optical memory.' In Dér, A. and Keszthelyi, L. (Eds.) *Bioelectronic Applications of Photochromic Pigments*, pp 30–43. Szeged, Hungary, IOS Press.

Stuart, J. A., Marcy, D. L., Wise, K. J. and Birge, R. R. (2002) Volumetric optical memory based on bacteriorhodopsin. *Synthetic Metals*, 127, pp. 3–15.

Stuart, J. A., Marcy, D. L., Wise, K. J. and Birge, R. R. (2003) 'Biomolecular Electronic Device Applications of Bacteriorhodopsin.' in Barsanti, L. *et al.* (Eds) *Molecular Electronics: Bio-Sensorys and Bio-Computers.* Netherlands, Kluwer Academic Publishers. pp. 265–299.

Stuart, J. A., Tallent, J. R., Tan, E. H. L. and Birge, R. R. (1996) Protein-based volumetric memory. *Proceedings of the IEEE Nonvolatile Memory Technologies*, 6, pp. 45–51.

Stuart, J. A (2008) *Binary optical compound and method of manufacture*. US Patent application 20080035897.

Sugihara, T., Blout, E. R. and Wallace, B. A. (1982) Hydrophobic oligopeptides in solution and in phospholipid vesicles: Synthetic fragments of bacteriorhodopsin. *Biochemistry*, 21, pp. 3444–3452.

Szundi, I. and Stoeckenius, W. (1987) Effect of lipid surface charges on the purple to blue transition of bacteriorhodopsin. A theoretical model of purple membrane surface. *Proceedings of the National Academy of Sciences, USA*, 84, pp. 3681–3684.

Szundi, I. and Stoeckenius, W. (1989) Surface pH controls purple-to-blue transition of bacteriorhodopsin: A theoretical model of purple membrane surface. *Biophysical Journal*, 56, pp. 369–383.

Takahashi, T., Yan, B., Mazur, P., Derguini, F., Nakanishi, K. and Spudich, J. L. (1990) Color regulation in the archaebacterial phototaxis receptor phoborhodopsin (sensory rhodopsin II). *Biochemistry*, 29, pp. 8467–8474.

Takeda, K., Matsui, Y., Kamiya, N., Adachi, S., Okumura, H. and Kouyama, T. (2004) Crystal structure of the M intermediate of bacteriorhodopsin: Allosteric structural changes mediated by sliding movement of a transmembrane helix. *Journal of Molecular Biology*, 341, pp. 1023–1037.

Takei, H., Lewis, A., Chen, Z. and Nebenzahl, I. (1992) Implementing receptive fields with excitatory and inhibitory optoelectrical responses of bacteriorhodopsin films. *Applied Optics*, 30, pp. 500–509.

Tallent, J., Song, Q. W., Li, Z., Stuart, J. and Birge, R. R. (1996) Effective photochromic non-linearity of dried blue-membrane bacteriorhodopsin films. *Optics Letters*, 21, pp. 1339–1341.

Tallent, J. R., Stuart, J. A., Song, Q. W., Schmidt, E. J., Martin, C. H. and Birge, R. R. (1998) Photochemistry in dried polymer films incorporating the deionized blue membrane form of bacteriorhodopsin. *Biophysical Journal*, 75, pp. 1619–1634.

Tamerler, C. and Sarikaya, M. (2009) Molecular biomimetics: Nanotechnology and bionanotechnology using genetically engineered peptides. *Philosophical Transactions of the Royal Society A*367, pp. 1705–1726.

Tamogami, J., Kikukawa, T., Miyauchi, S., Muneyuki, E. and Kamo, N. (2009) A tin oxide transparent electrode provides the means for rapid time-resolved pH measurements: Application to photoinduced proton transfer of bacteriorhodopsin and proteorhodopsin. *Photochemistry and Photobiology*, 85, pp. 578–589.

Taneva, S., Caaveiro, J. M., Petkanchin, I. B. and Goñi, F. M. (1995) Electrokinetic charge of the anesthetic-induced bR480 and bR380 spectral forms of bacteriorhodopsin. *Biochemica et Biophysica Acta,* p. 1236.

Terakita, A. (2005) The opsins. *Genome Biology*, 6, pp. 213–222.

Tian, X., Iriarte-Diaz, J., Middleton, K., Galvao, R., Israeli, E., Roemer, A., Sullivan, A., Song, A., Swartz, S. and Breuer, K. (2006) Direct measurements of the kinematics and dynamics of bat flight. *Bioinspiration and Biomimetics*, 1, pp. S10–18.

Tittor, J., Haupts, U., Haupts, C., Oesterhelt, D., Becker, A. and Bamberg, E. (1997) Chloride and proton transport in bacteriorhodopsin mutant D85T: different modes of ion translocation in a retinal protein. *Journal of Molecular Biology*, 271, pp. 405–416.

Tokaji, Z., Dér, A. and Keszthelyi, L. (1997) N-like intermediate in the photocycle of the acid purple form of bacteriorhodopsin. *FEBS Letters*, 405, pp. 125–127.

Tønnesen, J., Sørensen, A. T., Deisseroth, K., Lundberg, C. and Kokaia, M. (2009) Optogenetic control of epileptiform activity. *Proceedings of the National Academy of Sciences, USA*, 106, pp. 12162–12167.

Torres, J., Sepulcre, F. and Padros, E. (1995) Conformational changes in bacteriorhodopsin associated with protein-protein interactions: A functional alpha I-alpha II helix switch? *Biochemistry*, 34, pp. 16320–16326.

Trissl, H.-W. (1987) *Optoelectronics Magazine*, 3, pp. 105–115.

Trissl, H. W., Gärtner, W. and Leible, W. (1989) Reversed picosecond charge displacement from the photoproduct K of bacteriorhodopsin demonstrated photoelectrically. *Chemical Physics Letters*, 158, pp. 515–518.

Tukiainen, T., Lensu, L. and Parkkinen, J. (2007) 'Temporal characteristics of artificial retina based on bacteriorhodopsin and its variants.' In Mele, F., Ramella, G., Santillo, S. and Ventriglia, F. (Eds.) *Advances in Brain, Vision, and Artificial Intelligence.* Heidelberg, Springer.

Tuzi, S., Yamaguchi, S., Naito, A., Needleman, R., Lanyi, J. K. and Saito, H. (1996) Conformation and dynamics of [3-^{13}C]Ala-labeled bacteriorhodopsin and bacterioopsin, induced by interaction with retinal and its analogs, as studied by ^{13}C nuclear magnetic resonance. *Biochemistry*, 35, pp. 7520–7527.

Van Der Steen, R., Biesheuvel, P. L., Mathies, R. A. and Lugtenburg, J. (1986) Retinal analogues with locked 6–7 conformations show that bacteriorhodopsin requires the 6-s-*trans* conformation of the chromophore. *Journal of the American Chemical Society*, 108, pp. 6410–6411.

Váró, G. (1981) Dried oriented purple membrane samples. *Acta Biologica Academiae Scientiarum Hungaricae*, 32, pp. 301–310.

Váró, G., Brown, L. S., Lakatos, M. and Lanyi, J. K. (2003) Characterization of the photochemical reaction cycle of proteorhodopsin. *Biophysical Journal*, 84, pp. 1202–1207.

Váró, G., Brown, L. S., Needleman, R. and Lanyi, J. K. (1999) Binding of calcium ions to bacteriorhodopsin. *Biophysical Journal*, 76, pp. 3219–3226.

Váró, G. and Keszthelyi, L. (1983) Photoelectric signals from dried oriented purple membranes of *Halobacterium halobium*. *Biophysical Journal*, 43, pp. 47–51.

Váró, G. and Lanyi, J. K. (1989) Photoreactions of bacteriorhodopsin at acid pH. *Biophysical Journal*, 56, pp. 1143–1151.

Váró, G. and Lanyi, J. K. (1991) Kinetic and spectroscopic evidence for an irreversible step between deprotonation and reprotonation of the Schiff base in the bacteriorhodopsin photocycle. *Biochemistry*, 30, pp. 5008–5015.

Váró, G. and Lanyi, J. K. (1990) Protonation and deprotonation of the M, N, and O intermediates during the bacteriorhodopsin photocycle. *Biochemistry*, 29, pp. 6858–6865.

Vinchurkar, M. S., Bricarello, D. A., Lagerstedt, J. O., Buban, J. P., Stahlberg, H., Oda, M. N., Voss, J. C. and Parikh, A. N. (2008) Bridging across length scales: Multi-scale ordering of supported lipid bilayers via lipoprotein self-assembly and surface patterning. *Journal of the American Chemical Society*, 130, pp. 11164–11169.

Vsevolodov, N. N. (1998) *Biomolecular Electronics. An Introduction via Photosensitive Proteins.* Boston, Birkhauser.

Wang, J. and El-Sayed, M. (1999) Temperature jump-induced secondary structural change of the membrane protein bacteriorhodopsin in the premelting temperature region: a nanosecond time-resolved Fourier transform infrared study. *Biophysical Journal*, 76, pp. 2777–2783.

Wang, J. and El-Sayed, M. A. (2000) The effect of protein conformation change from alpha(II) to alpha(I) on the bacteriorhodopsin photocycle. *Biophysical Journal*, 78, pp. 2031–2036.

Wang, W. W., Sineshchekov, O. A., Spudich, E. N. and Spudich, J. L. (2003) Spectroscopic and photochemical characterization of a deep ocean proteorhodopsin. *The Journal of Biological Chemistry*, 278, pp. 33985–33991.

Weetall, H. H., Druzhko, A., De Lera, A. R., Alvarez, R. and Robertson, B. (2000) Measurement of proton release and uptake by analogs of bacteriorhodopsin. *Bioelectrochemistry*, 51, pp. 27–33.

Weetall, H. H., Druzhko, A., Samuelson, L. A., De Lera, A. R. and Alvarez, R. (1997) Optical and electrical properties of bacteriorhodopsin Langmuir-Blodgett films: II. D96N mutant and its 4-keto and 9-demethyl retinal analogs. *Bioelectrochemistry and Bioenergetics*, 44, pp. 37–43.

Wise, K. J. and Birge, R. R. (2004) 'Biomolecular photonics based on bacteriorhodopsin.' *CRC Handbook of Organic Photochemistry and Photobiology, 2nd Edition*, pp. 135(1)–135(15).

Wise, K. J., Gillespie, N. B., Stuart, J. A., Krebs, M. P. and Birge, R. R. (2002) Optimization of bacteriorhodopsin for bioelectronic devices. *TRENDS in Biotechnology*, 20, pp. 387–394.

Xi, B., Tetley, W. C., Marcy, D. L., Zhong, C., Whited, G., Birge, R. R. and Stuart, J. A. (2008) Evaluation of blue and green absorbing proteorhodopsins as holographic materials. *The Journal of Physical Chemistry B*, 112, pp. 2524–2532.

Xiao, Y., Partha, R., Krebs, R. and Braiman, M. (2005) Time-resolved FTIR spectroscopy of the photointermediates involved in fast transient H^+ release by proteorhodopsin. *The Journal of Physical Chemistry B*, 109, pp. 634–641.

Xu, J., Stickrath, A. B., Bhattacharya, P., Nees, J., Váró, G., Hillebrecht, J. R., Ren, L. and Birge, R. R. (2003) Direct measurement of the photoelectric response time of bacteriorhodopsin via electro-optic sampling. *Biophysical Journal*, 85, pp. 1128–1134.

Yamamoto, M., Hayakawa, N., Murakami, M. and Kouyama, T. (2009) Crystal structures of different substates of bacteriorhodopsin's M intermediate at various pH levels. *Journal of Molecular Biology*, 393, pp. 559–573.

Yan, B., Spudich, J. L., Mazur, P., Vunnam, S., Derguini, F. and Nakanishi, K. (1995) Spectral tuning in bacteriorhodopsin in the absence of counterion and coplanarization effects. *The Journal of Biological Chemistry*, 270, pp. 29668–29670.

Yang, J. and Wang, G. (1998) Image edge detecting by using the bacteriorhodopsin-based artificial ganglion cell receptive field. *Thin Solid Films*, 324, pp. 281–284.

Yao, B., Dalun, X. and Hou, X. (1997) Oriented bacteriorhodopsin film biomolecular devices and their photoelectric dynamics. *Proceedings of SPIE*, 2869, pp. 752–758.

Yoshizawa, T. (1994) Molecular basis for color vision. *Biophysical Chemistry*, 50, pp. 17–24.

Zhang, F., Aravanis, A. M., Adamantidis, A., De Lecea, L. and Deisseroth, K. (2007) Circuit-breakers: Optical technologies for probing neural signals and systems. *Nature Reviews Neuroscience*, 8, pp. 577–581.

Zhang, F., Wang, L.-P., Boyden, E. S. and Deisseroth, K. (2006) Channelrhodopsin-2 and optical control of excitable cells. *Nature Methods*, 3, pp. 785–792.

Zhang, F. S. and Deisseroth, K. P. (2009) *Systems, methods and compositions for optical stimulation of target cells*. US Patent application US2009/039949.

Zhang, L., Zeng, T., Cooper, K. and Claus, R. O. (2003) High-performance photovoltaic behavior of oriented purple membrane polymer composite films. *Biophysical Journal*, 84, pp. 2502–2507.

Zhang, Y., Ivanova, E., Bi, A. and Pan, Z.-H. (2009) Ectopic expression of multiple microbial rhodopsins restores ON and OFF light responses in retinas with photoreceptor degeneration. *Journal of Neuroscience*, 29, pp. 9186–9196.

Zimányi, L., Keszthelyi, L. and Lanyi, J. K. (1989) Transient spectroscopy of bacterial rhodopsin with an optical multichannel analyzer. 1. Comparison of the photocycles of bacteriorhodopsin and halorhodopsin. *Biochemistry*, 28, pp. 5165–5172.

Zimányi, L., Váró, G., Chang, B., Needleman, R. and Lanyi, J. (1992) Pathways of proton release in the bacteriorhodopsin photocycle. *Biochemistry*, 31, pp. 8535–8543.

Zingoni, J., OR, Y. S., Crouch, R. K., Chang, C.-H., Govindjee, R. and Ebrey, T. G. (1986) Effect of variation of retinal polyene side-chain length on formation and function of bacteriorhodopsin analogue pigments. *Biochemistry*, 25, pp. 2022–2027.

3

Polarisation effects in natural photonic structures and their applications

S. BERTHIER, Université Paris-Diderot, France and Facultés Universitaires Notre Dame de la Paix, Belgium

Abstract: Electromagnetic waves, unlike those which are generated mechanically, are transverse and introduce polarisation into optic phenomena. Many organisms have photosensitive sensors which enable them to respond to this. Most insects are sensitive to the polarisation of light and are capable of distinguishing two different states of polarisation. This is not the case in primates in general and in hominids in particular. This natural handicap has proved to be a source of bio-inspired innovation, as the use of simple instruments has made it possible to reveal what was previously hidden. Polarisation is used by many animals and may be adapted for use as a means of detecting forgeries. The first part of this chapter deals with fundamental theories on the polarisation of light and with experimental techniques. The photonic structures of insects are also considered. The chapter concludes with a presentation of a transfer bio-generated anti-counterfeiting mechanism used for documents or banknotes.

Key words: polarisation, bio-photonic, polarising structures, bio inspiration, insects.

3.1 Introduction

Vision is an important and dominant sense in human beings but is relatively atrophied when compared with other animal groups. Many of the electromagnetic signals received by humans escape due to the transverse nature of the wave. However, the light environment, even when restricted to the visible spectrum, is very strongly polarised. Many animals and insects in particular, are sensitive to linear polarisation of the electromagnetic wave and have developed means of making this phenomenon an integral part of their inter- and intra-specific communication systems (Gabor and Dezso, 2004).The basic principles of this development constitute an important source of ideas and are presented in this chapter.

3.2 Principles of polarisation

Following sections present outlines of the principles of polarisation. The topics that are covered are: the electromagnetic wave, Stokes vector and Mueller matrix representation, linear polarisation and circular polarisation.

3.2.1 The electromagnetic wave

In wave form, a beam of light consists of two fields: an electric field **E** and a magnetic field **H.** These are perpendicular to each other and to the direction of propagation. The spatio-temporal dependency of these fields was determined during the 19th century and summarised by J.C. Maxwell in the equations which bear his name (Jackson, 1975; Lipson *et al.*, 1995). All the examples include the handling of monochromatic plane waves. However, this is not restrictive as any electromagnetic wave may be decomposed into an infinite number of monochromatic fields of pulsation ω, each of which has an infinite number of plane waves with vector **k**. The Maxwell equations for macroscopic fields in an area with a macroscopic bound charge can be written as follows:

$$\begin{cases} \nabla \cdot \mathbf{E} = 0 \\ \nabla \times \mathbf{E} = -\dfrac{\delta \mathbf{B}}{\delta t} \end{cases} \quad [3.1]$$

and

$$\begin{cases} \nabla \cdot \mathbf{B} = 0 \\ \nabla \times \mathbf{H} = \dfrac{\delta \mathbf{D}}{\delta t} \end{cases} \quad [3.2]$$

E and **B** represent the electric field (in units of V/m) and magnetic induction (T) respectively, which are connected to the electric displacement **D** (V) and magnetic field **H** (A/m) by the following linear relations:

$$\begin{cases} \mathbf{D} = \varepsilon_0 \mathbf{E} + P \\ \mathbf{H} = \dfrac{\mathbf{B}}{\mu_0} - M \end{cases} \quad [3.3]$$

where **P** and **M** represent the electric and magnetic polarisation phenomena induced in matter by the wave. ∇ symbolises the vector operator 'nabla' and × the vector product. The rotationals indicate that **E** and **B** are perpendicular to each other. In looking for plane wave solutions, these fields may be represented as in Fig. 3.1.

By omitting details of the calculation which, although long, are simple and classic, and seeking monochromatic equations, these equations lead to the following pairs of solutions:

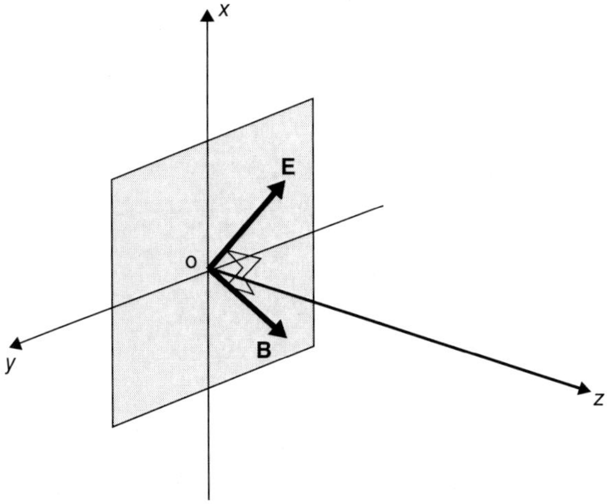

3.1 The electric field and magnetic induction of a plane wave are perpendicular to each other and to the direction of propagation.

$$\begin{cases} \frac{\delta^2 E_y}{\delta z^2} + \varepsilon_0 \mu_0 \omega^2 E_y = 0 \\ \frac{\delta^2 B_x}{\delta z^2} + \varepsilon_0 \mu_0 \omega^2 E_x = 0 \end{cases} \quad [3.4]$$

and

$$\begin{cases} \frac{\delta^2 E_x}{\delta z^2} + \varepsilon_0 \mu_0 \omega^2 E_x = 0 \\ \frac{\delta^2 B_y}{\delta z^2} + \varepsilon_0 \mu_0 \omega^2 E_y = 0 \end{cases} \quad [3.5]$$

where E_x, E_y, B_x and B_y are the components along o_x and o_y of fields **E** and **B** respectively. These equations show some of these components to be intimately linked: B_x and E_y and B_y and E_x form two plane waves totally independent of each other which can be separated spatially, creating polarised light. There is no imposed direction of vibration in natural light, so the wave is not defined as polarised. Fields **E** and **H,** which are always perpendicular to each other, may vibrate in any direction in the plane wave. However, it is possible to impose a direction of vibration to the fields, or to modify a pre-existing direction, making waves interact with various devices. The polariser and polarising films are the most common of such devices.

3.2.2 Stokes vector and Mueller matrix representation

The general polarisation state of a given wave may be determined by a set of parameters involving only measurable quantities as determined from the intensities of different components. These are known as the Stokes parameters (Gel'fand and Shapiro, 1956; Huard, 1994; Schalz *et al.*, 1999). These parameters are usually grouped in a vector quantity with four components and are known as the Stokes vector **S**. Natural light does not generally present in any particular state of polarisation and may be seen as the superposition of all possible states, both linear and circular. The wave may be decomposed on these bases to define the intensities of the different elements.

If I_x and I_y are the intensities of the components of the electric vector along the o_x and o_y directions, I_{+45} and I_{-45} those of the components at ±45° of the same directions, and I_G and I_D the respective intensities of the left and right circularly polarised components, the Stokes parameters will be defined by:

$$\begin{cases} P_0 = I_x + I_y \\ P_1 = I_x - I_y \\ P_2 = I_{+45} - I_{-45} \\ P_3 = I_G - I_D \end{cases} \quad \text{and} \quad \mathbf{S} = \begin{pmatrix} P_0 \\ P_1 \\ P_2 \\ P_3 \end{pmatrix} \qquad [3.6]$$

In any optical device which amends the optical polarisation state of the incident radiation, the emergent Stokes vector **S′** will differ from the incident vector **S**. The 4 × 4 matrix between these two vectors is called the Mueller matrix [**M**] and characterises the effect of the optical device.

The 16 elements of the matrix which link linearly measurable quantities, can be determined experimentally and represented by maps (Kusceri and Rubaric, 1959). Each element of the matrix is obtained by the linear combination of a number of measures of intensity (Fig. 3.2). VH, for example, corresponds to a vertically polarised (V) incident wave and a horizontally (H) detected wave. M_{11} is the total backscattered light intensity when the sample is illuminated with non-polarised light and is used as a standardisation for the other 15 elements. Symmetry in the M_{nm} and M_{mn} elements is obtained simply by reversing the optical incident or emerging polarisation devices and is the signature of an isotropic medium.

3.2.3 Linear polarisation

It should be noted that when a wave, even if un-polarised, falls on a surface, particular directions will appear ipso facto. The plane defined by the incident and emergent beam is termed the 'plane of incidence', regardless of

(OO)	(HO-VO)/2	(PO-MO)/2	(LO-RO)/2
(OH-OV)/2	(HH+VV)/4- (HV+VH)/4	(PH+MV)/4- (PV+MH)/4	(LH+RV)/4- (LV+RH)/4
(OP-OM)/2	(HP+VM)/4- (HM+VP)/4	(PP+MM)/4- (PM+MP)/4	(LP+RM)/4- (LM+RP)/4
(OL-OR)/2	(HL+VR)/4- (HR+VL)/4	(PL+MR)/4- (PR+ML)/4	(LL+RR)/4- (RL+LR)/4

3.2 16 elements of the Mueller matrix. 49 measures are needed to determine them. Each letter corresponds to an intensity of the wave: O: unpolarised; H and V polarised horizontally and vertically, P and M polarised to +45° and −45°; and L and R, circularly polarised left and right respectively.

the direction of the incident field and may be broken down into two vectors, one perpendicular to the plane of incidence, $\mathbf{E}_s$ (from the German *senkrecht*), and one $\mathbf{E}_p$ parallel to that plane (*parallel* in German). Note that other decompositions of the wave are possible, as for example, in transverse electric (TE) and transverse magnetic (TM) adapted to the study of gratings, which will be presented later.

These two fields define two perpendicular polarised waves which, if following the Snell–Descartes laws concerning their directions of propagation, may interact differently with matter. One of the most remarkable properties (often found in butterflies) is the angular dependence of reflected intensity. The s-wave and its electric field which is always in the level plane, regardless of the incidence, depends very little upon the latter, but the second, the p-wave, presents surprising variations in angle. There is a specific angle, known as the Brewster angle, for which the reflectance of the p-wave is equal to 0. If a surface is illuminated under the incidence of Brewster θ_B, the reflected light is strictly s-polarised (Abeles, 1967). A practical consequence of this, which is well known to photographers, is that this incidence makes it possible to remove the reflection of a surface. Under these conditions, the light reflected by certain butterflies may be very strongly polarised. The Brewster angle depends only on the refraction indices of materials which constitute the diopter:

$$\theta_B = tg\ n/n_0 \qquad [3.7]$$

When in contact with air of index $n_0 = 1$, biological materials with an index varying schematically from 1.5 to 1.8 will give Brewster angles between 50 and 60° (see Fig. 3.3).

3.2.4 Circular polarisation

Circular polarisation is not often encountered in nature, although it is generated by the carapace of beetles and merits specific attention. A wave is circularly polarised when the electric vector **E** rotates around the direction of propagation as it moves along this direction (Fig. 3.4).

It should be noted that any linear polarised wave may be seen as the sum of two circularly polarised waves turning in phase in opposite directions: a levorotatory and a dextrorotatory wave. These polarised waves can be created by introducing a phase equal to a quarter of the wavelength between two perpendicular polarised waves, due to a thin layer known as a 'quarter wave layer'. Many insects have developed a particular structure which produces the same effect in reflection as this transmission device. This imposes a stratified medium and it is necessary to consider how a given electromagnetic wave is propagated in such a medium.

Relation of dispersion

To understand how an electromagnetic wave or a photon propagates in a material, the relation of dispersion is calculated (Fox, 2001). This function establishes the relation between the energy of the particle E (or any quantity linked to the energy such as the frequency υ, the angular frequency ω) and its momentum **p** (or any quantity linked to it such as the wave vector **k**).

$$\begin{cases} E = h\nu = \hbar\omega \\ k = \dfrac{2\pi}{\lambda} = \dfrac{\omega}{\nu} \end{cases} \qquad [3.8]$$

Which leads to:

$$\omega(k) = \nu k. \qquad [3.9]$$

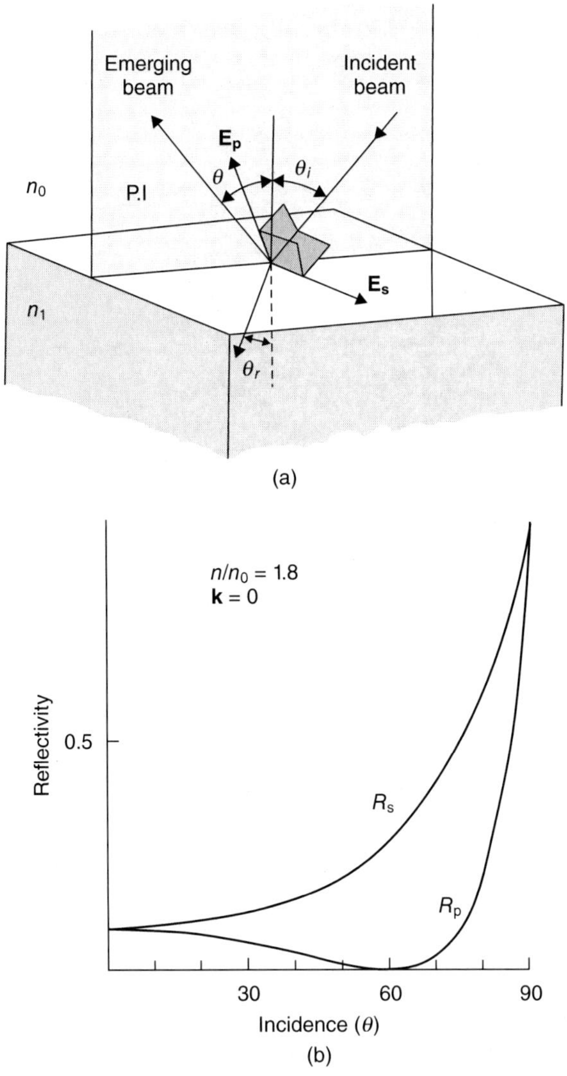

3.3 (a) The electric field of a non-polarised wave falling on a diopter can be decomposed into two perpendicular vectors $\mathbf{E}_s$ and $\mathbf{E}_p$ which are reflected differently depending on the angle of incidence (b).

In a homogeneous media with index n, this relation takes the form of a continuous straight line, indicating that any wave, whatever its frequency, can propagate in the medium with a velocity equal to the slope of the line (Fig. 3.5a). The situation is completely different in a periodically structured medium (Fig. 3.5b). It may be shown that each time the vector of the electromagnetic wave is equal to that of the structure, or to a subdivision

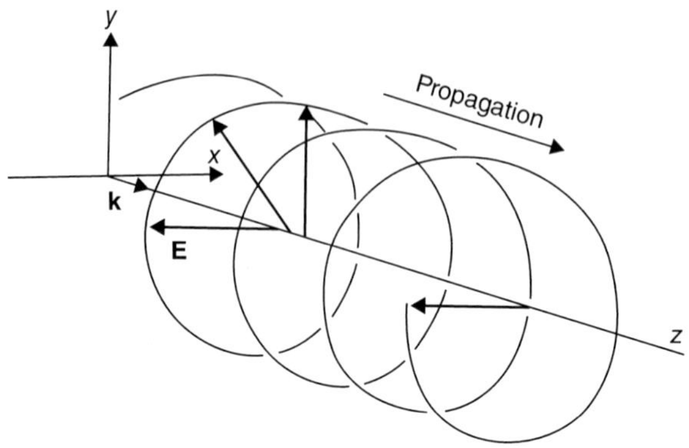

3.4 Schematic representation of a dextrorotatory wave.

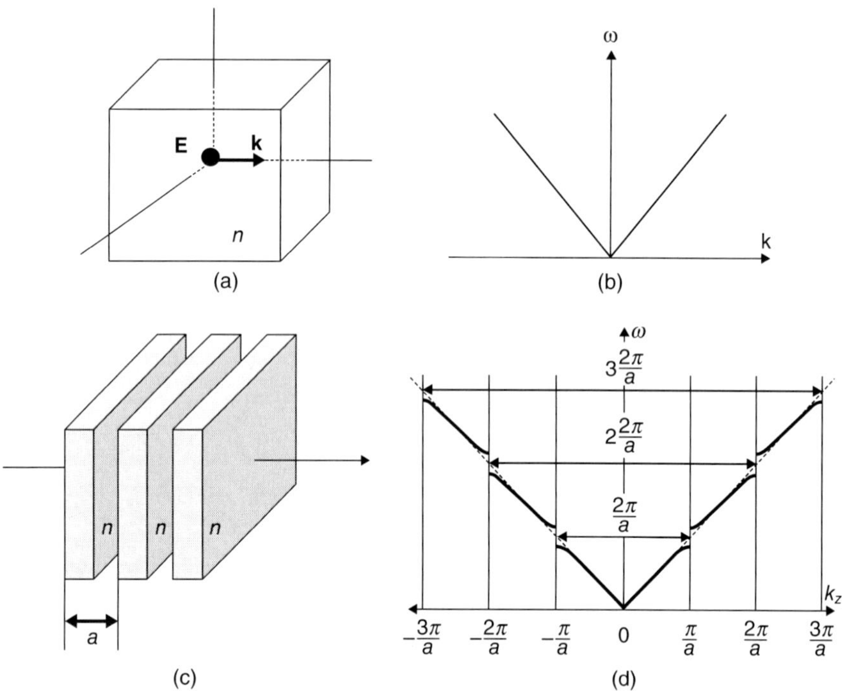

3.5 In a continuous medium with refractive index *n* (a), any electromagnetic wave, whatever its frequency, propagates with a velocity given by the slope of the relation (b). In a stratified medium (c), the straight line is broken when the frequency or the harmonics of the two waves – the electromagnetic and the material ones – match (d). At these frequencies, the electromagnetic wave cannot propagate and is reflected.

of it, the straight line is broken and frequency gaps are opened in the spectrum, indicating that waves of this frequency cannot propagate in the medium. They will therefore be reflected and cause a colour to appear (Ziman, 1979).

3.3 Experimental techniques to study polarisation

In addition to the classical spectroscopic techniques used to determine the reflectivity and transmissive qualities of insect wings and integuments, there are specific methods of polarimetric analysis such as various ellipsometric and polarimetric imaging techniques.

3.3.1 Ellipsometry

Ellipsometry has a long history of use in science and technology. However, it is only recently – with the exception of the precursory works of Michelson (Michelson, 1911) – that it has been used to characterise natural photonic structures. The spectra of these structures are extremely sensitive to surface conditions and are complex and difficult to interpret.

Principle

The principle of measurement is to determine the change of state in the polarisation of a wave reflected by the surface of a sample. This change can be directly related to the optical and geometrical characteristics of a surface, making ellipsometry a powerful tool for the optical analysis of surfaces.

It has been shown that the reflection coefficients R_p and R_s of a linearly polarised wave depends in differing ways on the angle of incidence of light: R_s is a monotonically increasing function, whereas R_p passes through a minimum for a particular angle of incidence (the Brewster angle). However these two waves may not be in synchrony with each other, thus the ratio of their amplitudes may be expressed in the following complex form:

$$\rho = \frac{r_p}{r_s} = tg\Psi e^{i\Delta} \qquad [3.10]$$

These are the angles Ψ and Δ, or more commonly $tg\Psi$ and $\cos\Delta$ which are directly measured.

The most general state of polarisation in a monochromatic wave is elliptical, i.e. the trace of the extremity of the electric vector field **E** in the plane wave is an ellipse. This field may be written in matrix form:

$$\mathbf{E}(t) = \begin{bmatrix} E_x(t) \\ E_y(t) \end{bmatrix} = \mathrm{Re}\left\{ \begin{bmatrix} Xe^{i\Delta} \\ Y \end{bmatrix} e^{-i\omega(t-t_0)} \right\} \qquad [3.11]$$

The state of polarisation is entirely determined by its Jones vector which takes the form:

$$\mathbf{V} = \begin{bmatrix} \sin\Psi e^{i\Delta} \\ \cos\Psi \end{bmatrix} \quad [3.12]$$

and thus by the two quantities Ψ and Δ. In the case of a semi-infinite media, the reflection coefficients R_p and R_s are expressed in terms of the complex refractive indices of the two media and their angles of incidence and refraction. These values may be obtained by inverting the equations and measuring the two angles. The same method may be applied to a thin layer deposited on a substrate and, as considered here, on a multilayer where the calculations are more complex. The multilayer is considered to be a composite medium and the characteristics of the component layers are determined. Analysis of the components can be made using models of actual environments and an effective theory of mediums such as the Maxwell Garnett theory (Cohen *et al.*, 1973; Berthier, 1993).

All types of ellipsometer are based on the principle illustrated in Fig. 3.6. A polariser and a compensator located on one arm of the ellipsometer determine a defined state of polarisation of the incident light. An analyser and a detector located on the other arm detect the change in polarisation produced by reflection on the surface of the sample. More complex procedures of generalised ellipsometry (GE) are used in the study of anisotropic materials, such as the wings of many butterflies and coleopteran (see Fig. 3.7).

3.3.2 Polarimetric imaging

As described in section 3.2.2, changes in the polarisation state of an incident wave induced by a given structure can be represented in a matrix form known as the Mueller matrix. New experimental devices allow the visualisation of the various elements of that matrix.

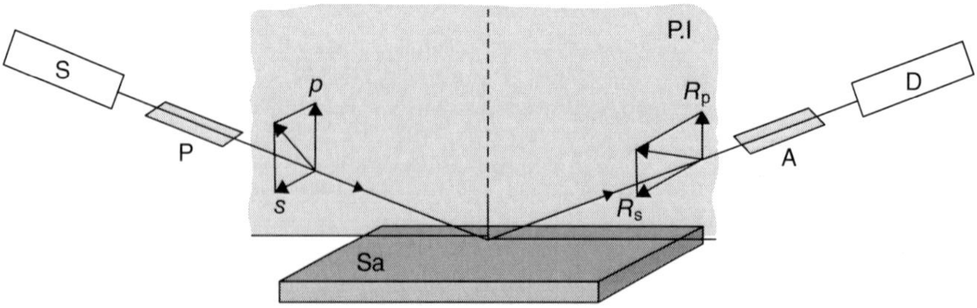

3.6 Schematic representation of an ellipsometer. (Sa: sample, S: source, D: detector, P: polariser, A: analyser)

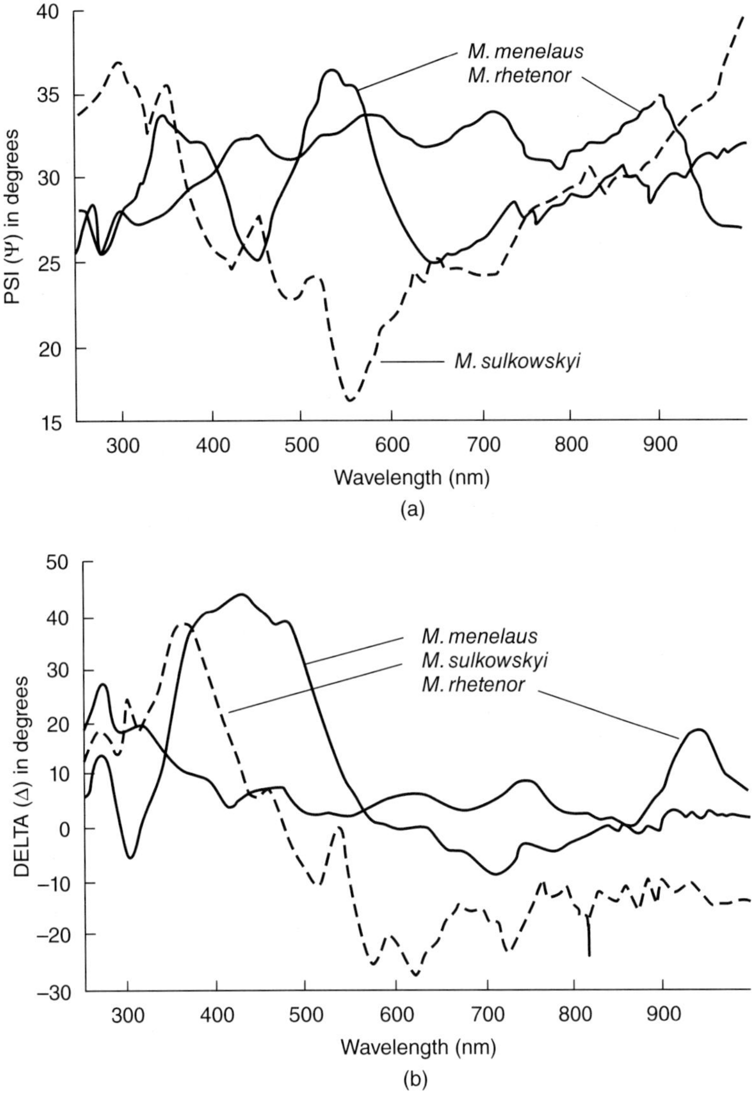

3.7 Variations of the angles Ψ (a) and Δ (b) in the ultraviolet, visible and near infrared for three Morphos belonging to three different sub genus (*M. menelaus*: Grasseia, *M. rhetenor*: Megamede and *M. sulkowskyi*: Cytheritis).

Monochromatic light from different lasers is passed through a rotating linear polariser or a polariser followed by a quarter wave plate (λ/4) which generates a circularly right or left polarised wave. The same device will be crossed by the light which is retro-diffused by the object before reaching a charge coupled device (CCD) camera measuring the intensity (Hielsher

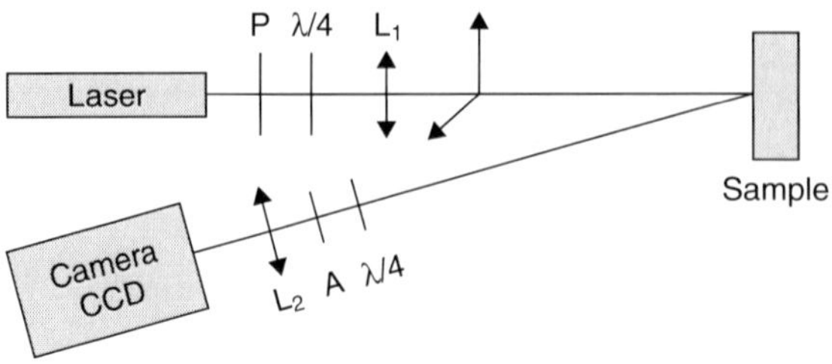

3.8 Schematic diagram of the polarimetric measurement installation.

et al., 1997). Thus, any state of polarisation can be sent to the object and measured after reflection, regardless of the incident state. It can therefore measure not only the reflectance of the structure for a given polarisation, but also any changes in polarisation and depolarisation phenomena. The camera is inclined 15° to the optical axis in order to remove specular reflections.

The 16 elements of the Mueller matrix are obtained by 49 measurements of intensity (Fig. 3.8). The matrix element M11 is the total backscattered light when the incident beam is not polarised. It is used as a reference for all other elements of the matrix.

Two modes of imaging are available: in real space or reciprocal space. It is obvious that each element of the matrix has more or less evident number of interpretations. See Plate II (see colour section between pages 96 and 97).

3.4 Polarisation structures in insects

In the following, details of polarisation structures in insects are given. Both linear and circular polarisation structures are included.

3.4.1 Linear polarisation

Polarisation at the Brewster angle

A one-dimensional structure composed of multi-layered films is considered here. In the natural world, this is found in the carapaces of many insects and in the scales of certain butterflies. Among these structures, which are generally of a plane or cylindrical form, some are constituted by concave, roughly hemispherical, multi-layered basin shapes. These concave structures are composed of a stack of thin films, which alternate solid (chitin embedded in proteins) with air, as for example, in the scales of butterflies of the genus *Papilio* (Vukusic and Samble, 2003), or solid films of various composition as in the elytrons of certain coleopteran of genus *Cicindela* (Berthier, 2007a, 2007b).

Papilio

The covering scales of most of the species of the genus *Papilio* present a unified bulk and surface structure. The upper membrane can be seen to consist of a multi-layered air/chitin film of about 5 to 10 layers having an undulating surface between the ridges which forms a regular set of concave basins (see Fig. 3.9). Depending upon the species, these basins are roughly spherical (*Papilio blumei*) or slightly elongated (*Papilio peranthus*) (see Plate V).

Cicindela

Similar structures can be found in the elytrons of the species *Cicindela* among others (Schultz and Rankin, 1985a, 1985b). They differ from the above by the solid/solid composition of the multi-layer and by the random disposition of basins on the surface (see Plate III).

Two different phenomena occur when light falls on such hemispherical structures at a given angle of incidence, according to the position in the basin. If the light falls on or near the centre, specular reflection and interferential phenomena occur, giving rise to a coloured effect which depends on the thickness **e** and the equivalent refractive index **n** of the layers, according to:

$$k\lambda_m = 2ne \cos \theta_r \qquad [3.13]$$

The extinguish wavelengh is denoted by λ_{min} and k an integer. At close-to-normal incidence, this reflected colour is either polarised or very slightly polarised. If the light falls at a distance from the centre, which causes the normal angles of incidence to the surface to intersect at right angles at this point, the light will emerge after a double reflection in the incident direction. The angle of incidence on the surface is exactly equal to $\pi/4$ which is not far from the polarising angle θ'_B. The wavelength which interferes is shifted towards lower values and the corresponding colour towards blue.

When observing the insect from above, a uniform colour is seen to be produced by the concentric combination of various colours, in a manner similar to the juxtaposed points which can be found in pointillist paintings. The particularity here is that the 'blue' component, at the periphery of the basins, is strongly polarised while the 'red' component at the centre, is not (see Plate VIII).

Depending upon the symmetry of the basin, if two opposite sides – for example right and left – are Transverse Electric (TE) polarised, the two others – up and down – are Transverse Magnetic (TM) polarised, thus no macroscopic polarisation effects can be observed unless the symmetry is broken. This is the case with (*Papilio paris*), whose basins are slightly elongated, in the scale axis direction.

3.9 (a) The scales of the dorsal side of *Papilio ulysses* (SEM). (b) Detail of the surface of a cover scale (SEM). (c) TEM view of a section of a cover scale showing the multi-layered structure of the upper membrane.

If this phenomenon is to be used to produce macroscopic effects for anti-counterfeiting devices, the ratio between S and P must be as great as possible, the ideal situation being infinite multi-layered grooves. This structure exists on a larger scale in a recently studied butterfly (Z. Bàlint, 2005): *Suneve coronata.*

Multi-scaled polarisation: Suneve coronata

Suneve coronata is a neotropical lycaenid butterfly distributed from Mesoamerica to NE South America (Berthier *et al.*, 2007b). In common with most butterflies, the lycaenid have two layers of scales on their wings. The inner ones which are directly in contact with the wing membrane are called 'ground scales' or 'basal scales' while the outer ones are the 'cover scales'. The scales of both sides of the wings have classical Urania-like structures with structural convex cover scales and pigmentary plane ground scales (see Plate IX).

The apexes of the cover scales of a given row cover the base of the neighbouring row perpendicular to its surface. The wing appears to be covered by long lines of alternate green and blue (Plate IX), corresponding respectively to the crests of the convex scales. These are barely polarised while the space between neighbouring rows is strongly polarised. The blue component can be almost extinguished with a linear polariser, thus changing the colour of the butterfly from blue to emerald green (Plate IV). This should provide the optimal situation for maximum polarisation contrast. However, the macroscopic measurements and direct observation do not show this to be the case (Fig. 3.10). This is attributable to the presence of another structure which polarises the reflected light in a direction perpendicular to the previous one. This structure is formed by the ridges of the scales which form a very regular periodic structure (Fig. 3.11).

Again, the mean angle is equal to $\pi/2$ but the axis of the ridges is perpendicular to the direction of the rows, which minimises the macroscopic polarisation ratio. This is a rare case where two structures with very different scales – a cover scale is about 200 μm × 70 μm, and a distance between two ridges which is equal to 2 μm – produce the same polarisation effect but in opposite directions, so that the macroscopic effect almost disappears.

Polarisation by diffraction

The Morphinae have developed a particular structure on their ground scales (and sometimes also on their cover scales) which is known as the 'Christmas tree' structure (Berthier, 2010) (Fig. 3.12). This acts as a complex grating which produces a blue colour by interference and diffracts it later-

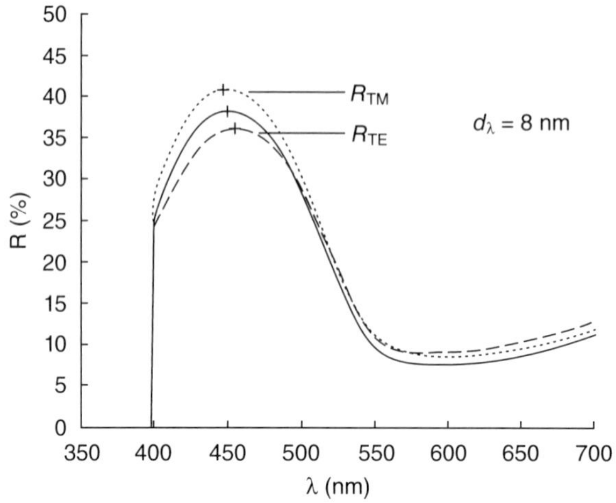

3.10 Hemispherical reflectivity of the wings of *Suneve coronata* for the two states of linear polarisation and for non-polarised light. The maximum reflectivity is slightly shifted by about 8 nm.

ally. This type of structure also produces a unique phenomenon in which the diffraction orders are differently polarised. It was first predicted by simulation (Berthier, 2006) and then confirmed by spectrometric measurement and polarimetric imaging (Plate VI).

Spectral analysis

As human eyes are not sensitive to the polarisation of light, its effects on the wings of Morphos appear primarily colorimetric (Berthier, 2010). The reflection spectra are shifted as the incident wave is polarised as transverse electric or magnetic. This can easily be observed under a light microscope or even on the insect itself. The effect appears to be regular, but varies in intensity according to the species (Fig. 3.13). A butterfly lit by a Transverse Electric polarised wave (TE) is always more blue in colour than the same illuminated by a Transverse Magnetic (TM) wave which will tend towards green (Fig. 3.14).

This observation is consistent with the theoretical predictions, which also reveal further information. It is significant that this diffracting structure slightly polarises the light, especially under quasi-normal incidence (Siewert, 1981, 1982), i.e. the diffraction orders have a quite different polarisation state when the structure is illuminated by natural non-polarised light. The order R1 is predominantly Transverse Magnetic polarised while the opposite order R-1 is predominantly in the Transverse Electric state. The specular order R0, is only minimally polarised, but still of very low intensity. All

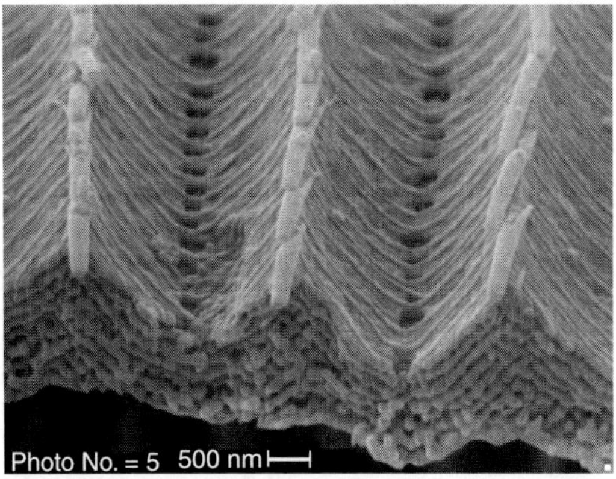

(a)

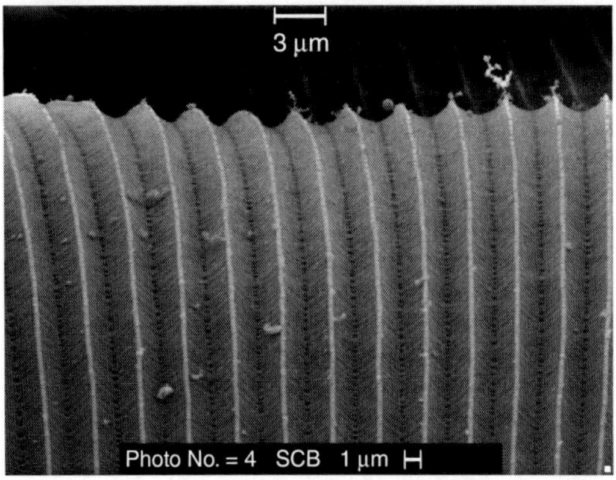

(b)

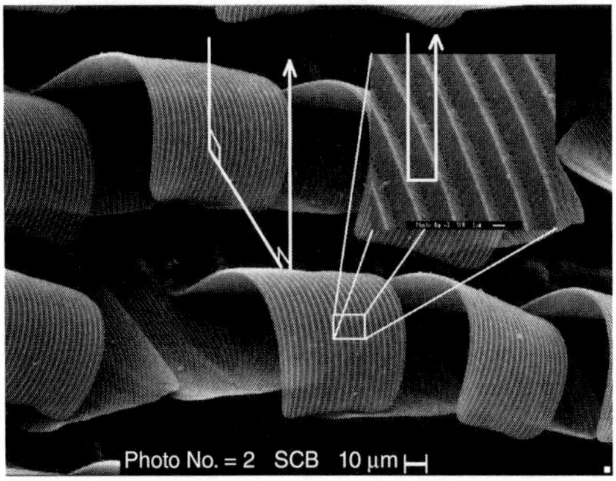

(c)

3.11 (a) SEM view of a cross-section of a cover scale of *Suneve coronata*, showing the multi-layered structure of the membrane and a front view of the scale showing the regular disposition of the right angles.

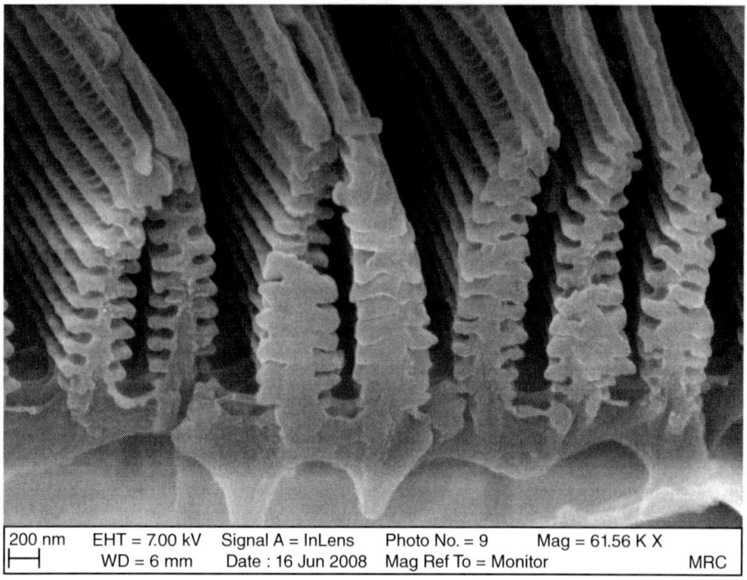

3.12 SEM view of a cut of a ground scale of *Morpho rhetenor cassica* showing the 'Christmas tree' structure of the scales.

the properties of an electromagnetic wave reflected by a surface can be represented by the 'Bidirectional Reflectivity Distribution Function' (BRDF). This function is generally too complex to be completely determined and can only be ascertained for a limited number of parameters (e.g. a given wavelength, or a given direction). The calculations and maps of BRDF present a different picture. Illuminated under a non-polarised light, the diffracted wave is TE polarised on one side and TM polarised on the other. Although two polarisation states can be distinguished in *Morphos*, the advantage of this to the species is not obvious. Although highly attenuated, the effect may be verified experimentally. A division along the axis Φ at an angle of about 45° clearly shows the imbalance of the polarisation states between the two diffraction orders. (Fig. 3.15).

3.4.2 Circular polarisation

The *Cetonia* and *Plusiotis* genus are well known for the extraordinary bright metallic appearance of many species (Moron, 1990). The epicuticle shows a helicoidal structure which generates both interference and circular polarisation. It consists of an anisotropic solid/solid multi-layer structure with a direction rotating from one layer to the other, creating a periodic, though weak, gradient of index in the epicuticle (Bouligan, 1969; Neville 1969, 1975;

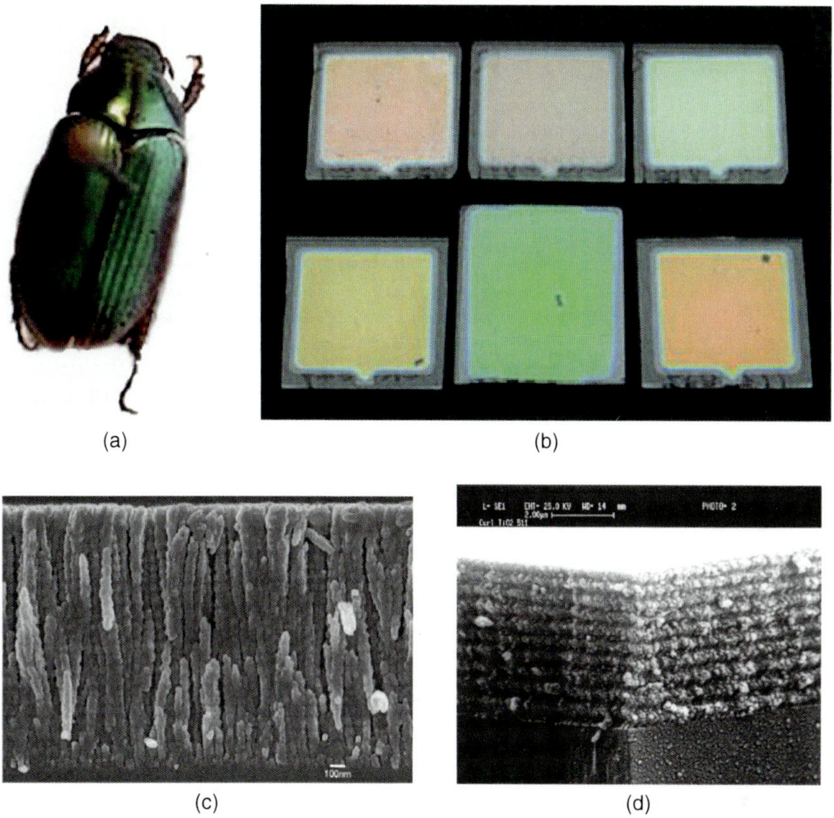

Plate I (a) A Manuka (scarab) beetle with (b) titania mimetic films of slightly different pitches. (c) Scanning electron micrograph of the chiral reflector in the beetle's cuticle. (d) Scanning electron micrograph of the titania mimetic film. Images by L. DeSilva and I. Hodgkinson, reproduced with permission.

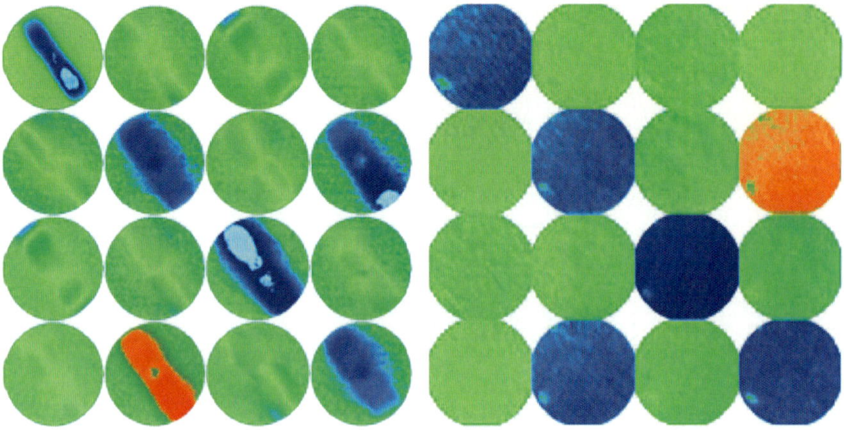

(a) (b)

Plate II (a) Mueller matrix in Fourier space, measured on a scale of *M. rhetenor* (normal incidence, λ = 450 nm). Mueller matrix of the same area, image mode (b).

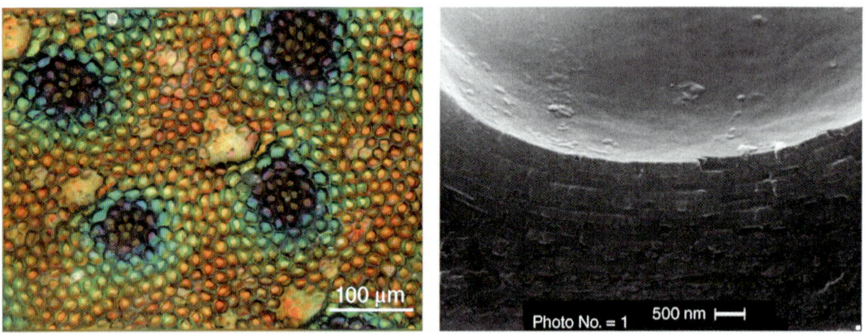

(a) (b)

Plate III (a) Surface of the elytrons of *Cicindela hybrida* showing the coloured basins and (b) an SEM view of a cross-section of a single basin.

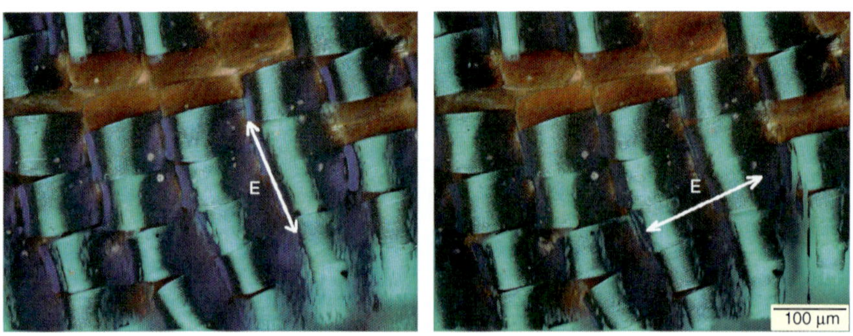

(a) (b)

Plate IV The scales of *Suneve coronata* under TM (a) and TE (b) polarisation light. TE and TM are relative to the direction of the rows. The blue component, which is strongly TM polarised, nearly disappears when the incident light is TE polarised.

(a)

(b)

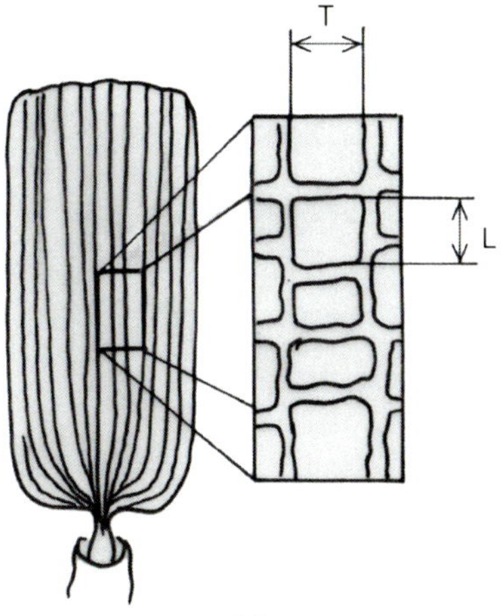

(c)

Plate V Cover scales of two papilio. (a) *Papilio blumei*: the basins are roughly spherical. (b) They are slightly elongated for *Papilio peranthus*. (c) Schematic representation of the scales.

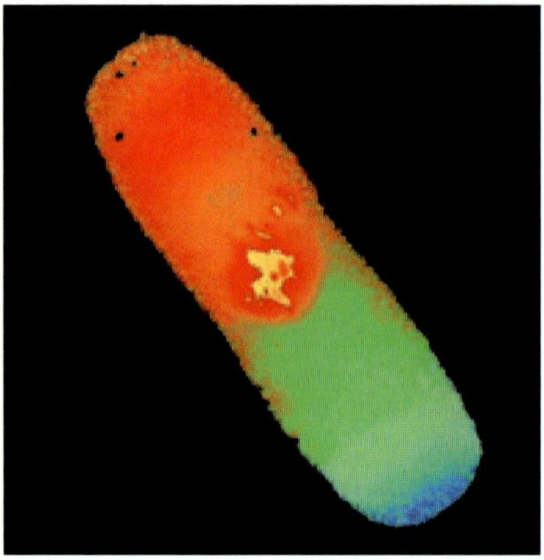

Plate VI Element Mxx of the Mueller matrix of *Morpho rhetenor* showing a great contrast in the polarisation of the diffraction orders.

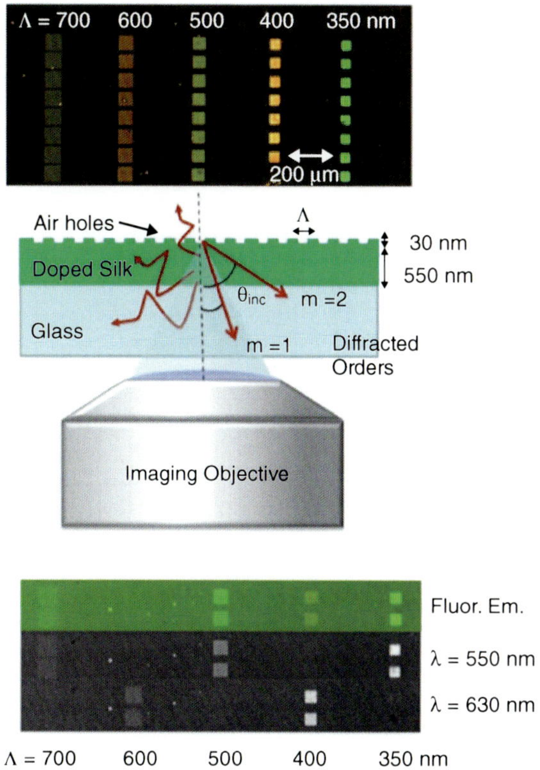

Plate VII (top) Image of the patterned, undoped silk film illuminated at grazing angle under white light illumination. Different colours appear because of the pitch of the individual patterns. The scheme of the experimental layout is illustrated in the figure as well, whereas the bottom image shows the nanopatterned doped silk film. The enhancement in emission corresponds to specific lattice constants that are matched to the fluorescence spectrum from different pitch nanopatterns. Shown underneath the fluorescent image are images collected at λ = 550 nm and λ = 630 nm, illustrating the specific nanopatterned squares that give rise to enhancement at those wavelengths.

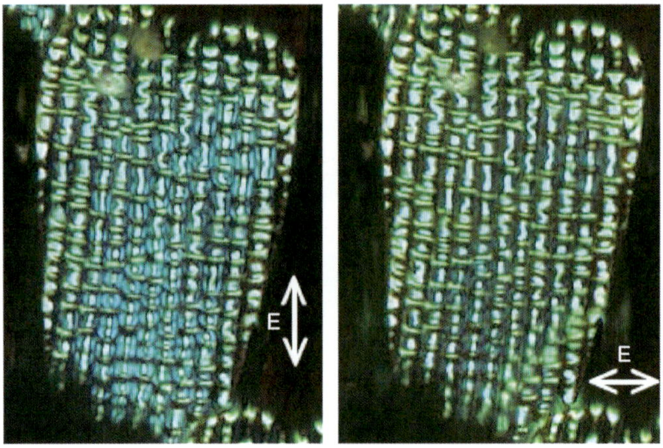

(a)

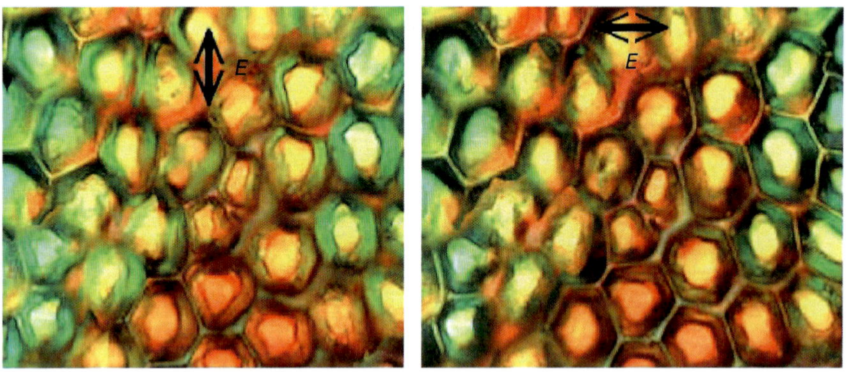

(b)

Plate VIII (a) Polarisation effect on *Papilio paris* scales. On the left, the polarisation of the incident wave is parallel to striae. The high edges of the basin illuminated under high incidence, send back a blue polarised light in this direction. On the right, polarisation is perpendicular to striae, the small edges of the basin now reflect the polarised wave. Under non-polarised light, the elongated shape of basins leads to a reflected light that is slightly polarised in the direction of striae. (b) Polarisation effect resulting from reflection in the basins of *Cincidela hybrida* elytrons. On the left, the polarisation direction is vertical, while it is horizontal on the right. The yellow background is not polarised, contrary to the green edges of basins.

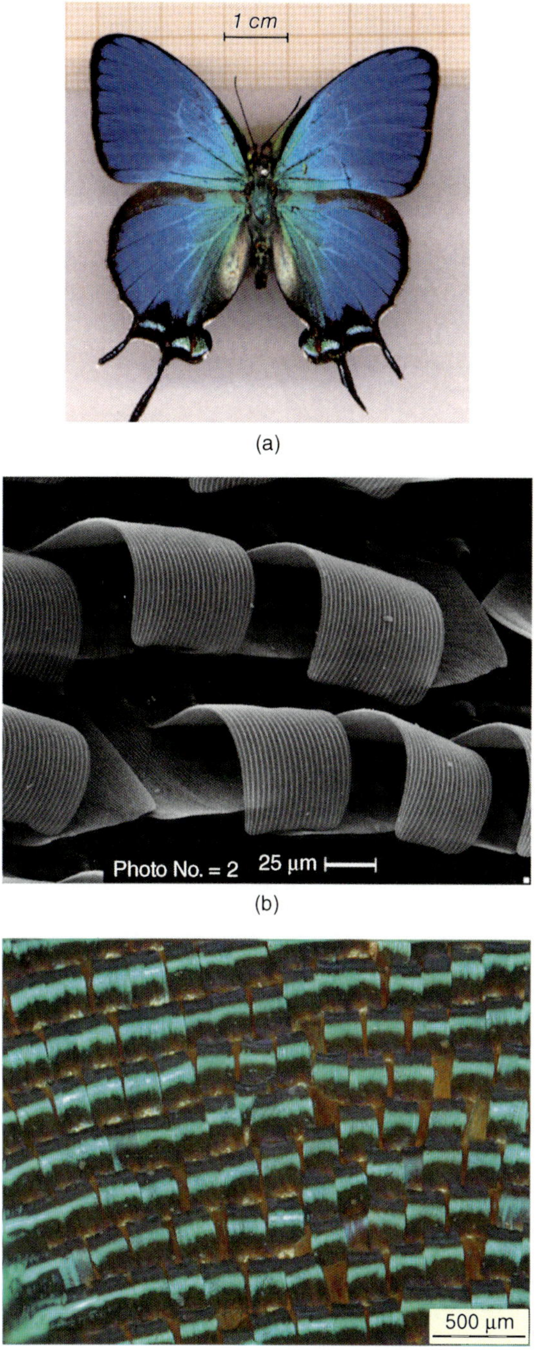

(a)

(b)

(c)

Plate IX (a) *Suneve coronata*. (b) SEM view of the dorsal scales. Cover scales are convex and intersect at right angle. The ground scales are plane. (c) Photonic microscope view of the wing under unpolarised light.

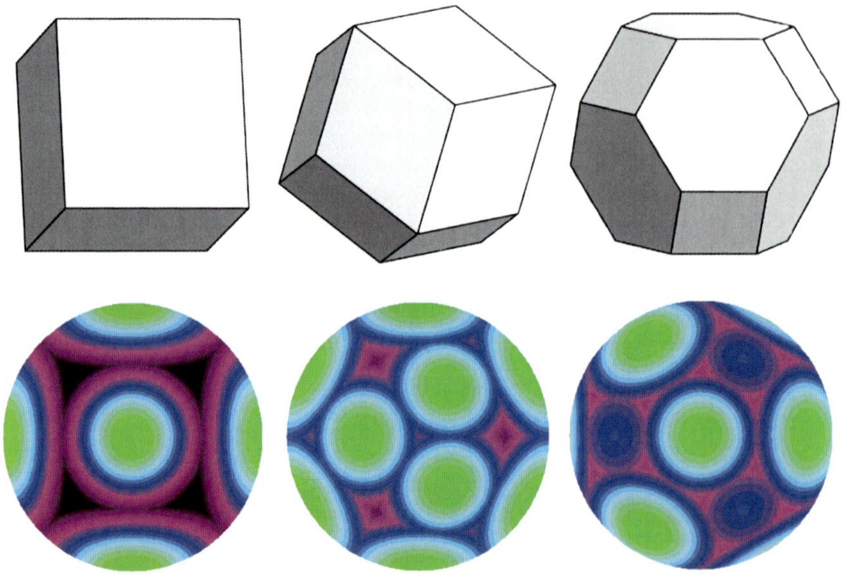

SC (a) BCC (b) FCC (c)

Plate X The Brillouin Zones (BZ) for the following lattice structures: (a) simple cubic SC, (b) body-centred cubic BCC, and (c) face-centred cubic FCC. Below each BZ is the indicative changes in colour with angle.

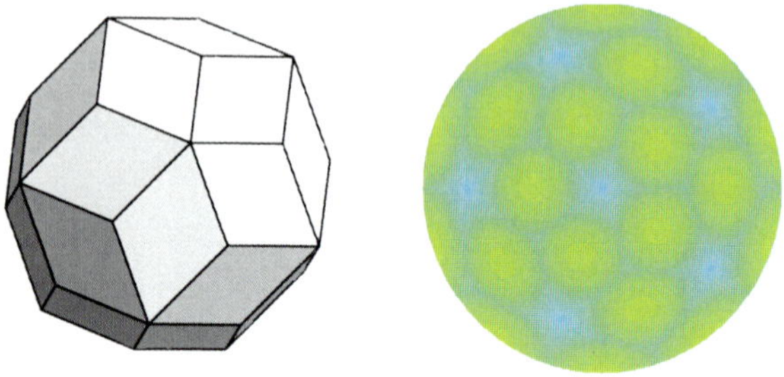

Plate XI The Brillouin Zone for a quasi-crystal with five-fold symmetry and the corresponding iridescence map showing the indicative changes in colour with angle.

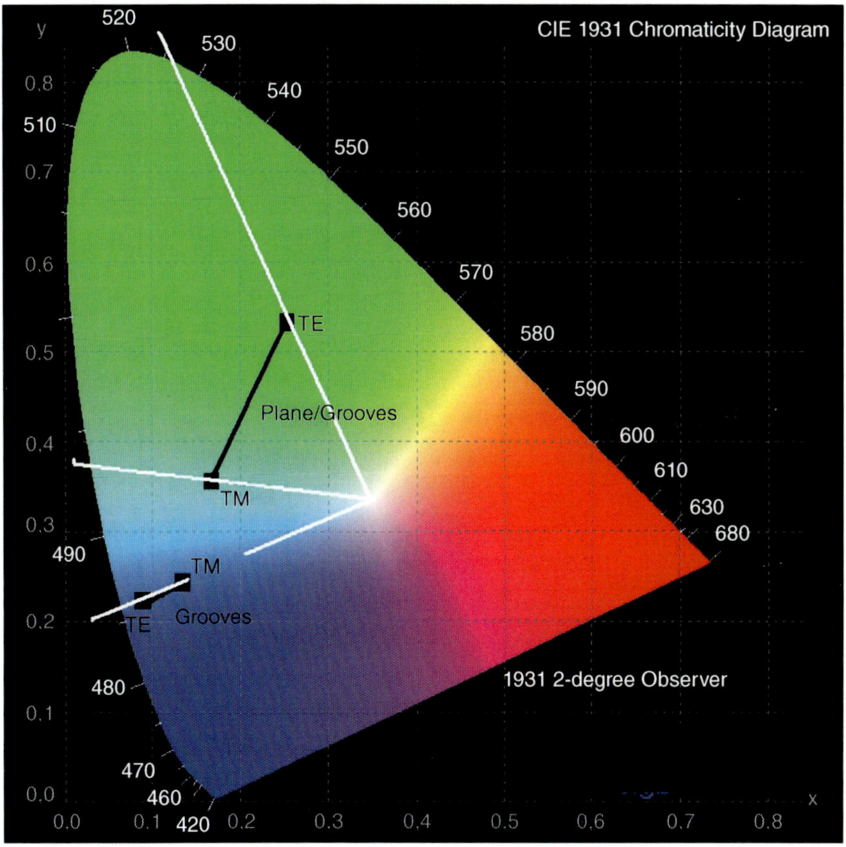

Plate XII Changes in colouration with the polarisation of the two systems: plane/grooves and perpendicular grooves. In the first one, the colour change but not the luminosity, in contrast to the perpendicular grooves system.

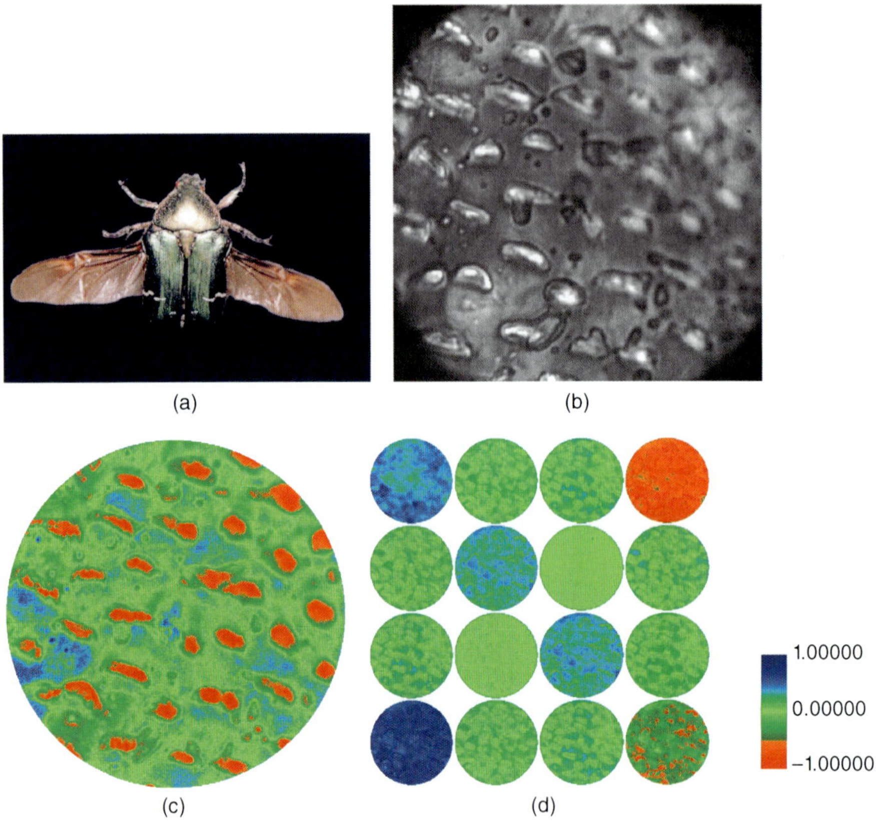

Plate XIII *Cetonia aurata* in flying position (a). Optical microscopic view of the elytron (×100) (b) and its Mueller matrix element 1 × 4 in image mode showing that the circular polarisation arises from very local part of the elytron (c). The complete Mueller matrix of the elytron in image mode (×10) (d).

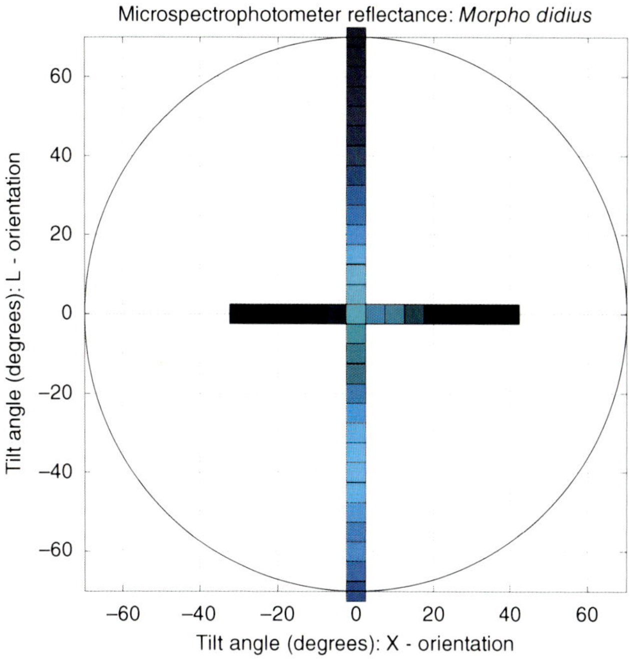

Plate XIV Perceived colour as a function of scale orientation for *Morpho didius*. The colour of each square is based on the measured spectrum and converted to a colour using the CIE co-ordinate system.

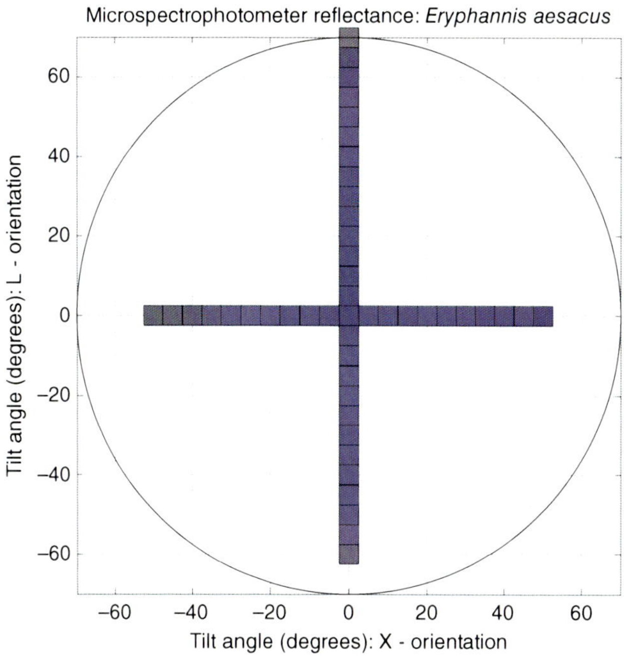

Plate XV Angle-dependent spectral measurement for a micro-rib butterfly, *Eryphannis aesacus*.

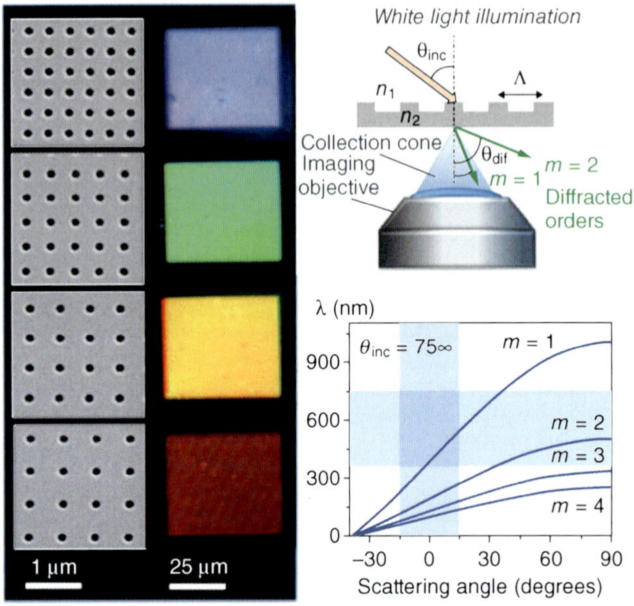

Plate XVI Various periodic lattices imprinted in silk and the associated diffracted spectral signatures as observed under dark-field illumination microscopy.

(a) (b)

Plate XVII Scales of *P. sesostris* viewed from above (a) and below (b), finally revealing their iridescence.

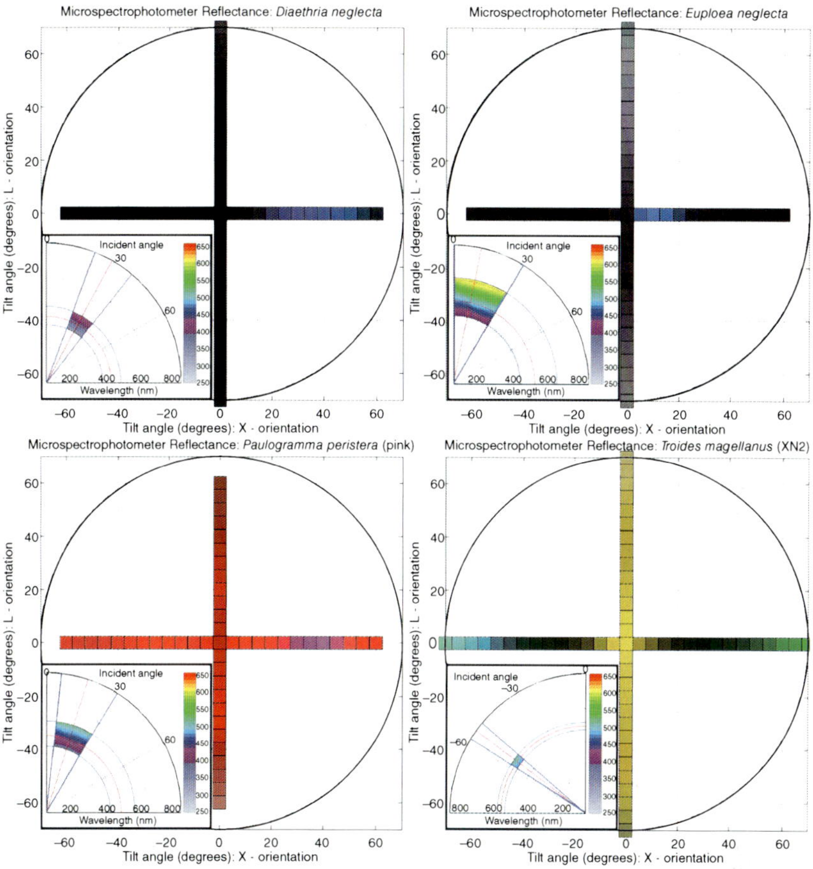

Plate XVIII Angle-dependent spectral measurements of the four sloped ridge lamellae butterflies. The inset shows the theoretical prediction of the wavelength and X-tilt angle range of the strongest reflection as predicted by the bi-grating analysis (artificial colour).

Plate XIX Bent scales *P. sesostris* viewed from below, finally revealing their iridescence.

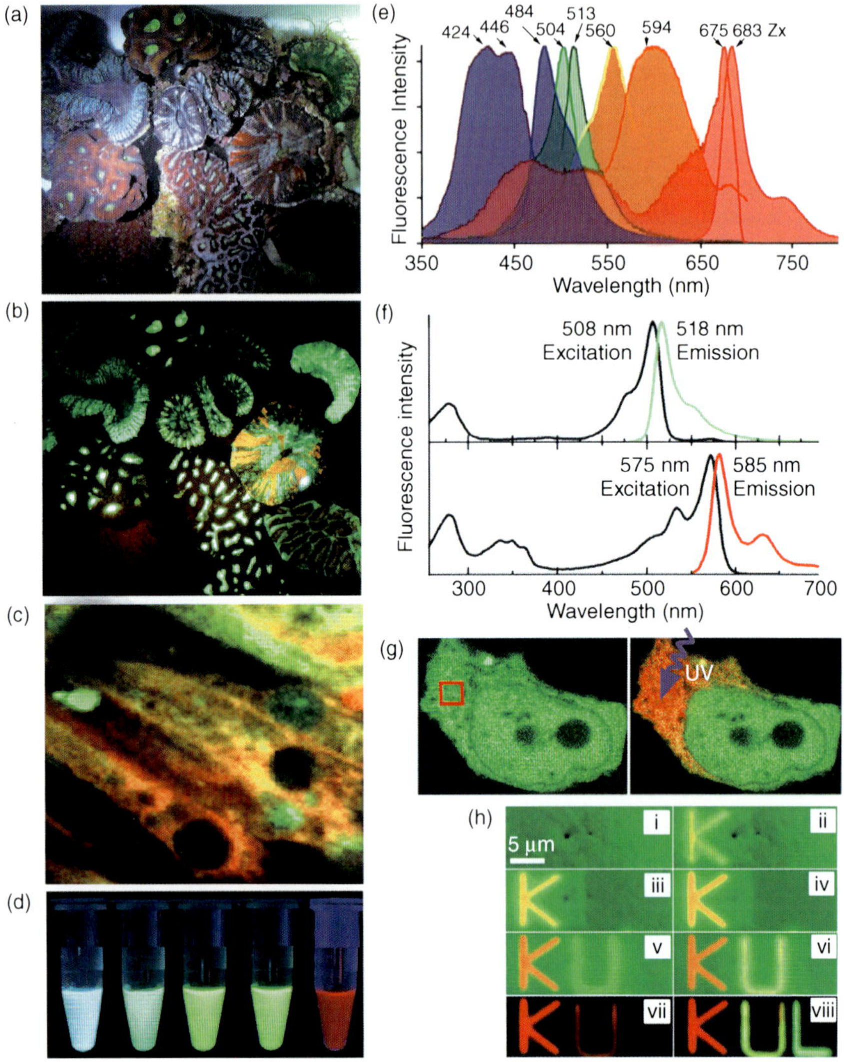
(a)
(b)
(c)
(d)
(e)
424
446
484
504
513
560
594
675
683 Zx
Fluorescence Intensity
350
450
550
650
750
Wavelength (nm)
(f)
508 nm
Excitation
518 nm
Emission
575 nm
Excitation
585 nm
Emission
Fluorescence intensity
300
400
500
600
700
Wavelength (nm)
(g)
UV
(h)
5 μm
i
ii
iii
iv
v
vi
vii
viii

Plate XX Diverse colours of GFP-like proteins in reef corals and in biophotonic applications. (a) Patterning of tissues pigmented with GFP-like proteins in different coral species observed in white light and (b) by blue light excitation in the dark, revealing tissue fluorescence. (c) Multicoloured GFPs, including EosFP, in ectodermal cells of the coral *Lobophyllia hemprichii.* (d) Cloned and purified cyan, green and red GFPs from the coral *Acropora millepora.* (e) The excitation and emission spectra (only 3 of 5 GFPs are shown) of these cyan, green and red proteins in coral tissues exhibit characteristics of donor–acceptor pairs tuned for fluorescence energy transfer, with emissions from cyan spectrally overlapping the excitation of the green emitters and the emission of greens overlapping the excitation of reds. Red spectra and shading in (e) show excitation and emission of photosynthetic pigments of coral's symbiotic microalgae. Energy transfer via GFPs channel excess energy towards peridinin, the accessory photosynthetic pigment of microalgae and into the gap of photosynthetic absorption (adapted from Salih *et al.* 2004). (f) Green-to-red photoconversion of EosFP from *L. hemprichii* showing absorption and emission spectra of the green and photoconverted red forms (adapted from Wiedenmann *et al.* 2004). (g) mEosFP expressed in a mammalian cell and photoconverted to red by a brief irradiation by 405 nm (Salih, unpublished). (h) Encoding in four colours using IrisFP in a surface of a large agarose-Ni-NTA bead (117 mm diameter) imaged at 488 nm (52 W/cm^2) and 561 nm (125 W/cm^2) for green and red channels (sub-panels i–viii). Using a 405-nm laser light, the letters were written, fully or partially photoconverting the initially green molecules to their red emitting states and creating red, yellow and green colours (adapted from Adam *et al.* 2010).

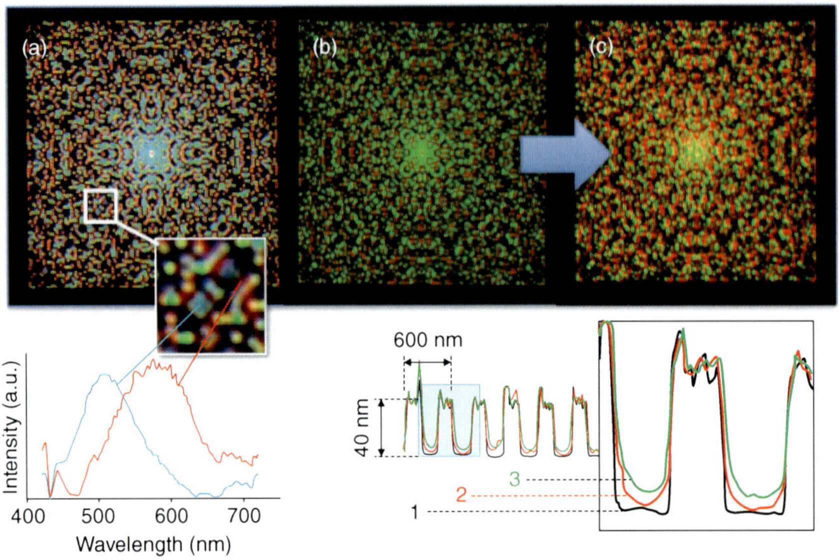

Plate XXI (a) Nanopatterned aperiodic array of gold nanoparticles and its colorimetric response. The patterned surface occupies a 50 μm × 50 μm area. The inset shows a 5 μm × 5 μm area and the associated spectra extracted from a multispectral image of the lattice. By displaying only the red and green spectral challenge a colour transition is overtly observed (image b to image c). This transition corresponds to a change in topography of a silk monolayer, which is spin-coated on the structure. The atomic force microscope profiles below the images show the change in thickness corresponding to the different (and successive) spin-coating process.

(a)

(b)

3.13 Chromatic variations of the wings of *Morpho zephyritis* for two perpendicular linear polarisation states of the incident light: (a) Transverse Magnetic and (b) Transverse Electric.

Caveney 1971; Filshie 1988; Hadley 1982; Beireither-Hahn *et al.*, 1984). The small difference in index is offset by the large number of layers which cause a bright appearance. It is important to note that there is no alternation of materials with low and high indices in the multi-layer. This is the result of the rotation of a layer composed of a single material having two different indices in the perpendicular direction. Interference occurs between layers which have the same direction.

The layers are parallel in the exocuticles of most coleoptera, whereas the chitin micro fibrils which constitute the elytra bend at a relatively constant

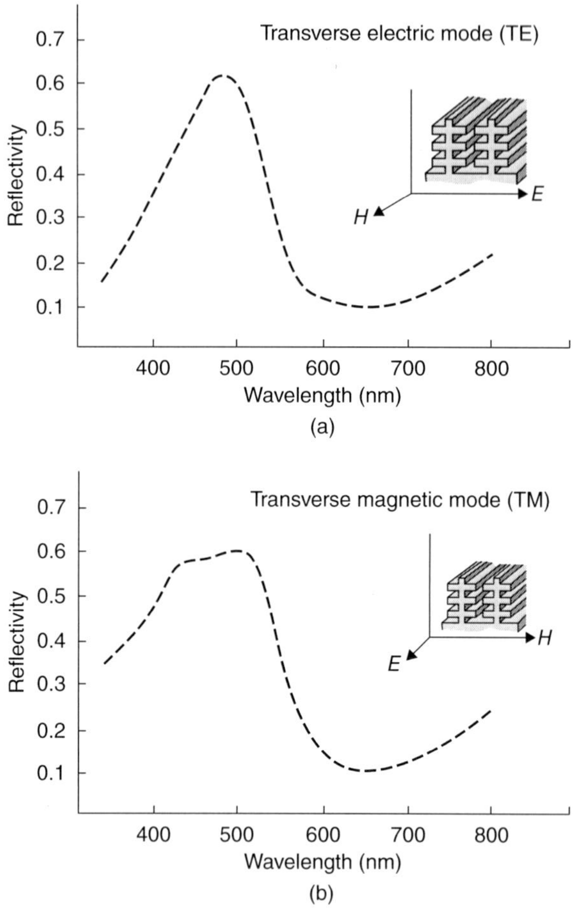

3.14 Coefficient of reflection of the wing of *Morpho menelaus* for both the TE (a) and TM (b) polarisation. In the TM mode, the spectrum consists of two peaks and is shifted towards green.

angle which is called the rotator, resulting in a helicoidal structure similar to that of some liquid crystals in the cholesteric phase (Fig. 3.16).

Determining the rotator is not an easy task. The classical method consists of making an oblique section of the elytra at a specific angle. A complex arrangement of arches then appears which allows the thickness of the lamella to be measured. The coefficient of rotation α is given by $\theta = \alpha z$. Another technique consists in polishing the surface of the elytra and to proceed by ion bonding of the surface which causes the micro fibrils to appear (Fig. 3.17).

When circular polarised light falls on such a structure, two different situations may arise according to the respective rotation. Under conditions in

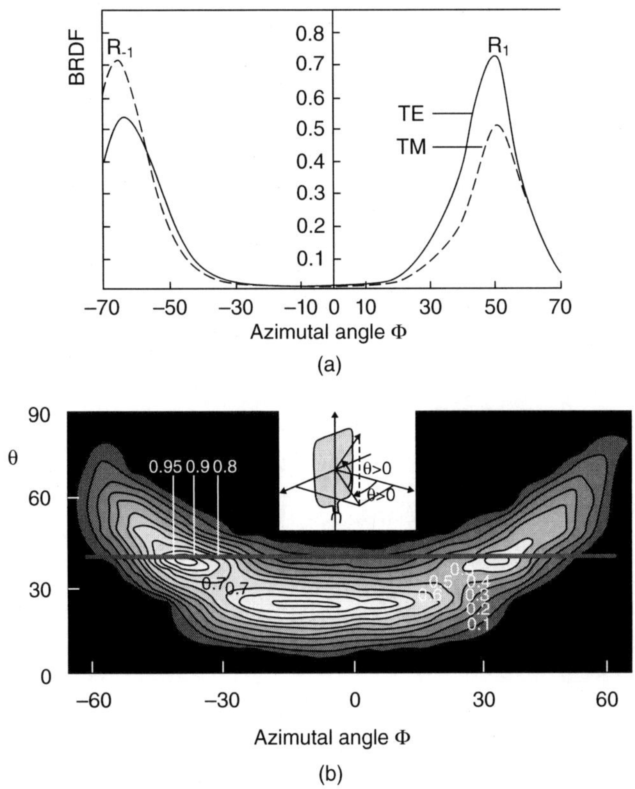

3.15 Cup in Bidirectional Reflectivity Distribution Function (BRDF) of *M. menelaus* for $\phi = 45°$ and $\lambda = 450$ nm (a). The order R1 is predominantly TE polarised, the R-1 order TM polarised. The BRDF in intensity, and the axis of cut (b).

which the wave and the structure turn in the same direction, the electric field **E** will always remain parallel to the chitin fibrils. The wave therefore experiences a homogeneous medium in which the direction of dispersion is a straight line and travels freely within it. However, when the wave and the structure turn in opposing directions, **E** is alternatively parallel and perpendicular to the fibrils. The effect is then that of a stratified medium of alternatively high refractive index when **E** is perpendicular to the fibrils, and is of low index when it is parallel. Gaps open in the direction of dispersion and the wave is reflected (Fig 3.18).

This phenomenon is clearly visible in most of the genus *Plusiotis* and in certain Scarabaeidae (*Cetonia aurata*). Plate XIII shows the Mueller matrix of the elytra of *Cetonia aurata*. The strong circular polarisation effect is evidenced by the M_{14} and M_{41} elements.

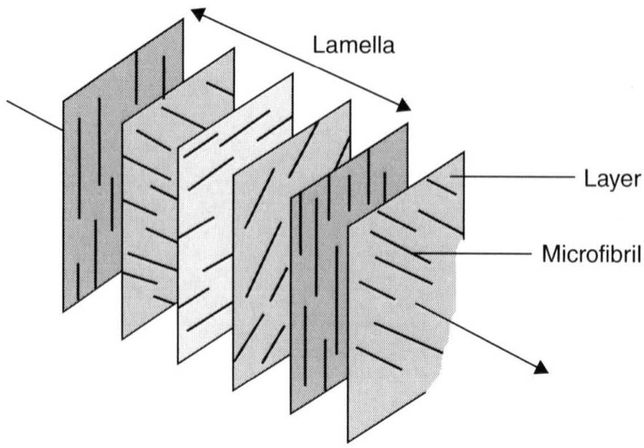

(a)

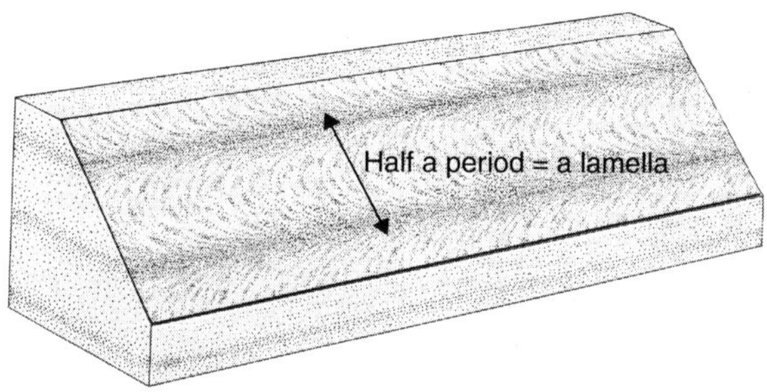

(b)

3.16 In a cholesteric stage, the director orientation regularly varies along the axis. A period shows up when the elements undergo a 180° rotation, determining a lamella (a). A cross-section through the endocuticle points out arched structures characterising cholesteric-type arrangement. Lamellae are formed by a 180° rotation of sticks (b).

3.5 Bio-inspired applications: anti-counterfeiting patterns

These structures can be adapted to provide an original solution to the problem of protecting against the forging of bank notes. Each bank note consists of a certain number of elements which ensure protection at three different levels and correspond to three types of users. Level-1 protection requires any individual to be able to recognise the authenticity of a bank note without the use of equipment. This involves various motifs, colours, papers, watermarks and holograms. Level-2 requires simple equipment which may be used by shopkeepers and cashiers to detect concealed properties through using ultraviolet light or polarisation. Level-3 protection

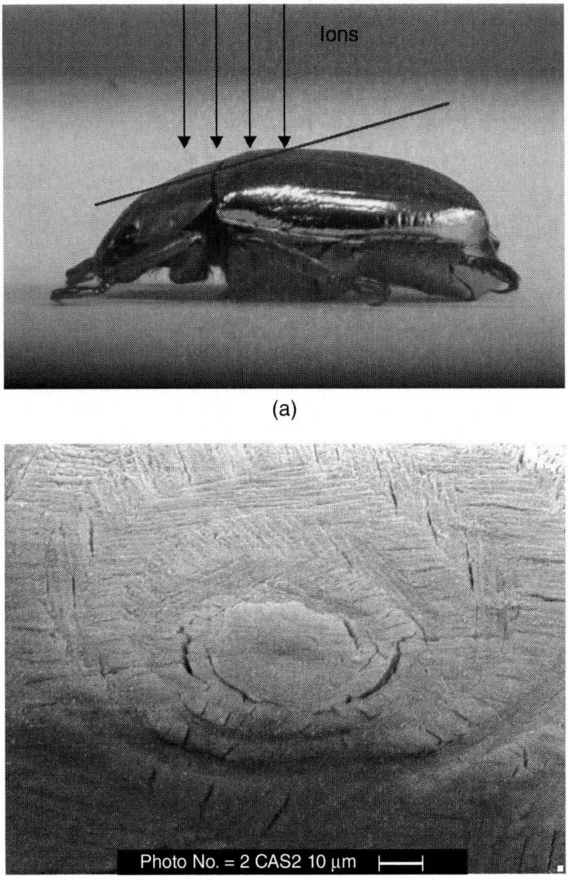

(a)

(b)

3.17 (a) Analysis of the cholesteric structure of the elytra of *Plusiotis chrysargirea*. The top of the elytra is polished and attacked by an ion beam. (b) The ion etching reveals the orientation of the chitin fibrils in each layer.

involves all the physico-chemical properties of bank notes which can only be determined in a laboratory by the use of cumbersome equipment.

Once their symmetry is broken, the effects produced by *Papilio* and *Cicindeles* structures can ensure the two first levels of protection. These generate bright and iridescent colours which fulfil the level-1 protection requirements and are also capable of creating chromatic effects dependent upon polarisation which meets the needs of level-2 protection (Berthier *et al.*, 2007a, b).

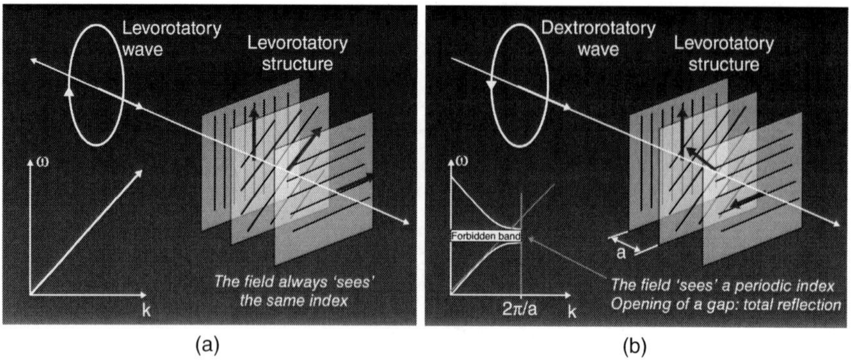

3.18 The relations of dispersion in a cholesteric medium for two circular waves turning in opposite directions. When the wave and the structure turn in the same direction, and if the rotators are equal, the layered medium appears homogeneous (a). When they turn in opposite directions, the refractive index changes periodically: the relation of dispersion is broken (b).

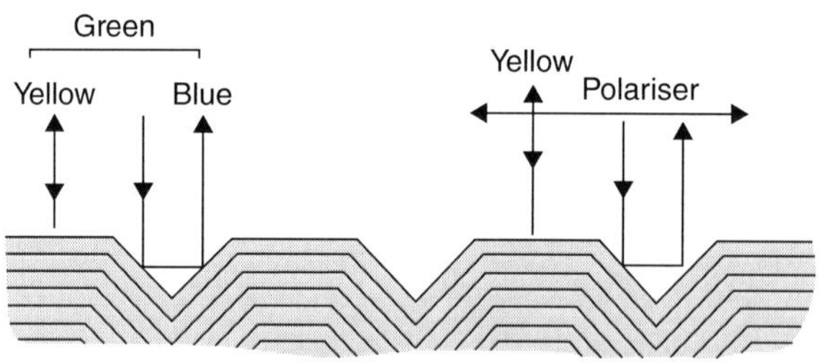

3.19 A coloured effect linked to the polarisation of one of the components. By suppressing a large part of the blue component with a polariser, one can make the general green colour turn yellow.

3.5.1 Change of colour

Several different processes can be created which are based on two main effects. The one most closely resembling *Papilios* and *Ciccindeles* is a succession of grooves and alternated planes of the same width. This produces a compound colour, one of the components of which can be suppressed by using a polariser to modify the colour. The colour of a motif may thus be varied from green to yellow for example, either by suppressing or permitting the blue component reflected by the multi-layer grooves which reflect yellow under normal incidence (see Fig. 3.19).

The calculated spectra of these different elements are presented in Fig. 3.20. The system consists of a multi-layer of alternatively high (n_h =

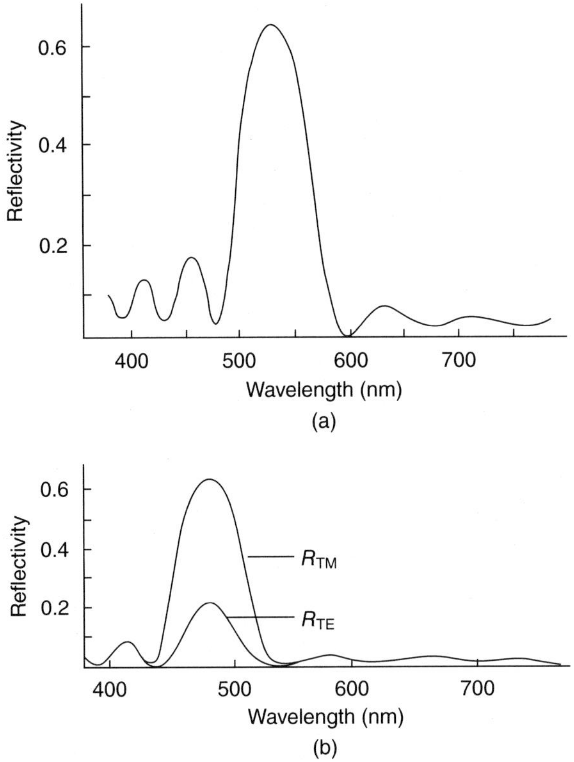

3.20 (a) Calculated reflectivity of a plane multi-layer under normal incidence. (b) Reflectivity of the structured film for the two polarisation states (TE and TM are relative to the axes of the grooves).

1.94) and low ($n_l = 1.51$) indices and thicknesses e_h and e_l respectively which produce a green colour under normal incidence which is very similar to that of the butterfly *Sunive coronata* : $e_h = 185$ nm and $e_l = 117$ nm (see Fig. 3.21).

The polarisation contrast is relatively important (P ~ 0.43) and leads to a significant change in colour. The luminosity remains almost constant (Tables 3.1 and 3.2). The coloured effects are presented on a CIE diagram in Plate XII.

3.5.2 Change of luminosity

Another method consists of using only grooved areas perpendicular to each other. Without a polariser, these appear to be exactly the same colour, but when perpendicularly polarised, only one will appear, thus causing concealed motifs to emerge (Table 3.2) (see Fig. 3.22).

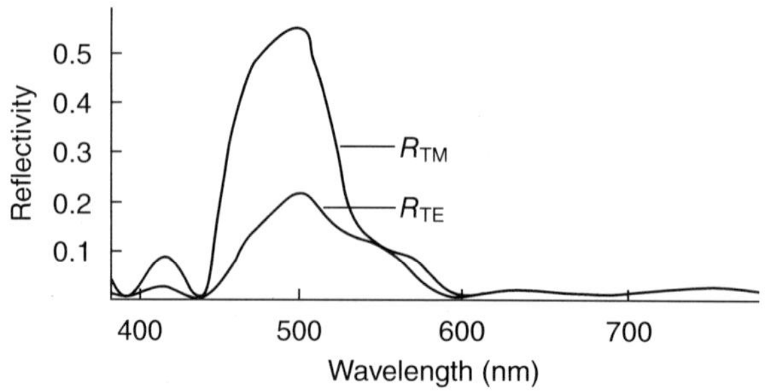

3.21 Calculated reflectivity of a succession of multi-layered planes and grooves of the same width and for the two polarisation states.

Table 3.1 Chromaticity coordinates of a motif composed of multi-layered planes and grooves; the luminosity is nearly constant but the colour changes

	CIE coordinates (x', y')	Luminosity
TE	0.23, 0.52	16.9
TM	0.16, 0.36	6.5

Table 3.2 Chromaticity coordinates of a motif composed of multi-layered perpendicular grooves; the colour stays unchanged but the contrast of luminosity is important

	CIE coordinates (x', y')	Luminosity
TE	0.09, 0.23	13
TM	0.12, 0.24	33

3.6 Conclusion

Insects have long been known to be sensitive to the polarisation of light. It has been shown that polarised light increases the sensitivity of compound eyes by about 15–30% when compared to unpolarised light of the same intensity (Rossel, 1988). Exhibiting a polarising structure could therefore be advantageous for the insect as it enhances its visibility. Many insects, including most of the Morphinae and many coleoptera (*Cetonia, Plusiotis*) reflect a fairly strong polarised light, in either a linear or circular manner. Many others have developed surface structures that depolarise reflected light. This is the case in many butterflies of the genus *Papilio* (Papilionidae) and certain coleoptera (Cicindelidae). Except under normal incidence, the light reflected by a given surface is always polarised to some degree. To

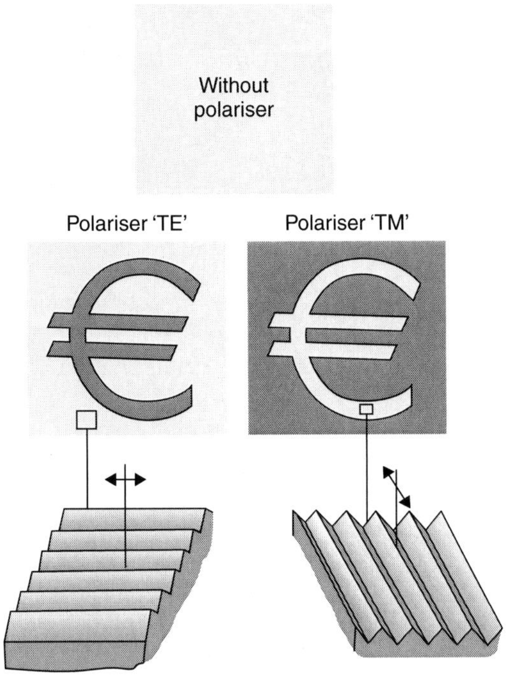

3.22 When one uses ridged areas only, they present the same colour under natural light and it is impossible to distinguish any pattern. They can become apparent by using a polariser letting the reflected component penetrate. TE and TM are relative to the central motif structures.

avoid this phenomenon, the surface has to present facets perpendicular to each other in the same proportions. This is generally achieved by concave structures on the wing scales of butterflies or on the surface of the elytra. In some rare cases, as for example, in some Lycaenidae, the depolarising structure is multi-scaled. One polarising structure is formed by the scales themselves, while the other is formed by the grooves of these scales. The exposed surfaces are roughly equivalent and as they reflect light polarised in perpendicular directions, they appear unpolarised from the macroscopic point of observation.

Significant local changes in colour or luminosity may be made to appear by the use of a polariser and these effects can be used for the first two levels of protection for banknotes and credit cards. By combining multi-layered planes and grooves, an infinite number of devices may be posited that will produce coloured effects linked to polarisation. Two examples are presented, one producing a change in colouration, the other giving a variation of intensity. Circular polarisation may also be envisaged.

3.7 References

Abeles F (1967), *Optics of thin films*. In *Advanced Optical Techniques*, Amsterdam, North Holland Publishing, pp. 145–187.

Beireither-Hahn J, Matoltsy A G and Richards K S (1984), *Biology of the integument, vol. 1 : Invertebrates*, Berlin, Springer-Verlag.

Berthier S (1993), *Omtique des milieux composites*, Paris, Polytechnica, ISBN 2-84054-015-0.

Berthier S (2006), 'Structure and optical properties of the wings of Morphidae', *Insect Sciences* **13**, 145.

Berthier S (2007a), *Iridescence, The Physical Colors of Insects*, Springer Science + Business Media.

Berthier S, Boulenguez J and Balint Z (2007b), 'Multiscaled polarization effects in Suneve coronata (Lepidoptera) and other insects: application to anti-counterfeiting of banknotes', *Appl. Phys.* A86, 123–130.

Berthier S (2010), *Photonique des Morphos*, Paris, Springer France.

Bouligand Y (1969), 'Sur l'existence de pseudomorphoses cholesteriques chez divers organismes vivants' *J. de Physique* **30**, C4-90, 103.

Caveney S (1971), 'Cuticule reflectivity and optical activity in scarab beetles: the role of uric acid', *Proc. R. Soc. London B* **178**, 205–225.

Cohen RW, Cody GD, Couts MD and Abeles B (1973), *Phys. Rev*. B8, 3689.

Filshie BK (1988), '*Fine structure of the cuticule of insects and other arthropods*', *Insect Ultrastructure* **1**, 281–312.

Fox M (2001), *Optical Properties of Solids*, New York, Oxford University Press.

Gabor H and Dezso V (2004), *Polarized Light in Animal Vision*, Heidelberg, Springer Verlag Berlin.

Gel'fand IM and Shapiro ZY (1956), 'Representation of the group of rotations of 3-dimensional space and their application', *Amer. Math. Soc. Translation* **2,** 207.

Hadley NF (1982), 'Cuticule ultrastructure with respect of the lipid waterproofing barrier', *J. Exp. Zoology* **222–223**, 239–248.

Hielsher AH, Eick AA and Mourant JR (1997), 'Diffuse backscattering Mueller matrices of highly scattering media', *Optics Express* **1**, 441.

Huard S (1994), '*Polarisation de la Lumière*', Paris, Masson.

Hutley MC (1982), *Diffraction Gratings,* London, Academic Press.

Kusceri I and Rubaric M (1959), 'Matrix formalism in the theory of diffusion of light', *Optica Acta* **6**, 42.

Jackson JD (1975), *Classical Electrodynamics*, New York, Wiley.

Lipson SG, Lipson H and Tannhauser DS (1995), *Optical Physics*, Cambridge, Cambridge University Press.

Michelson AA (1911), 'On the metallic colouring in birds and insects', *Phil. Mag*. **21**.

Moron MA (1990), *The Beetles of the World*, 10, France, Sciences Nat.

Neville AC and Caveney S (1969), 'Scarabaeide beetle exocuticule as an optical analogue of the cholesteric liquid crystals', *Biol. Rev*. **44**, 531–562.

Neville AC (1975), *Biology of the Arthropod cuticule*, Berlin, Springer-Verlag.

Rossel S (1988), Polarization sensitivity in compound eyes. In *Facets of the Vision* (Stavenga DG and Hardie RC eds.), Springer-Verlag, Berlin.

Schultz TD and Rankin MA (1985a), 'The ultrastructure of the epicular interference reflectors of tiger beetles (Cicindela)', *J. Exp. Biol.* **117,** 111–117.

Schultz TD and Rankin MA (1985b), 'Developmental changes in the interference reflectors and colorations of tiger beetles (Cicindela)', *J. Exp. Biol.* **117,** 87–110.

Schulz FM, Stammes K and Weng F (1999), 'VIDISORT: an improved and generalized discrete ordinate method for polarized radiative vector', *J. Quant. Spectrosc. Radiat. Transfer* **61**, 1005.

Siewert CE (1981), 'On the equation of transfer relevant to the scattering of polarized light', *The Astrophys. Journal* **245**, 1080.

Siewert CE (1982), 'On the phase matrix relevant to the scattering of polarized light', *The Astrophys. Journal* **109**, 195.

Vukusic P and Samble J (2003), *Nature* **424**, 852.

Ziman JM (1979), *Principle of the Theory of Solids*, Cambridge, Cambridge University Press, 2nd edition, ISBN 978-0521297332.

4 Biomimetic nanostructures for anti-reflection (AR) devices

S. CHATTOPADHYAY, National Yang Ming University, Taiwan, Y. F. HUANG, National Yang Ming University, Taiwan and Institute of Atomic and Molecular Sciences, Taiwan, K-H. CHEN, Institute of Atomic and Molecular Sciences; and L-C. CHEN, National Taiwan University, Taiwan

Abstract: This chapter highlights natural designs for controlling light and structural coloration in general and anti-reflection (AR) in particular. It is now possible to impart natural functionalities, such as AR from moth eyes or cicada wings, to biomimetic structures. Following these natural clues, we have been able to produce superior AR structures and designs, often in the sub-wavelength scale, to reduce reflectances below 1% over the broad ultraviolet and visible region and achieve an enhanced AR regime in the long wavelength, infrared (IR) and terahertz (THz), part of the spectrum. The chapter will discuss briefly the incorporation of these designs into real applications such as solar cells.

Key words: anti-reflection, biomimetic, photonic, nanostructures, sub-wavelength.

4.1 Introduction

Nature offers efficient designs for special functions (Fig. 4.1). Biomimicry is hence a natural choice. Plants, insects and animals around us offer many such designs which are often hidden from view and which only reveal themselves through careful observation. This is because most natural designs are on the micro- or nano-scale. Hydrophobicity of the lotus leaf, low hydrodynamic drag of the skin of aquatic animals, thermal insulation of animals, for example polar bear fur, and anti-reflection (AR) from moth eyes are few examples. In this chapter, we will introduce biomimetic AR structures in detail.

Reflection (R) controls the appearance and colour of an object since it is this fraction of the incident light that reaches the observer's eye. However, reflection is not independent and is associated with absorption (A) and transmission (T) by the simple relation $A + R + T = 1$, where the sum adds to 100% of the incident light. R contains, in addition to the specular part, a diffuse component, and interference effects. Diffraction, interference, scattering, absorption and emission processes are the basic colour production mechanisms. For example, the colour produced in thin films [1], soap

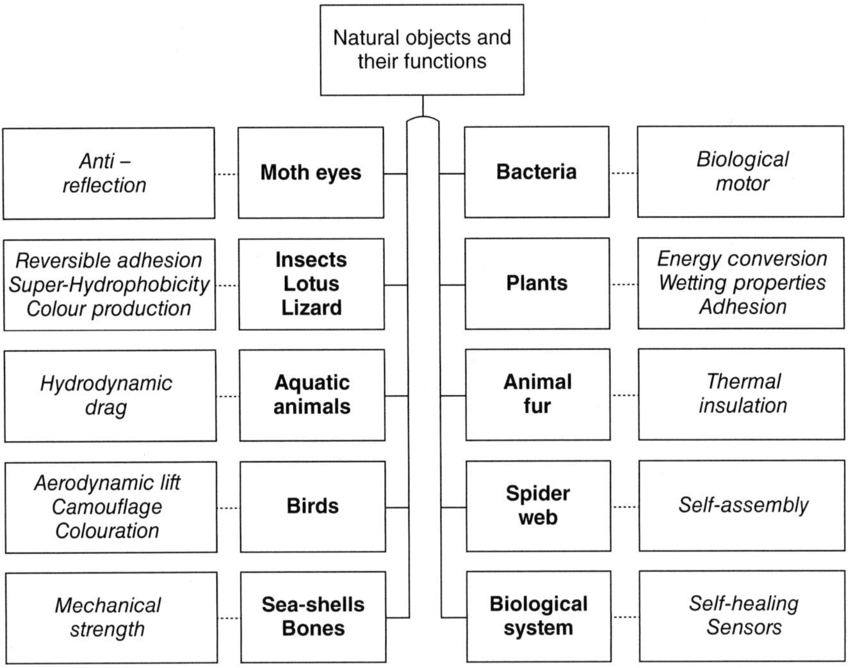

4.1 Natural designs (in bold) for various special functions (italicized).

bubbles or peacock's tail feathers is due to interference, the colour of the sky due to Rayleigh scattering and the rainbow due to refraction. In most cases a 'non-motion structural (or physical)' [2,3] green object (for example, peacock's feather) appears such because of the anti-reflecting property of it to other non-green wavelengths (λ). Such physical colours in nature are found in butterfly wings [4], the cuticle of arthropods such as beetles [5–7], feathers of peacock or hummingbird [8,9], Indigo snake skin [10] or crystals [11–14] such as opal. On the other hand, the colours of the chameleon (family *Chamaeleonidae*) appear owing to the presence of skin chromatophores (pigment-containing cells), which can be stimulated physiologically. Such colours or colour changes that are taxonomical or have behaviouristic nature are 'motion structural' [3] and is outside the scope of this chapter together with colours emitted by electronic transitions.

It is nature's strategy to produce colours by surface texturization [15,16], which controls the reflection from it. Refractive index (RI, denoted by 'n') characterizes a material medium, according to the velocity of light in it with respect to that in air, and is a function of λ. Any change in the velocity of light, or optical disturbance or reflection, will be zero if the RI of the propagating medium does not change. Since air is the all-pervading medium (n = 1.0), for most anti-reflection (AR) purposes of human interest, it is desirable to have an object of low RI as close to 1.0 as possible. Few examples of such materials are MgF_2, CaF_2 and SiO_2 having n = 1.39, 1.44, 1.46, respectively. In the organic world, guanine is the material with highest transparency having n = 1.8. These materials will work as potential AR objects in water (n = 1.33) also. Nature has its own strategy of developing low RI materials by combining a 'material' with 'air' in a definite ratio. For example, a certain volume of a porous material 'A', say silicon (n = 3.8), will have a certain fraction (f) of silicon and (1–f) fraction of air with n = 1.0. Effectively, porous silicon will have an RI in between 1.0 and 3.8 depending on the volume fraction 'f'. The lower the 'f' the lower is the effective RI; however, the variation of effective RI with 'f' is not linear.

The naturally occurring surface microstructures are special ways of imparting roughness to achieve selective AR. As RI is a function of λ, structures have different degrees of RI matching with ambient, say air, in the ultraviolet (UV), visible (VIS) or infrared (IR) part of the spectrum resulting in low reflectance. Hence, it is difficult to obtain broadband, over large wavelength regimes, AR using a uniquely designed material. Nature taught us to design the AR surfaces by engineering the effective RI between air (n = 1.0) and the surface (n > 1), known as the RI profile, especially at the plane of incidence. Lord Rayleigh [17] demonstrated, theoretically, that layers with gradual variation of density, hence RI, can work as AR coatings. However, the AR properties of these coating design fail at grazing angles of incidence (AOI) [18–20] indicating the angle-dependence of AR properties. Evidently, the interaction of a structured surface with electromagnetic radiation will depend on the thickness of the surface and polarization of the radiation also. In short, the colour production and AR effects are in fact related terms controlled by RIs of materials, which have a wavelength, viewing angle and polarization dependence.

4.2 Anti-reflection (AR)

The following sections outline the basic principles of anti-reflective coatings. Details are also given for single, double and multi-layer coatings.

4.2.1 Anti-reflective coatings

The idea behind an AR coating can be realized from a single-layer dielectric thin film with a low RI (n) on a substrate with different RI, n_s, where $n_s > n$ [21]. The thin film configuration creates double interfaces, resulting in two reflected waves. When these two waves are out of phase, destructive interference occurs, cancelling both beams partially or totally before they exit the surface.

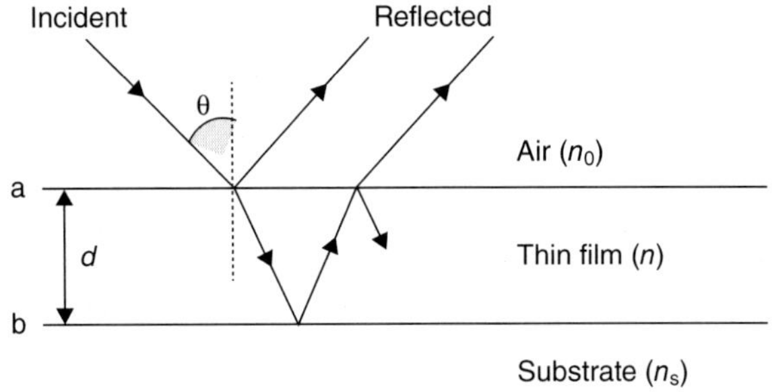

4.2 A schematic of a single thin film anti-reflection coating, wherein n is the index of refraction of the object of interest and n_0 and n_S are the indexes of refraction of the ambient and substrate, respectively.

As presented in Fig. 4.2, the incident energy is equal to the sum of reflected and transmitted energy, ignoring the contribution of scattering and absorption. The AR is achieved as the reflected waves, from the two optical interfaces 'a' (air–film) and 'b' (film–substrate), are out of phase and consequently cancel each other by destructive interference, assuming their amplitudes being identical, as depicted in Fig. 4.2.

Two requirements should satisfy for the exact cancellation of the reflected waves: (1) the two reflected waves are exactly 180° (π radians) out of phase, just before they exit the film, and they have the same intensity, (2) the optical thickness of the thin film coating must be an odd number of quarter wavelengths ($\lambda/4$). At normal incidence ($\theta = 0$), we will obtain

$$R = \left|\frac{(n_0 n_s - n^2)}{(n_0 n_s + n^2)}\right|^2 \qquad [4.1]$$

Interestingly, the AR phenomenon ($R = 0$) must occur at a particular λ, if

$$n = \sqrt{n_s n_0}. \qquad [4.2]$$

For a double-layer thin film coating, with each layer-thickness of quarter-wavelength, at normal incidence, reflectivity becomes

$$R = \left|\frac{(n_2^2 n_0 - n_s n_1^2)}{(n_2^2 n_0 + n_s n_1^2)}\right|^2, \qquad [4.3]$$

where n_1 and n_2 are the RI of the two layers. Hence, R will be exactly zero at a particular wavelength, if

$$\left(\frac{n_2}{n_1}\right)^2 = \frac{n_s}{n_0}. \qquad [4.4]$$

4.2.2 Single, double and multi-layer coatings

Availability of proper materials limits the fabrication and application of single-layer AR coatings. For example, to minimize reflection from coatings on glass or transparent plastic substrates (n_s ~ 1.5), the thin film coating, with quarter–wavelength thickness, must have n ~ 1.22 that is rare. Materials with low RI, such as magnesium fluoride (MgF_2, n = 1.38), can reduce reflectance to 1.5% (Fig. 4.3a) when coated on plain glass. It is virtually impossible to achieve near zero reflectance from a single-layer AR coating.

The application of a double-layer, each layer with quarter-wavelength thickness, can improve the performance of AR coating if a high-index-layer (e.g. Al_2O_3, n_1 = 1.69), while $n_1 > n_s$, is coated prior to the coating of low-index-layer (MgF_2, $n_2 = 1.38 < n_s$) onto the glass substrate (n_s ~ 1.5). This approach leads to a better design for AR effect, as shown in Fig. 4.3b. It is observed that the effective reflectance (R_{eff}) of double-layer AR coatings decrease significantly, approaching zero in the region of target wavelength ~ 500 nm. The reflectance curve of the double-layer AR coating exhibits a V-shaped spectrum; hence, they are often termed as V-coatings [22].

The AR efficiency of a double-layer coating minimizes R_{eff} at a particular target wavelength (λ_t), but simultaneously increases the R_{eff} on both sides

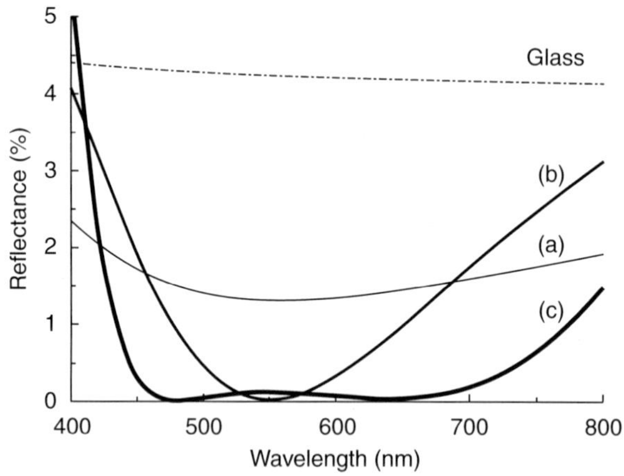

4.3 Reflectance as a function of wavelength for anti-reflection coatings: (a) single-layer AR coating: *Air* (n_0 = 1.0)/0.25λ_t –MgF_2 (n = 1.38)/*Glass* (n_S = 1.52); (b) two-layer AR coating: *Air*/0.25λ_t –MgF_2/0.25λ_t –Al_2O_3 (n_1 = 1.69)/*Glass*; (c) three-layer AR coating: *Air*/0.25λ_t –MgF_2/0.5λ_t –ZrO_2 (n_m = 2.05)/0.25λ_t –CeF_3 (n_1 = 1.64)/*Glass*. Here, λ_t represents the target wavelength, at which the effective reflectance will be minimum, following Eq. [2]. Reprinted with permission from Ref. [22]; copyright 2010: Elsevier.

of λ_t, even higher than that of a single-layer AR coating (Fig. 4.3 a, b). Thus, the number of layers has to be increased. The three-layer design of AR coating (quarter-half-quarter wavelength design), involves a thicker layer (M, $\lambda_t/2$ thickness) sandwiched between two $\lambda_t/4$ layers (H, L). The structure of the three-layer AR device is: *substrate/H-2M-L/Air*, where H and L are the high- and low-index layers, respectively. Simultaneously, the RI (n_m) of the middle layer, M, should satisfy $n_m > (n_1^2 / n_s)$, in order to achieve the broadband AR effect. Fig. 4.3c shows the AR effect of the 3-layer structure.

4.3 Gradient refractive index structures

Efficacy of layered AR coatings is limited to a single λ and small AOI by either or both of design and material. Hence, the gradient-RI AR structure becomes a desirable next generation. The previous sections dealt with AR achieved with external coatings on a surface. There is an alternative method to fabricate the AR surfaces focusing on the surface itself and its topography (Fig. 4.4). Changing or grading the RI reduces the optical reflection at the boundary of two medium having contrasting RI (Fig. 4.4a). The RI changes from that of the light propagating medium (say air, RI~1.0, lower limit) to that of the reflecting medium (say Si, RI~3.8, upper limit) (Fig. 4.4b). Figure 4 (c, e) shows several ways of achieving such grading of RI. These AR surfaces must satisfy a critical criterion: *substrate material is mixed with air on a sub-wavelength scale* [23], such as porous or textured surface. Interestingly, the porous (Fig. 4.4d, type I) and textured surfaces (Fig. 4.4d, type II–IV) are man-made via chemical or physical processes. Observation of natural moth eyes resulted in the first sub-wavelength structures (SWS) [24, 25]. Here we will interpret the different optical properties and optical analytical methods of AR surfaces, including effective medium theory (EMT) and gradient refractive index (GRIN) theory.

Simple EMT [26–28] can describe the interaction of light with SWS by representing the heterogeneity in AR surfaces in terms of a homogeneous medium possessing a single effective optical constant. For the specific case of a multilevel (stepped or discontinuous) surface-relief profile, the EMT assumes a film stack where each layer corresponds to a distinct level of the surface-relief profile (Fig. 4.4c, left). For a continuous profile, the equivalent effective medium is a gradient film (Fig. 4.4c, right). Since EMT rests on the weighted spatial averaging of the profile's optical properties in a particular direction, say along the depth z, the RI of the effective medium increases layer by layer from the value at the ambient-layer interface to that of the substrate ($n_{air} < n_1 < n_2 < n_s$). On the other hand, the RI increases along the thickness ($n_{air} < n(z) < n_s$), for a continuous profile (Fig. 4.4d) [29].

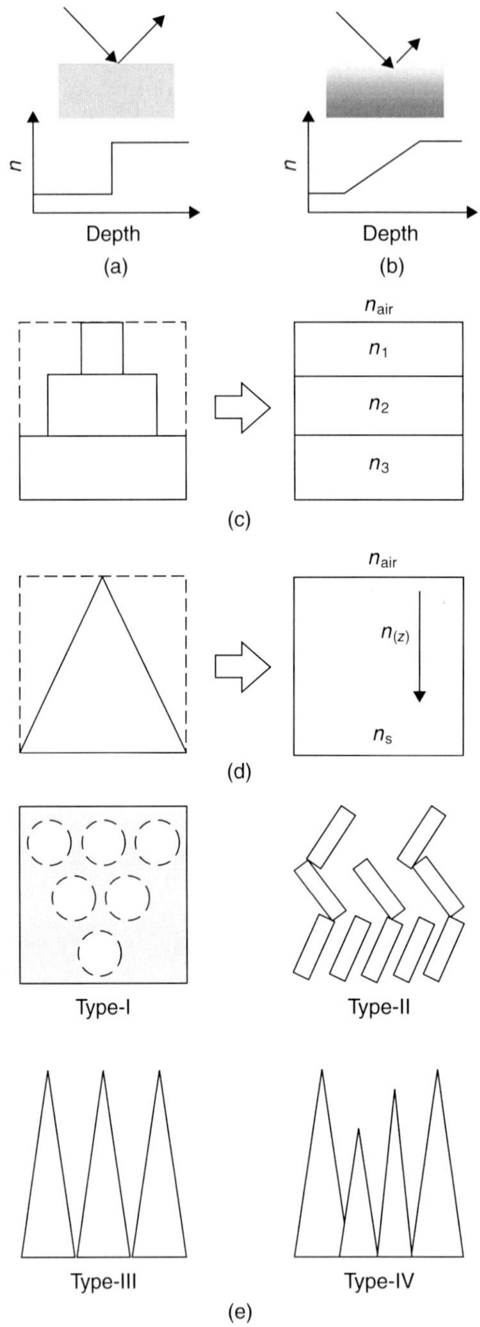
n
Depth
(a)
n
Depth
(b)
n_air
n_1
n_2
n_3
(c)
n_air
n_(z)
n_s
(d)
Type-I
Type-II
Type-III
Type-IV
(e)

4.4 Schematic of strong and reduced optical reflection (top panel) from a solid with (bottom panel) (a) ungraded (fixed), and (b) graded (depth varying) RI, respectively. Different architectures result in the graded RI solid (c–e). (c) Multilayer film with increasing RI (from n_1 to n_s); (d) tapered AR structure with a continuously varying RI (as a function of the solid volume fraction according to the effective medium theory); RI is small at the apex and increases towards the base. (e) Four kinds (Type I–IV) of inhomogeneous media producing gradient RI: (Type I) Cross-section view of a porous layer, with porosity (pore volume fraction, shown by white circles) decreasing, hence RI increasing, with increasing depth; (Type II) Cross-section view of tilted arrays of nanorods, grown by glancing angle deposition, with nanorod number densities, hence RI, increasing with increasing depth of the layer; array of (Type III) periodic nanotips, and (Type IV) random nanotips, with RI scaling as the solid volume fraction at different depths within the array. RI is near 1.0 at the apex of the array where solid volume fraction is ~zero.

4.3.1 Gradient refractive index profiles

Different profiles investigated for gradient-RI layers include linear, parabolic, cubic, Gaussian, quintic, exponential, exponential-sine, and Klopfenstein, for broadband and omnidirectional AR coatings [18, 30–35]. Linear RI profiles, for AR and increased transmission in the visible and near IR for several AOI, are generated on different substrates such as silicon, quartz, or silicon nitride (SiN_x) [36].

Presented below are the typical expressions for the continuous gradient-RI with linear, cubic and quintic profiles [31]:

$$\text{Linear index profile: } n = n_0 + (n_s - n_0)t,\ 0 \le t \le 1 \quad [4.5]$$

$$\text{Cubic index profile: } n = n_0 + (n_s - n_0)(3t^2 - 2t^3) \quad [4.6]$$

$$\text{Quintic index profile: } n = n_0 + (n_s - n_0)t\ (10t^3 - 15t^4 + 6t^5) \quad [4.7]$$

where n_0 and n_s, are the RIs of the incident and substrate media, t is thickness of graded interface region, respectively. The calculated spectral and angular dependence of the reflectivity can be found for linear, cubic and quintic profiles exhibiting low reflectivity for both transverse electric (TE) and transverse magnetic (TM) polarisations over a broad spectral range, for a substrate with $n_s = 2.05$ [37].

Quintic- and exponential sine (ES) index profiles are generally accepted as nearly optimum profiles for gradient-RI AR coatings. The ES profile [38] is expressed as:

$$n(x) = \eta_{\max} \exp\left(\frac{1}{2}\ln\left(\eta_{\max}/\eta_{\min}\right) \times \left\{\sin\left[\pi\left(\frac{x}{x_{tot}}\right) + \pi/2\right] - \sin\left(\pi/2\right)\right\}\right), \quad [4.8]$$

where $x = \int_0^z n(z')dz'$ and $x_{tot} = \int_0^d n(z')dz'$ are the optical distance from the substrate and the total optical thickness, respectively. Here z, the metric thickness, has its origin at the substrate–film interface and d is the total metric thickness of the inhomogeneous layer. $n(z)$ represents the RI of the inhomogeneous layer at a position z within it, while η represents the effective admittance at oblique AOI. Poitras and Dobrowolski proposed modified forms of both quintic and ES profiles to generate the perfect AR coating [38]. For both the modified profiles, the solid adjacent to the ambient affects the AR performance significantly for almost all AOI as high as 89°, over a large bandwidth.

4.3.2 Porous layers

The application of nanoporous AR films (Fig. 4.4e, type I) became commonplace after the development of the etching techniques [39, 40]. Here, the pore sizes (smaller than 300 nm), and the effective-RI of the nanoporous medium, are averaged over the entire film-surface.

Chemical etching/leaching processes applied to glass surfaces have achieved successful fabrication of single-layer gradient-RI porous AR films [41]. Reflectance can be reduced effectively from ~ 8 to 0.5% in the wavelength region between 0.35–2.5 μm. The thin film, considered as an inhomogeneous layer whose RI changed exponentially with the film thickness (d), is surrounded by a homogeneous dielectric media where light is incident normally on the surface [42]. The porous AR film can suppress the reflectance entirely for AOI lower than 40°. The reflectance is reduced, from 16.4 to 5.7% , for the blank glass at higher AOI ~ 70° [43].

By solution processing one may achieve a porous layer with low RIs to satisfy the AR condition stated in Eq. [2]. Transparent substrates, such as glass or plastic, when coated with low RI polymer materials may have similar transmission levels as when uncoated. However, when the coating gets porous, the reflection does go down as discussed before. The challenge, however, is to maximize a pore volume ratio, keeping the pore size smaller than that of visible wavelength [44].

Physical vapour deposition technique can grow porous films as well. Thin films will grow with decreasing densities, compared to that of the bulk, when deposited at increasingly oblique or non-normal AOI. Films with low density, or high porosity, and hence low RI can be grown by the glancing angle deposition (GLAD) [45] technique. The density (hence the RI) can be controlled at each surface or section of the growing film via the obliqueness of the growth flux (Fig. 4.4e, type II). Hence, such a porous film can also have a graded refractive index (GRIN) as a function of its thickness

(z). A graded RI of the film ($n(z)$), from $n_{m(air)}$ to n_S, is achieved by controlling the density, to match even a theoretical profile, say a Gaussian type, increasing the efficiency of the AR structure [46].

One mask-less wet etching product is porous silicon (PSi) which is prepared by electrochemical etching of silicon in hydrofluoric acid solution [47]. Here the pore volume fraction, hence density or effective RI, is controlled by simply the etching current and time of etching. Active solar cells devices have used AR coatings made of such PSi [48]. A dynamic etching procedure, where increasing or decreasing current densities, with time, controls the pore volume fraction at different depths of the PSi layer [49], can replace the conventional constant current etching method. A single-layer PSi, with a gradient-RI, could replace a single homogeneous PSi layer, with a single effective RI.

4.3.3 Rough surface and moth eye structures

To realize the concept of AR from SWS, we have to distinguish between light reflecting from macroscopic and microscopically rough surfaces characterized by the mean surface level of irregularities (vertical) and the correlation length between these irregularities (horizontal) [50]. If the root-mean-square irregularity is larger than the λ of light (macroscopically rough surface) the reflected light will diffuse via scattering to degrade the transmitted wave-front. Whereas, for microscopically rough surfaces, both the mean value of irregularity and the corresponding correlation lengths are smaller than λ, and the light interacts with the whole surface rather than interacting with individual rough spots (Fig. 4.4e, type III-IV).

The periodic SWS as AR surfaces were first discovered in nature while inspecting the eyes of night-flying moths by Bernhard in 1967 [24]. Scientists have successfully reproduced the moth eye structures with interference lithographic techniques, precisely fabricating a regular array on glass substrate, as shown in Fig. 4.5a [25]. The basic principle of 'moth eye' AR surface is that the outer surface consists of nipple arrays, typically of submicron height and spacing. The RI of the surface layer varies gradually from air to substrate, and effectively suppresses the specular reflectance at the interface of the two media.

The dependence of the reflectance on the effective height (h) and spacing (s) of the array, and the radiation λ, may lead to an optimum design of 'moth eye' AR surface. Representing the graded interface by many layers of equal thickness, having progressively increasing RI [25], is a modification of the original approach by Rayleigh [17]. If $s < \lambda < 2.5\ h$ at normal incidence, the reflectance is essentially zero, as shown in Fig. 4.5b, c. The widest bandwidth, in an ideal case, can be achieved by tuning the spacing as fine as possible and the height as long as possible.

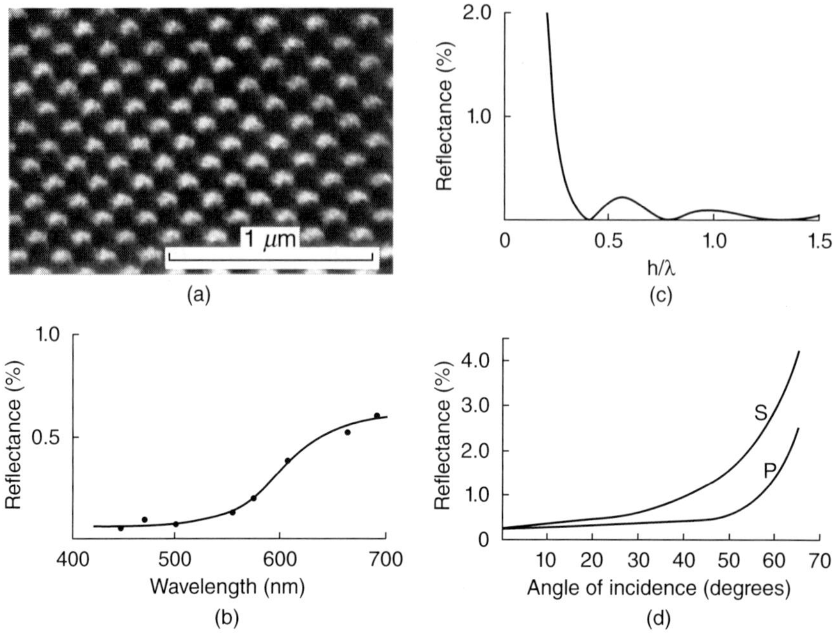

4.5 (a) Stereoscan picture of photo-fabricated array. (b) Measured specular reflectance as a function of wavelength at normal incidence. (c) Computed dependence of reflectance on h/λ. Reprinted with permission from Ref. [25]; copyright 1973: Nature Publishing Group. (d) Reflectance as a function of angle of incidence and polarization of light for a moth eye with 3300 grooves per millimetre at λ = 647.1 nm. Reprinted with permission from Ref. [51]; copyright 1982: Taylor & Francis.

It is worth mentioning that the reduction of specular reflectance in the moth eye is not due to the increase in diffuse scattering or the degradation of the transmitted wave-front, but solely the consequence of enhanced transmission; hence suitable for the purpose of camouflage. Another remarkable phenomenon is the insensitivity of the reflectance value on the AOI over a wide range (Fig. 4.5d). Hutley's group also showed that the moth eye effect could be widely applied for various wavelengths using different substrates such as gold, copper and nickel [51].

Recent progress in the research on AR surfaces of insects provides precise estimation of structure dimensions, and enables suitable optical modeling of them [52, 53]. Three models for the nipple profile, having conical, paraboloidal and Gaussian-bell shape, were proposed (Fig. 4.6a–c). The optical multi-layer model, assuming 100 layers with thickness $h/100$, where h (height of nipples) varies from 50 to 250 nm, revealed that the nipples of paraboloidal-shape showed better AR performance compared to the other two, for light incident normally. The reflection is progressively diminished with increased h, and can be suppressed significantly in the visible region for h = 250 nm (Fig. 4.6d, e). The strongest reduction in reflectance occurs for nipples with larger width (0.53 nm), where the nipples touch each other at their base (Fig. 4.6e), in comparison with the non-touching nipples (width = 0.4 nm, Fig. 4.6d). Similar strong reduction in the angle dependent reflectance results, for TE (s-) and TM (p-) polarized light with increasing h, at AOI < 50° (Fig. 4.6 f–g).

In 1991, Southwell, using numerical methods, showed that the surface–relief structures having a modified pyramidal configuration (having curved rather than flat triangular sides) possess the optimum gradient-RI profile [54]. Such SWS, with the basic pyramidal shape differing slightly with a change in the RI of the material (Fig. 4.7), was approximated by the quintic gradient-RI profile (Eq. [7]) [54]. The surface reflection can be reduced by several orders of magnitude over the broadest bandwidth. Later, based on two-dimensional EMT, Grann *et al.* also pointed out that the shape of continuously tapered SW grating structures is the best design for broadband AR surface [35].

In a recent work [55], simulated biomimetic profiles and their corresponding reflectances, at 1000 nm wavelength, were compared (Fig. 4.7c–e) as a function of height and period of those structures. With a surface profile, as shown in Fig. 4.7c, reflectance of less than 1% could be obtained for height > 400 nm and period > 300 nm. However, for large height (> 1200 nm), a wider window of period exists for which reflectance is below 1%. This window is extremely narrow for the other profiles (Fig. 4.7d, e) indicating practical difficulty in achieving sub-1% reflectivities. Calculated reflectance contours for all the surface profiles shown (Fig. 4.7c–e) demonstrate superiority of that shown in Fig. 4.7c.

4.4 Biomimetic photonic and anti-reflecting nanostructures

Biomimetics is about mimicking natural designs in the laboratory to achieve specific functions, such as AR. Another example will be the wings of butterflies [4, 56–61] that use photonic structures to bring about the iridescence (Fig. 4.8). Several other biological species, for example beetle, peacock, hummingbird, Indigo snake skin, with similar micro-features, also manipulate light.

Certain butterflies and moths [51–53, 62–65] which benefit from their corneal microstructures, to see in low-light conditions (Fig. 4.9a–g), demonstrate the AR function. These are GRIN structures, discussed before, that match the optical impedance of two adjacent media at their interface, minimizing reflection or improving the photon collection efficiency. Nature

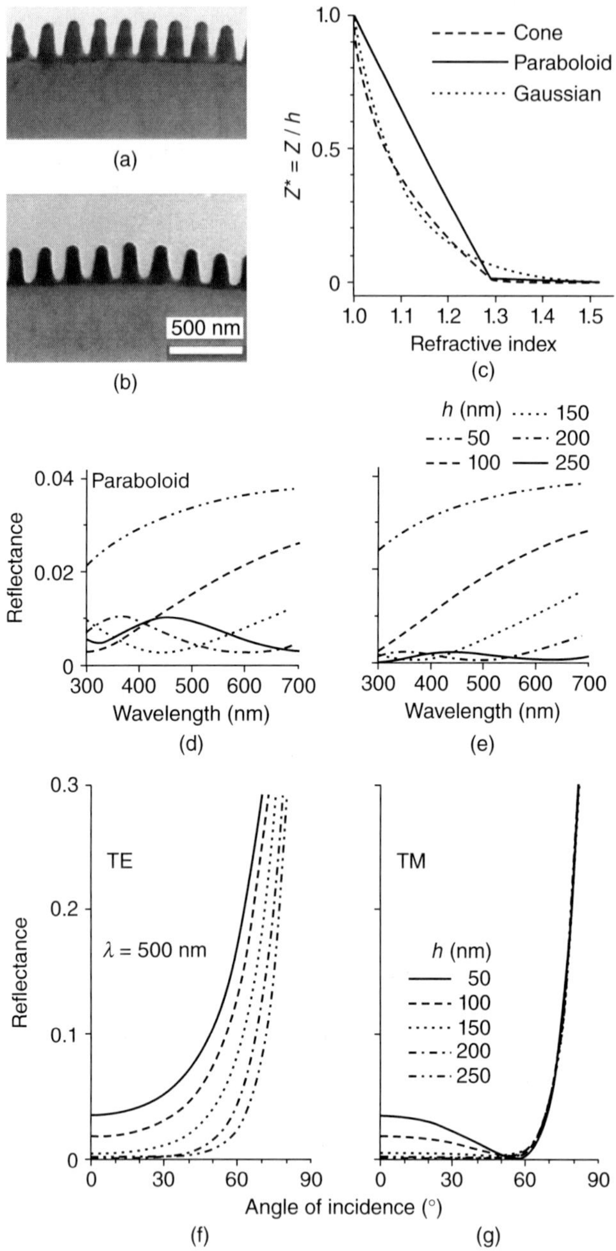

(a)
500 nm
(b)
Cone
Paraboloid
Gaussian
Z* = Z / h
Refractive index
(c)
h (nm)
50
100
150
200
250
Paraboloid
Reflectance
Wavelength (nm)
(d)
(e)
TE
λ = 500 nm
TM
Angle of incidence (°)
(f)
(g)

4.6 Transmission electron microscope (TEM) image of corneal nipple arrays in the nymphalids (a) *Bicyclus anynana* and (b) *Polygonia c-aureum.* (c) Effective refractive index values, at level z^*, for three model nipple types with cone, paraboloid and Gaussian-bell shape. $z^* = z/h$, where z represents Z-axis coordinate, perpendicular to the corneal surface, so that the nipple array troughs were at $z^* = 0$ and the nipple peaks at $z^* = 1$. (d–e) Reflectance of nipple arrays with paraboloidal-shape, for normally incident light, as a function of the height of nipples; for two nipple-widths 0.40 nm (d) and 0.53 nm (e). Line symbols are common for d and e. (f–g) Dependence of the reflectance on the angle of incidence for different nipple heights, for TE (s-) polarized and TM (p-) polarized light, respectively. The nipples were taken here to be touching paraboloids (width = 0.53 nm, Fig. 4.6e). The light wavelength was 500 nm. Line symbols are common for f and g. Reprinted with permission from Ref. [52]; copyright 2006: Royal Society Publishing.

achieves this function by an array of tapered elements (Fig. 4.9b, e), known as nipple arrays, on these arthropodal ommatidial surfaces. These structures can have varied shapes and sizes (Fig. 4.9c, d) controlling the extent of AR they can deliver. These nipple arrays can also be found on other parts of the insects, such as the wings (Fig. 4.9h), to produce a transparency useful for camouflaging [66]. However, the transparency in the wings of hawk-moths requires these structures to be present on both surfaces of the wing. Conventional engineering techniques could replicate these structures in the laboratory (Fig. 4.9i). In some cases, the natural structures serve as a template themselves to create an artificial copy [66, 67] rendering the complicated laboratory processes redundant. For example, the array structures found on cicada wings (Fig. 4.10a, b) could be easily replicated using the real wing as a template [66–68].

Figure 4.10c–h shows the use of a cicada wing as a bio-template to replicate the wing surface structure on gold (Fig. 4.10f) as well as PMMA (Fig. 4.10j). In a similar fashion, the cicada wing can be used as a stamp to replicate its surface features on a soft PMMA (at 190 °C) film [66]. This process is similar to nano-imprint lithography. The functionality of the cicada wing structure is to produce a high degree of AR (Fig. 4.10i). The actual reflectance values in these wings are around 2% in the visible region (Fig. 4.11a). They may increase to 8% maximum in the visible, if the nipple arrays or the protuberances are crushed (Fig. 4.11a). The ratio of the reflectance in the smooth wing (nipples crushed) to the actual wing is shown in Fig. 4.11b. The reflectance from a biomimetic PMMA structure, shown in Fig. 4.10j, was found to be nearly six times lower (Fig. 4.11c) than that of a planar (without the nipple arrays) PMMA film in the wavelength regime of 250–800 nm [67].

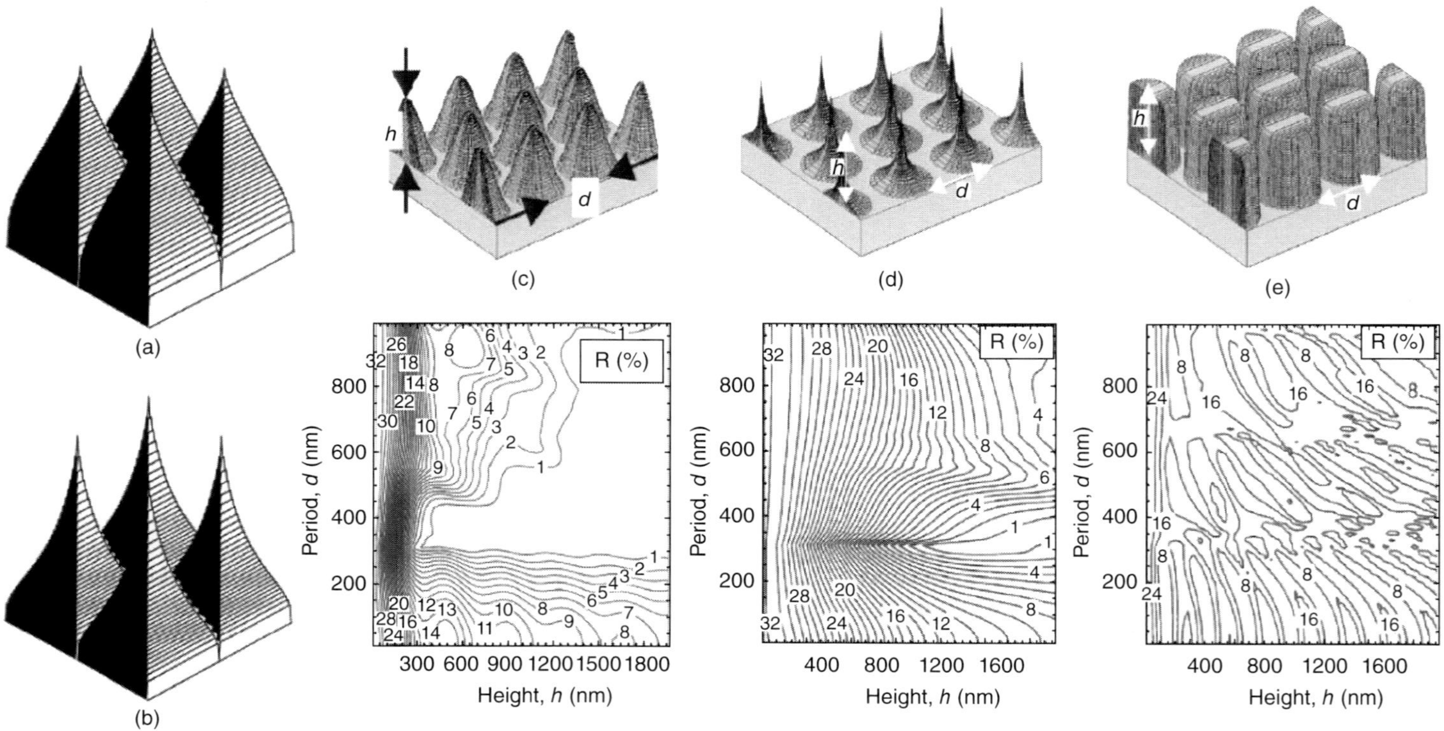

4.7 Surface relief pattern for an equivalent quintic on a substrate with RI (a) 1.45 and (b) 3.42. Reprinted with permission from Ref. [54]; copyright 1991: Optical Society of America. (c–e) Biomimetic moth eye profiles defined by simulation software and corresponding parameter scan of reflectance of light with a wavelength of 1000 nm vs. height (*h*) and period (*d*) for the hexagonal array of pillars shown alongside. The silicon refractive index used for a wavelength of 1000 nm is 3.617 + 0.0041i. Reprinted with permission from Ref. [55]; copyright 2008: American Institute of Physics.

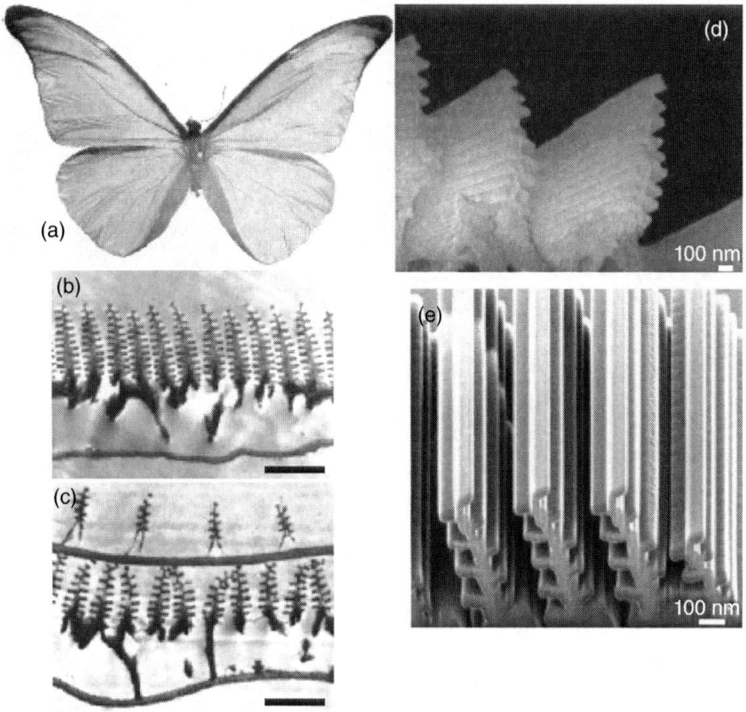

4.8 (a) The iridescent wings of a *Morpho* butterfly. TEM showing wing scale cross-sections of (b) *M. rhetenor* and (c) *M. didius* butterfly. The high occupancy and high layer number in *M. rhetenor* creates an intense reflectivity that contrasts with a more diffusely coloured appearance of the *M. didius*. Scale bar in (b) and (c) is 1.3 μm and 1.8 μm, respectively. (d) Scanning electron micrographs (SEM) of the structures in the wings that reflect blue light. (e) A mimic fabricated with focus ion beam-chemical vapour deposition (FIB-CVD) method. (a), (d) and (e) are reprinted from Ref. [61] with permission; copyright 2005: Japan Society of Applied Physics. (b) and (c) are reprinted from Ref. [60] with permission; copyright 2003: Nature Publishing Group.

In the field of AR, these biomimetic structures did play a big role. The apparently simple concept of AR becomes an experimental challenge when one tries to suppress the reflectance over a wide band of wavelengths, or AOI or using both polarizations of the incident light. The (nano)-patterned SWS used for AR are optically active to shorter wavelengths (ultraviolet and visible) affecting the reflectance, whereas they fail to do so for radiation with longer wavelengths (infrared and above) to which they appear featureless. In other words, it is easier to suppress reflectance at shorter wavelengths than longer ones, and in single SWS, the reflectance increases with increasing λ. Again, as discussed in the introduction, most AR structures

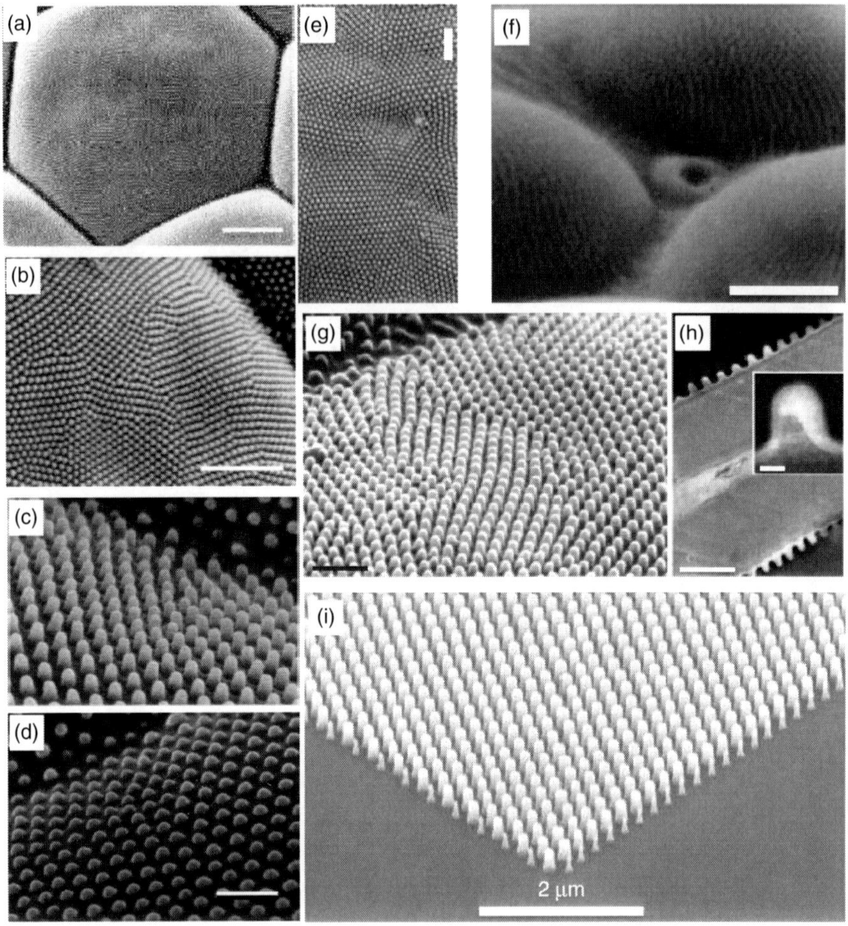

4.9 (a) Corneal nipple arrays in one facet lens and (b) detail showing the local arrangement of domains with highly ordered nipple arrays in the peacock (*Inachis io*), a nymphalid butterfly. Scale bar in (a) and (b) are 5 and 2 μm, respectively. Corneal nipple arrays in (c) nymphalid *Polygonia c-aureum*, and (d) lycaenid *Pseudozizeeria maha*, showing difference in nipple height and shape. Scale bar, 500 nm. Reprinted with permission from Ref. [52]; copyright 2006: Royal Society Publishing. (e) SEM image of an anti-reflective surface from the eye of a moth. Scale bar, 1 μm. (f) Anti-reflector with ridges on three facets on the eye of a 45 million year old fly (*dolichopodid*). Scale bar, 3 μm. (g) Nipple arrays on ommatidial surfaces of a lepidopteran eye appear identical to those found (h) on the transparent wings of certain hawkmoths. This image (h) shows a transparent wing section with a nipple array on both surfaces. Scale bar is 1 μm. Inset in (h) shows the magnified profile of a single nipple. Scale bar is 100 nm. (i) The biomimetic replica of a moth eye fabricated using ion-beam etching. (e), (f) and (i) are reprinted from Ref. [62] with permission; copyright 2007: Nature Publishing Group. (g) and (h) are reprinted from Ref. [60] with permission; copyright 2003: Nature Publishing Group.

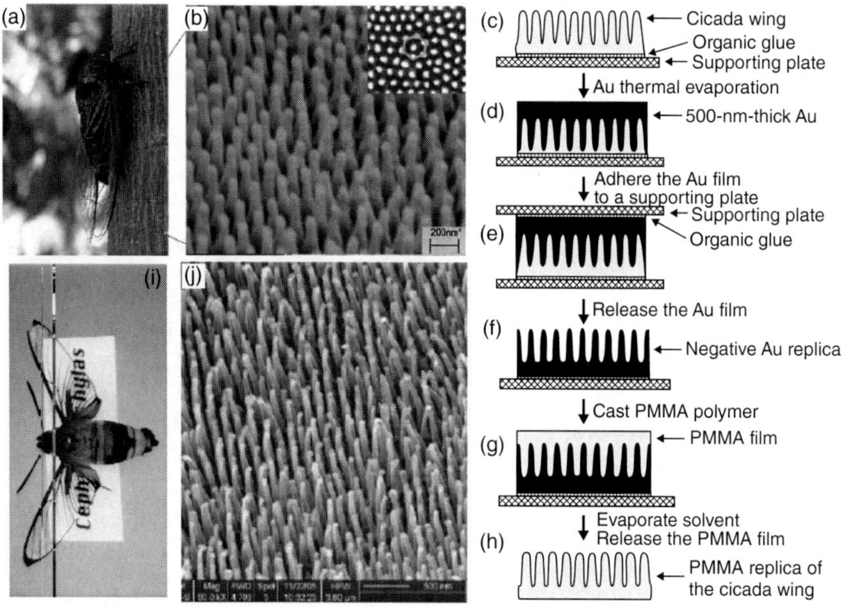

4.10 (a) Photograph of a *Cryptympana atrata Fabricius*. Reprinted with permission from Ref. [66]; copyright 2006: Wiley-VCH Verlag GmbH & Co. KGaA. (b) Tilted top view SEM image of the surface of a cicada wing. The pillars are arranged in a hexagonal array (inset) with 190 nm inter-pillar spacing. The inset in (b) shows a top view SEM image of the cicada wing. (c)–(h) Schematic diagram of the replica moulding procedure using the cicada wing as a bio-template to fabricate a sub-wavelength anti-reflective nanostructure on a PMMA polymer film. Reprinted with permission from Ref. [67]; copyright 2008: IOP Publishing Ltd. (i) Photograph of a *Cephonodes hylas* demonstrating the wing transparency. Reprinted with permission from Ref. [68]; copyright 2002: Society for Science on Form. (j) SEM image of the replicated PMMA film (as in (h)) using the negative Au-mould (as in (f)). Reprinted with permission from Ref. [67]; copyright 2008: Institute of Physics Publishing Ltd.

fail when it goes to grazing AOI. Hence if we treat a normal incidence as AOI = 0°, then an AR structure is superior if it can suppress reflectance over longer λ and AOI. Researchers have been trying to extend the limits of the wavelength span and AOI and have been successful largely.

4.4.1 Biomimetic silicon nanotips for anti-reflection

Silicon nanotip (SiNTs) array, adapted from the natural moth eye structure, are fabricated monolithically on planar silicon wafers using the self-masked dry etching technique [69–71]. A detail review on nanotip structures is

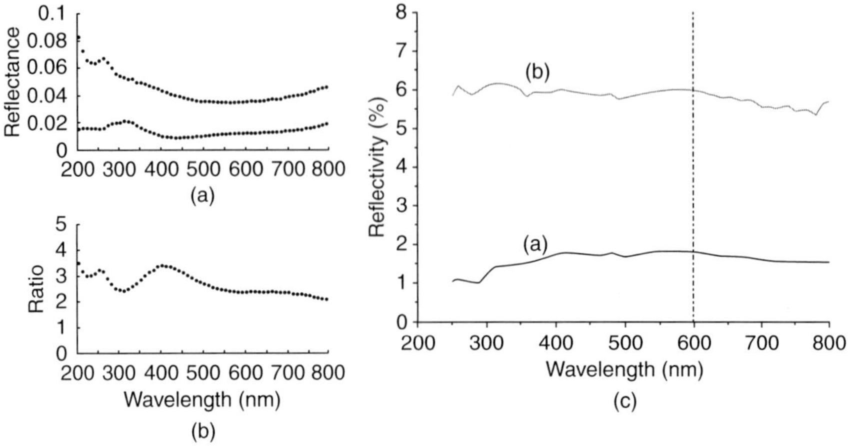

4.11 (a) Representative reflectance spectra of the intact wing of *Cephonodes hylas* (as in Fig. 4.10i) with the protuberances (lower trace) and the smooth wing with the protuberances crushed (upper trace). (b) Spectrum of the reflectance ratio of the smooth wing to the intact one. Reprinted with permission from Ref. [68]; copyright 2002: Society for Science on Form. (c) Wavelength dependence of the measured reflectivity of the (curve a) replicated PMMA film with the nano-nipple arrays on the surface (as in Fig. 4.10j), (curve b) unpatterned planar PMMA film. Reprinted with permission from Ref. [67]; copyright 2008: IOP Publishing Ltd.

available [72]. Several microns of these SiNTs (Fig. 4.12a), with base and apex diameters of 200 nm and 10 nm, respectively, when formed on the Si wafer renders it black (Fig. 4.12b) and shine-less.

The black and shine-less colour of the SiNTs reveals the AR property inherent in it. Hemispherical (specular + diffuse) and specular reflectances were measured on the SiNTs, having different lengths and compared to that of planar commercial silicon wafers, over a broad band of wavelengths spanning UV–VIS–NIR (Fig. 4.13a), mid-IR (Fig. 4.13b), far-IR (Fig. 4.13c), and terahertz (THz) (Fig. 4.13d) [73]. A significant suppression of reflectance results in the SiNTs, especially for the long (16 μm) ones, in comparison to planar silicon wafers. In the UV–VIS–NIR region, the 16 μm long SiNTs showed highest reflectance of only 0.7% (Fig. 4.13a). Expectedly, the absolute reflectance values increased, with increasing wavelengths, independent of the length of the SiNTs.

Even in the THz regime the 16 μm long SiNTs showed lower reflectance than planar wafers at least up to 750 μm of wavelength (Fig. 4.13d). This broadband AR efficiency in the SiNTs is even superior to natural moth eyes that are most effective in the visible 300–800 nm range [73]. The differences between the nipples in moth eye and SiNTs are their height, which can be

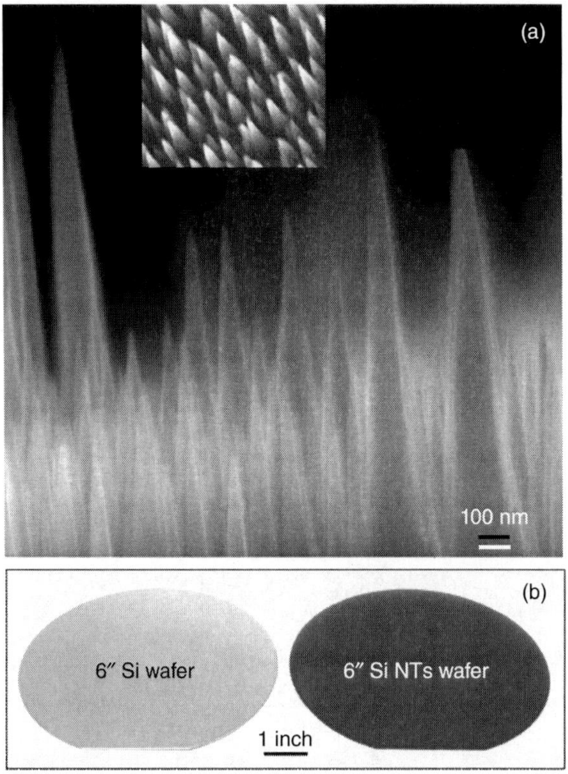

4.12 (a) A cross-section SEM image of the SiNTs. Inset in (a) shows a tilted top view SEM image of the SiNTs. Common scale bar. (b) Optical photograph comparing the shiny polished silicon wafer (left) and the black SiNTs wafer (right). Reprinted with permission from Ref. [73]; copyright 2007: Nature Publishing Group.

higher in the latter, and the shape, which is rounded pillar-like for the nipples and tapered conical for the SiNTs. In addition, another difference lay in their arrangement, where the moth eye nipples demonstrated a hexagonal packing, but the SiNTs are random. Later we will discuss the effect of the shape on the RI profile and hence the reflectance.

Reflection is also a function of the AOI and polarization, and a good anti-reflector must show low reflectance over a wide range of AOI values for both forms of light polarization, *s* and *p* [21]. The specular reflectance from planar silicon and 16 μm-long SiNTs surfaces were compared using unpolarized (Fig. 4.14a) and polarized light (Fig. 4.13b, c) at wavelength ranges of 0.3–2.0 μm (Fig. 4.14a–c) and 2.5–20 μm (Fig. 4.14d–f), respectively [73]. The strong dependence of specular reflectance with AOI, for both the *s*- and *p*-polarized light in polished silicon (Fig. 4.14b, e), could be

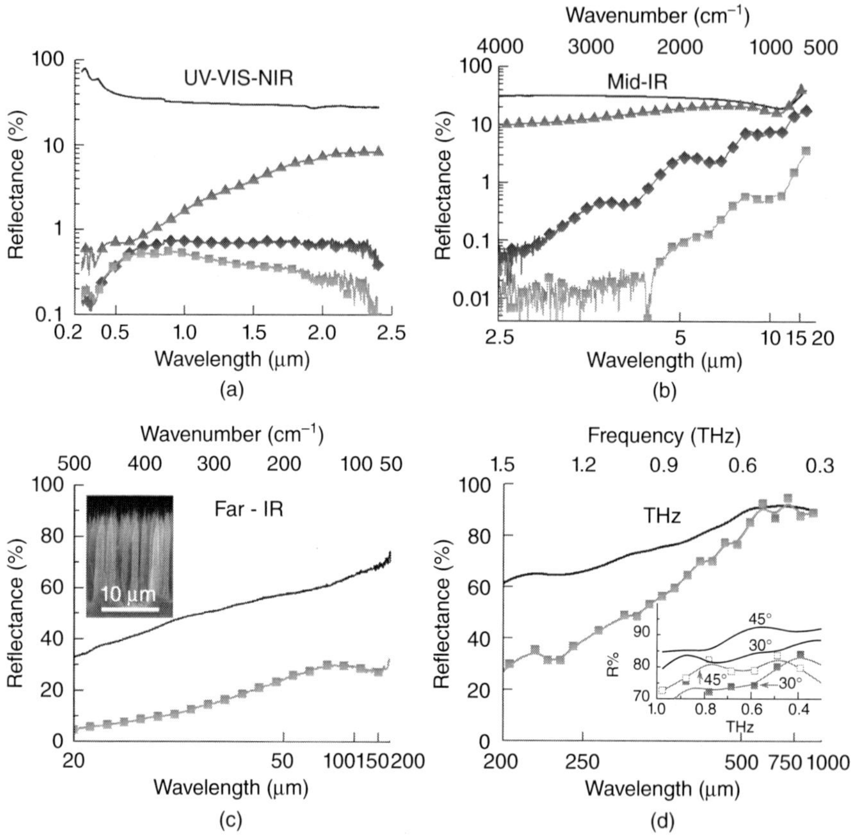

4.13 (a) Comparison of the hemispherical reflectance (using IS) as a function of wavelength for a planar Si wafer (solid line, black) and SiNTs (symbols) with L = 1.6 μm (▲), 5.5 μm (♦) and 16 μm (■) at UV, VIS and NIR wavelengths. (b) Comparison of specular reflectance (without IS) as a function of the wavelength in the mid-infrared region for an AOI of 30°. (c), (d) Comparison of specular reflectance as a function of wavelength for planar silicon wafer (solid line, black) and SiNTs with L = 16 μm (■) in the far-infrared (c) and terahertz (d) regions for an AOI of 30°. Inset in (c) shows the cross-sectional SEM image of the L = 16 μm SiNTs. Inset in (d) compares the reflectance in planar silicon (solid line, black) and SiNTs (symbols, ■) with unpolarized light and an AOI of 30° (filled squares) and 45° (open squares). The solid lines in (d) are guides to the eye. Reprinted with permission from Ref. [73]; copyright 2007: Nature Publishing Group.

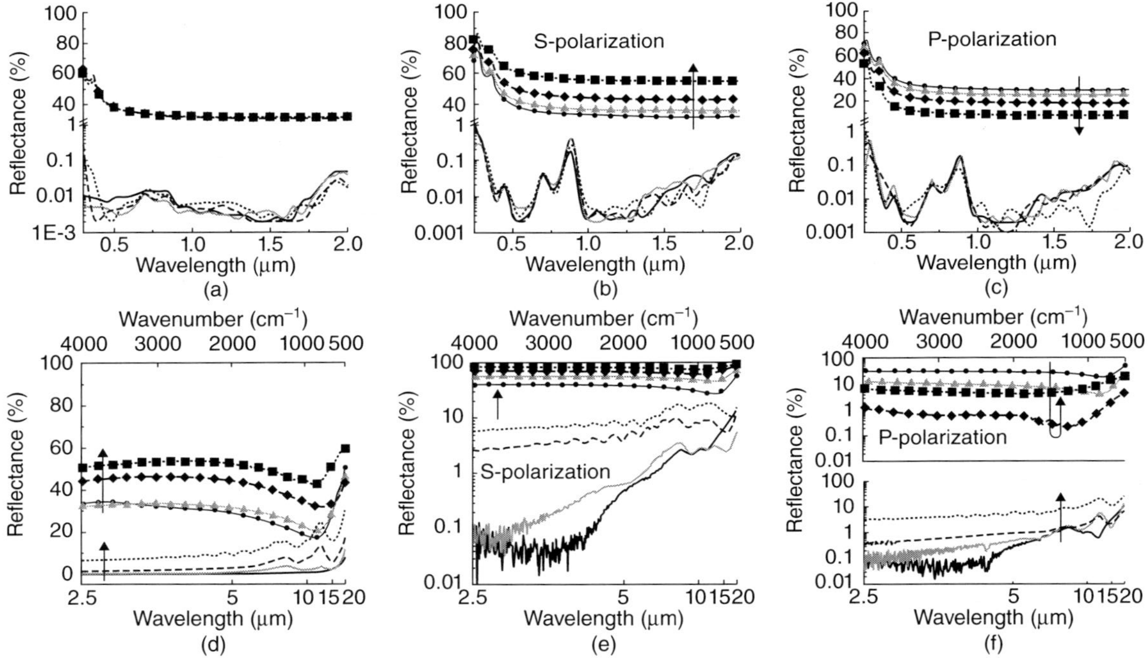

4.14 (a–c) Comparison of the specular reflectance from planar Si wafer (symbols) and SiNTs (lines), with L = 16 μm, as a function of wavelength for unpolarized (a), *s*-polarized (b), and *p*-polarized (c) light at wavelengths below 2 μm for four different values of AOI: 5° (●,—), 30° (▲,—), 45°(◆,– – ·) and 60° (■,……). Note that the *y*-axis is linear for planar Si wafer data and logarithmic for the SiNT data. (d–f) Comparison of the specular reflectance from a planar Si wafer (solid lines) and SiNTs (symbols), with L = 16 μm, as a function of wavelength for unpolarized (d), *s*-polarized (e), and *p*-polarized (f) light at wavelengths between 2.5–20 μm for four different values of AOI: 30° (●,—), 60° (▲,—), 75°(◆,– – ·) and 80° (■, ……). Note that the *y*-axis for each panel is different, and that the planar Si and SiNT data in (f) are drawn in separate graphs for clarity. The direction of the arrows indicates the variation in reflectance as the AOI were increased. Reprinted with permission from Ref. [73]; copyright 2007: Nature Publishing Group.

lifted upon nanostructuring of the surface. The specular reflectance of the 16 μm-long SiNTs remains below 1% at AOI values between 5° and 60° for both *s*- and *p*-polarized light (Fig. 4.14b, c) in the UV–VIS–NIR region [73]. For the same sample, with *s*- and *p*-polarized light within the AOI range of 30 to 80°, the reflectance is below 20% in the mid-IR range (Fig. 4.14e, f). Barring monochromatic measurements, most SW or quarter wavelength structures, independent of the material used, show a suppression of reflectance below an AOI of 60° (in an extended VIS to IR region) [35, 43, 74–77]. In this case, the suppression of reflectance is easily observed up to 70°. The reflectance from the aperiodic SiNTs demonstrates a near independence from the AOI compared to the periodic multi-layer coatings and SWS, as well as non-periodic porous materials.

The best RI profile of the SiNTs had to be determined by the theoretical analyses [38] and simulated fitting (using WVASE 32, by J. A. Woolam) [78] of the experimental data to interpret the wide angle and polarization independent AR properties. The RI profile (Fig. 4.15a) that correctly simulates the measured data, for the 1.6 μm long SiNT array, assumes three layers of Si, instead of one, having gradient-RI. These three regions, I, II and III,

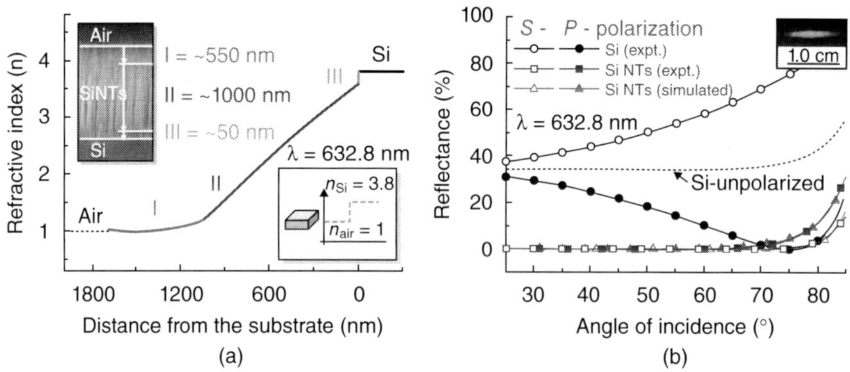

4.15 (a) The refractive-index profile of the SiNT surface. Three distinct regions (described in the text), region I with RI close to 1.0, region II with a graded index, and region III with RI close to 3.8, were observed. The top left inset shows a cross-section SEM image of the SiNT showing the three distinct regions. The bottom right inset shows the refractive index profile of the polished Si wafer, for reference. (b) The reflectance of the Si wafer and SiNTs (L = 1.6 μm) as a function of angle of incidence using *s*-, *p*- polarized light of 632.8 nm wavelength; line joining the data points represent the simulated (calculated) reflectance data for SiNTs (L = 1.6 μm) using the profile in (a). The simulated data match exactly with the real data shown in the plot. Top right inset in (b) shows the optical image of the laser spot at an AOI of 85°. The measured reflectance data for planar Si with unpolarized light is also shown. Reprinted with permission from Ref. [73]; copyright 2007: Nature Publishing Group.

could be distinguished in the RI profile (Fig. 4.15a) to fit the reflectance spectra as shown in Fig. 4.15b. Region I is around 550 nm from the apex of the SiNTs that serves as the air–SiNT interface (inset, Fig. 4.15a). Region II follows region I with a depth of ~ 1000 nm and constitutes the bulk of the SiNTs (inset, Fig. 4.15a). Region III extends about 50 nm from the base of the SiNTs and serves as the SiNTs–Si interface (inset, Fig. 4.15a). Region I and region III, having an RI close to that of air (1.0), and Si (3.8 at 632.8 nm), respectively, behave as the two limits of the RI profile of the SiNTs. Region II have an RI graded with the volume fraction of Si as one goes deeper into the SiNTs.

In contrast to the multi-layered coatings or multi-material structures [37, 79], having stepped gradient-RI, the SiNTs result in a smooth profile only with a single element. The average volume fraction of Si (ratio of volume of silicon to total volume of the SiNTs), for SiNTs array having different NT densities, was estimated to be < 0.5. The Bruggeman EMA [80], leads to a reasonable effective RI of the SiNTs at 1.6–1.8. The improved flatness of the RI profile near the air–silicon interface [38] is the key for the AR performance. Longer apexes (region I) of the SiNTs have shown to reduce the reflection both as a function of the AOI and wavelength. Hence, these SiNT structures are not optimal. Grating structures, with groove depths exceeding 30 μm, conventionally used in the THz regime [81, 82], indicate that longer nanotips may perform better in the THz regime since they would have a finer RI gradient at the air–solid interface. However, the AR performance at the short wavelengths does not deteriorate due to increasing lengths of the SiNTs (Figs. 4.13a and 4.16). We note here that a longer region I (at the solid–ambient interface) (Fig. 4.16a, b) actually results in a better AR performance at all AOI (Fig. 4.16c) and all wavelengths (Fig. 4.16d) [83]. Lee *et al.* also supported the importance of this interface layer [84]. ZnO nanorods, with tapered ends (Fig. 4.16e, f), had much better AR properties (black line, Fig. 4.16g) if they marginally differed in length (region I in Fig. 4.16f) resulting in a rough profile of the apexes compared with an array of identical length nanorods resulting in a planar profile of apexes [84]. A rough apex profile is another way of decreasing the volume fraction (f) and the resultant effective RI. From the results of this work it appears that arrays of long cylindrical structures with tapered apexes and preferably unequal lengths is the near-ideal [38, 54] AR design to work over a broadband and larger AOIs.

The principle of the self-masked dry etching technique [69–72], described above to produce the biomimetic SiNTs [73], has also been used to produce silicon AR structures through reactive ion etching (RIE). In this process [85], silicon is subjected to a radio frequency plasma of carbon tetrafluoride (CF_4) and oxygen (O_2) that creates random masks of fluorine-based compounds. In the second step the silicon is alternatively etched through these

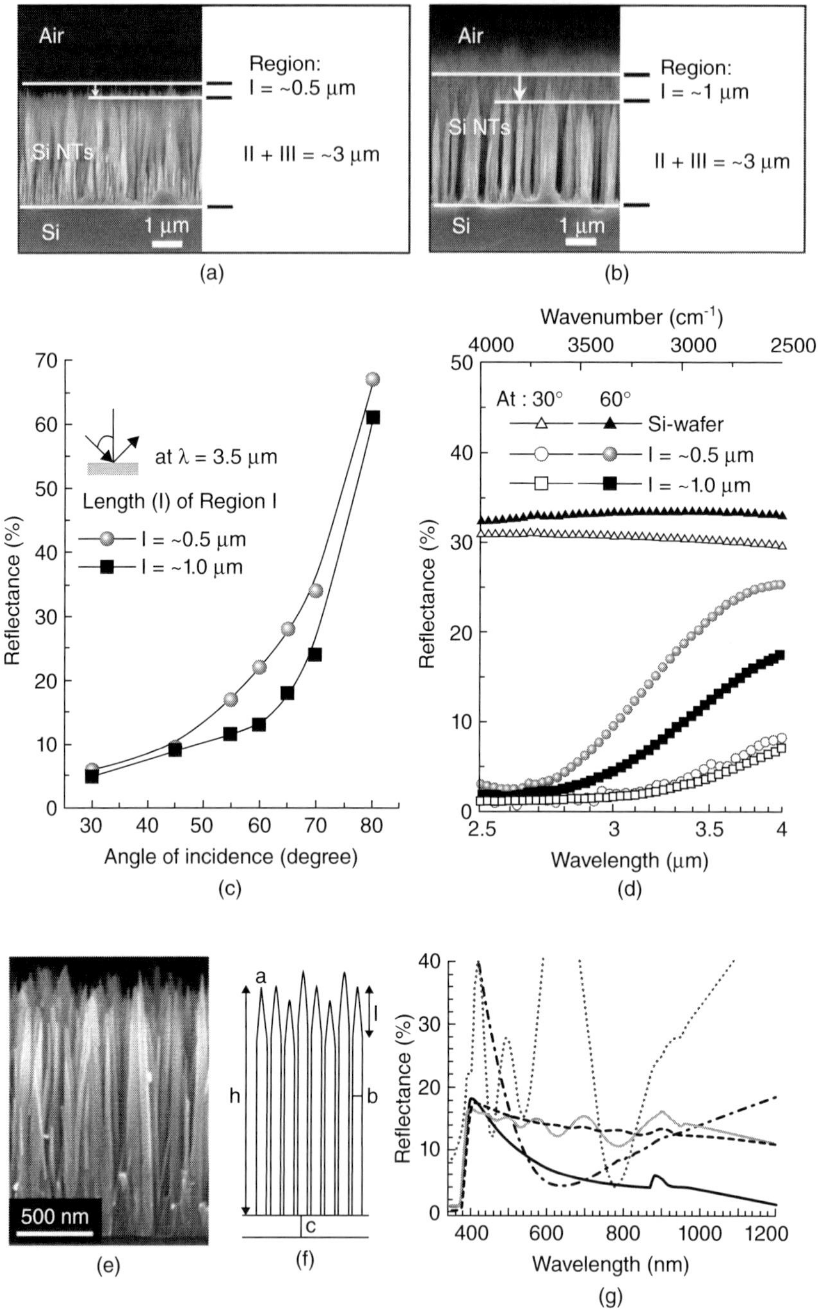
Air
Region:
l = ~0.5 μm
Si NTs
II + III = ~3 μm
Si
1 μm
(a)
Air
Region:
l = ~1 μm
Si NTs
II + III = ~3 μm
Si
1 μm
(b)
at λ = 3.5 μm
Length (l) of Region I
l = ~0.5 μm
l = ~1.0 μm
Reflectance (%)
Angle of incidence (degree)
(c)
Wavenumber (cm^{-1})
At : 30° 60°
Si-wafer
l = ~0.5 μm
l = ~1.0 μm
Reflectance (%)
Wavelength (μm)
(d)
500 nm
(e)
a
l
h
b
c
(f)
Reflectance (%)
Wavelength (nm)
(g)

4.16 Geometry-dependent reflectance data for SiNTs as a function of angle of incidence and wavelength. SiNTs with region I (defined in text) equal to (a) 0.5 μm, and (b) 1.0 μm. (c) Angle-dependent and (d) wavelength-dependent reflectance data at 30° and 60° AOI, clearly showing better performance for the sample shown in (b) compared to that in (a). In (d), the reflectance data for planar silicon wafer are shown for reference. Reprinted with permission from Ref. [73]; copyright 2007: Nature Publishing Group. (e) Cross-section SEM image of a highly tapered ZnO nanorod array (NRA). (f) Schematic representation of the corresponding nanorod parameters extracted via software analysis. (g) Front reflectance spectra of flat top NRA (—), tapered NRA (– – ·), sol-gel film (······), optimized SiN single-layer anti-reflection coating on Si photovoltaic cell with metallic contact (– · –), and highly tapered NRA (—). The bump at 900 nm is due to a detector change during data collection. Reprinted with permission from Ref. [84]; copyright 2008: American Chemical Society.

masks with sulfur hexafluoride (SF_6) using a deep RIE, and passivated with octafluorocyclobutane (C_4F_8). This process, commonly known as the Bosch process or time-multiplexed etching process, creates nearly vertical structures. The RIE-produced silicon nanotips, although not categorized as biomimetic by the authors, are larger (20 nm apex) than the single step self-masked dry-etching (SMDE)-produced biomimetic nanotips (5 nm apex) and have a corkscrew-like surface feature. These silicon nanotips also show impressive AR characteristics with less than 8% hemispherical reflectance in the entire 0.2–2.5 μm wavelength regime [85].

Both the nanotip [73, 85] structures demonstrate superior AR performance over broadband owing to the gradient index achieved in the tapered structure. Considering the low absorption coefficient of 5200 and 70 cm^{-1} for crystalline silicon at 600 and 1100 nm, respectively, the AR performance of the SMDE-produced 16 μm long silicon nanotips is commendable [73]. This, however, implies a better AR performance from longer (say > 30 μm) silicon nanotips. It would be a challenge to fabricate amorphous silicon nanotips having higher absorption coefficient and even lower reflectance than these silicon nanotips.

4.4.2 Biomimetic non-silicon structures for anti-reflection

The efficiency of the gradient index dictates the overall reflectance. The option of using starting materials with lower RI still exists. Let us describe the performance of non-silicon AR structures here. The published reports can broadly be classified into five major groups: (i) silicon oxide based, such as glass ($n \sim 1.52$–1.80), quartz ($n \sim 1.54$), and fused silica ($n \sim 1.459$), (ii) polymers ($n \sim 1.46$–1.55), (iii) diamond ($n \sim 2.417$) or diamond-like coatings,

(iv) metals (gold $n \sim 0.47$, Ag $n \sim 1.35$) and (v) Group III–V semiconductors such as GaAs, GaN, GaSb or InP.

Silicon oxide based

Lithographic principles and dry etching technique, such as RIE or fast atom beam (FAB), produce conical structures on fused silica [86], quartz [87] and glass [88] substrates. For SWS on fused silica [86] and quartz, a fluorocarbon plasma has been used whereas an FAB of SF_6 has been used for SWS on glass [88]. Minimum reflectance values of 0.5–1% and maximum transmittance values of 99% were obtained, in the visible, for glass and fused silica structures [86, 88, 89]. Double-side sub-wavelength grating structures on quartz resulted in reflectance values of ~ 0.1% in the 400–800 nm range. Optimized grating structures on quartz showed a reflectance value of only 0.035% [87] in the 450–525 nm range under normal incidence compared to 3.5% in planar fused quartz ($n \sim 1.46$). This result is significant in the sense that it is damage resistant and is transparent birefringent in the UV–IR region.

Polymer based

Moth eye like structures have been fabricated using plasma treatment on PMMA surfaces [90]. 80–120 eV Ar^+ ion beams impinged the PMMA surface, in the presence of reactive O_2 gas at 3×10^{-4} mbar pressures, for 200–400 s duration. The average reflectance in single- and double-side treated PMMA surface is below 4 and 1%, respectively, in the 300–600 nm region. Similar radio-frequency (rf) plasma (CF_4+O_2/N_2) treatment on 200–300 nm thick evaporated metallophthalocyanine (MPc) films, such as copper phthalocyanine (CuPc) or lead Pc (PbPc), on glass also demonstrate AR characteristics [91].

Two-dimensional array of silver ($n \sim 1.35$) coated polystyrene spheres have also been measured for its reflectance in the 1–2 μm wavelength region. However, it only produced a strong modulation of the reflectance compared to the uncoated spheres [92].

Diamond based

Diamond is a wide band gap ($E_g \sim 6$ eV) material and a suitable candidate for infrared (IR) optics. Polycrystalline diamond was grown (bottom up approach), by microwave CVD on photolithographically patterned germanium (Ge) substrates, for AR applications [93]. In this way, a moth eye type surface-relief pattern of diamond was achieved on Ge. The transmittance, of the structure, increased by approximately 10%, compared with pure Ge, in the 4–12 μm region. Again, the graded index theory explained this increase in transmittance.

On the other hand, a top-down approach through laser patterning of CVD diamond films (thickness of 300–400 μm) using laser (1078 nm) scanning has produced relief structures of rather large period of 3 μm and groove depths of ~ 2 μm [94]. Such patterning on both sides of the CVD diamond films resulted in an increase of the IR transmission by ~ 10% above 9.6 μm wavelengths, compared to polished CVD diamond [95].

Metal based

Metals are generally reflecting but grating structures can reduce the specular reflectance in them. Metallic gratings can also absorb a particular band of frequency in the incident radiation instead of diffracting them [95]. A design developed for an AR metallic grating used effective index method and impedance matching [96]. The grating structure was prepared on a chromium-coated photolithographic plate having a period of 1.0 μm, groove depth of 147.5 nm and filling factor of 50%. This metallic grating demonstrates a sharp dip in reflectance, approaching ~ 1–4% at ~ 480 nm, on both sides of which the reflectance increases significantly [97]. This might be associated with plasmon absorption in gold that occur around 500 nm and red shifts with size.

Group III–V based

Group III–V materials are known for having tunable band gaps (E_g) from the ultraviolet (GaN, AlN), visible (GaAs), to the infrared (InN, InP, GaSb). Efficient AR properties resulted above the band gap energy.

Vapour phase epitaxy grown GaN films (thickness ~ 3 μm) on sapphire substrates (thickness ~ 0.3 mm) could be etched, using a chlorine and Ar plasma, to generate GaN nanotips [98]. These nanotips, with lengths of 0.1–2 μm and densities of 10^9–10^{11}/cm^2, can reduce reflectance and increase transmittance compared to planar GaN. The GaN nanotips show reflectance below 10% in the 400–900 nm regime compared to 25% for planar GaN. Below 360 nm, the reflectance approaches a minimum due to enhanced absorption. Similarly, for GaN pillars, of different height (300–720 nm), patterned on a GaN film grown on quartz substrate, also shows reduced reflectivity with increasing height, compared with planar bulk GaN, over a wide AOI and *s*- and *p*-polarized light [99]. Cone shaped sub-wavelength AR gratings have also been directly generated on GaAs substrates. These SWS GaAs can reduce the reflectivity from 30 to 3% in the wavelength range of 0.4–1.0 μm. Such broadband anti-reflective properties of the AR grating match the rigorous coupled-wave analysis model's prediction [100].

Sub-wavelength gratings (SWG), having 350 nm period and ~ 1300 nm deep grooves [95], were fabricated on lithographically (e-beam) patterned GaSb substrates using fast atom beam etching. The GaSb SWG, which is similar in feature to the moth eye, suppresses the reflectance in the 0.5–2.0 μm spectral region below 10% [101]. Although conventional ZnS and MgF_2 AR coatings degrade upon heating, limiting its applicability in solar cells, these GaSb SWG do not show pronounced degradation.

Non-linear optical compound based

Non-linear optical materials exhibit interesting phenomena such as, second harmonic generation, sum and difference frequency generation, optical parameter amplification and wave mixing and can be used in optical computing, switching, data storage, and parametric oscillators. These materials are used under intense irradiation condition and maintenance of their AR properties is crucial for their device performance. To conclude this section, we select the report of AR properties from a non-linear optical material called zinc germanium phosphide (ZGP) having an RI of 3.17. This crystal can be pumped with 2.09 μm (λ_{in}) laser on one side and emits at longer wavelengths of 3–6 μm (λ_{out}), which is broadband tunable, on the other. The surface of the crystal can be patterned with moth eye structures, the period (Λ) and depth of which could be calculated depending on the target λ_{out} using the relation

$$\Lambda \leq \frac{\lambda_{out}}{\{n_{ZGP} + n_0 \sin\theta_{\max}\}}, \qquad [4.9]$$

where n_{ZGP} (= 3.17) and n_0 (= 1.0) are the RI for ZGP and air, respectively, θ_{max} ~ 10° is the acceptance angle. For a target output of 4.18 μm, Λ is ~ 1.17 μm, assuming the depth of the pattern is 40% of λ_{out}, i.e. 1.7 μm. These moth eye pattern dimensions are achievable over a large area using a combination of interference lithography for pattern definition and mask assisted chlorine RIE [102]. The transmission spectra for these structures, over 2.0–5.0 μm wavelength, show an enhancement above λ_{out} ~ 3.8 μm for both convex and straight tapered structures compared to the planar (bare) ZGP crystal. A double-side patterned ZGP crystal is estimated to have a transmission of less than 87% above 4.0 μm wavelength assuming 67% transmission for the straight tapered structures, which is close to the largest theoretical prediction of 73% [102].

The non-silicon AR structures performed as well as their silicon counterparts, although the advanced knowledge of silicon processing results in better control of the latter structures. Compared to crystalline materials, there is scope for further development in amorphous starting materials such as glass and polymers.

4.5 Future trends and conclusions

Research on efficient AR devices has shifted focus from coatings (Generation I), such as multiple thin films with alternating high and low refractive index materials, to gradient index structures (Generation II), such as biomimetic moth eye or cicada wing structures [73], in pursuit of lower reflectance values over a broad band. Such AR structures are not limited to some preferred materials. For Si-based materials, researchers could suppress reflectance, in the visible region, up to ~ 4% and to ~0.1% for Si-based micro- and nano-AR structures, respectively. Careful observations reveal that there is a trend that random structures, with no specific feature dimension or spacing, perform better than the periodic ones even when hemispherical and specular reflectance were compared. Figure 4.17 shows a representative broad band applicability of the AR structures indicating that random GRIN structures are superior [73, 75, 85, 103].

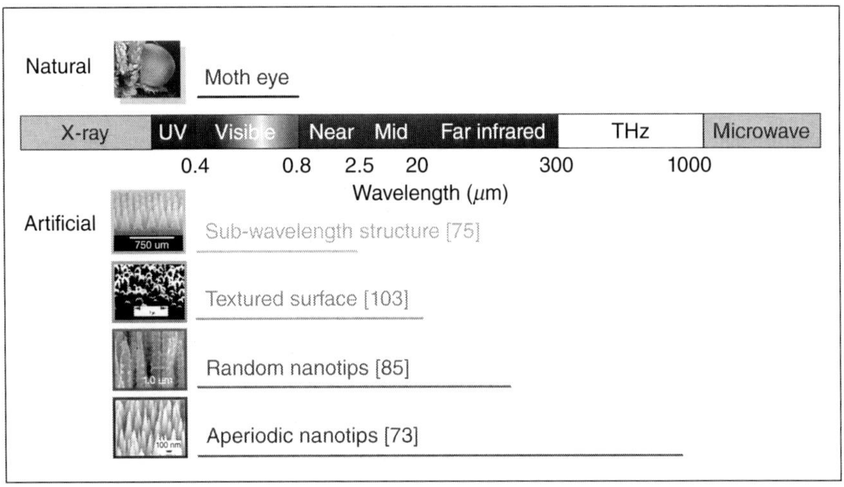

4.17 Representative broad-band applicability of natural and artificial AR structures. The extent of the lines indicates the spectral band (wavelength) over which those AR structures have demonstrated performance. Representative SEM images, shown with source reference numbers (in square brackets), have been reprinted with permission from corresponding publishers: Ref. [75], copyright 1999: Optical Society of America; Ref. [103], copyright 1979: American Institute of Physics; Ref. [85], copyright 2005: American Chemical Society; Ref. [73], copyright 2007: Nature Publishing Group. Reprinted with permission from Ref. [22]; copyright 2010: Elsevier.

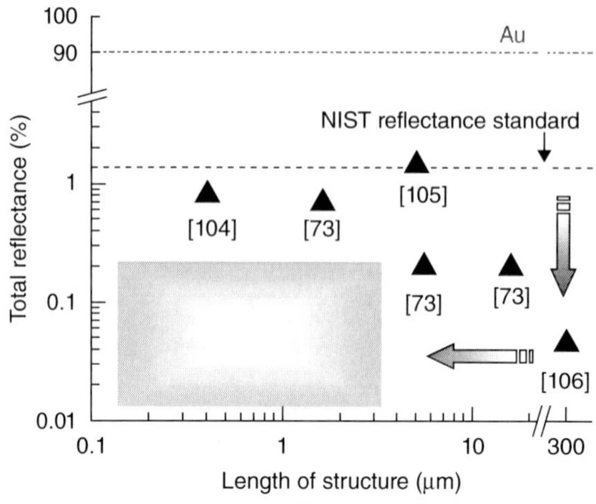

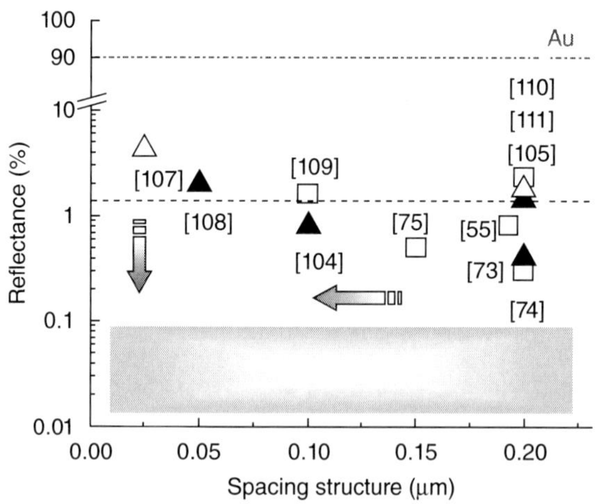

4.18 Reflectance as a function of (a) length and (b) spacing of AR nanostructures. Solid and hollow symbols indicate hemispherical and specular reflectance, respectively. The top dashed line indicates the reflectance (~90%) of polished bulk gold (Au). The lower dashed line indicates the NIST reflectance standard (Reflectance = 1.4%) [106]. The arrows in the figure indicate future directions for better AR efficiency and design. The shaded region indicates the future target regime for improved AR performance. Representative data points have been shown with source reference numbers, in square brackets. Reprinted with permission from Ref. [22]; copyright 2010: Elsevier.

Almost all the gradient index structures in the Generation-II technology used tip-like or tapered structures [72] to achieve the AR purpose. The newer, Generation III, of the AR structures is focusing on the refinement of the gradient index technology, especially at the air–material interface, to mimic not only natural but also theoretical AR structures that promised ideal AR properties. The total reflectance values of AR structures show strong dependence on the structure height/length (Fig. 4.18a) [73, 104–106] and spacing (Fig. 4.18b) [55, 73–75, 104–105, 107–111]. Longer length and optimized spacing lowers the reflectance of these structures. The former is to achieve better gradient index and the latter for multiple scattering.

Future research will mostly address the optimization of the length (Fig. 4.19a) and the profile of the apexes (Fig. 4.19b, e). The absolute value of reflectivity/reflectance will depend on the length/height (*H*, Fig. 4.19a), and density (determined by *S*, Fig. 4.19a) of these structures. Deep cylindrical structures, exceeding 300 μm, which have been used for AR purposes [112], should be tailored at their apex for broadband efficiency. A possible design for an optimal broadband AR structure may consist of a deep vertical grating with tapered ends (Fig. 4.19a). Increasing length (h') and density (decreased *S*) of the grating will increase the absorption of the structure that will lower the absolute reflectance value, while the tapered end will take care of the required gradient index at the air interface [73]. The ratio of the length of the vertical part (h') to the tapered part (h''), however, has to be optimized for particular wavelength bands. Hence, stress will be on

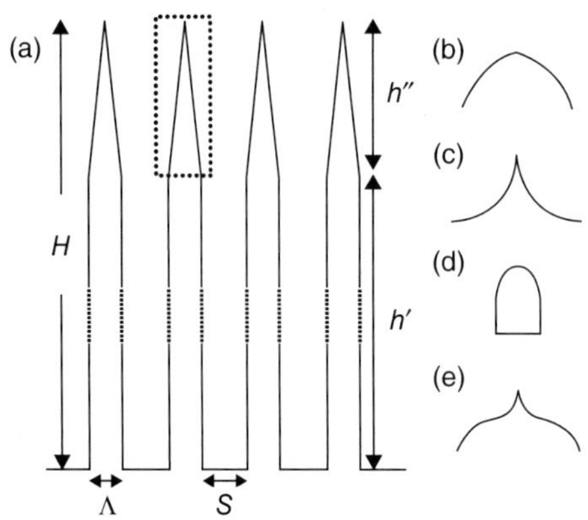

4.19 (a) Schematic of optimal gradient index AR structure, where *h*′, *h*″ and *H* denote the height of the cylindrical part, tapered part (shown in the dashed rectangle) and total structure, respectively. '*S*' is the inter-structure spacing that controls its density, and 'Λ' is the process controlled diameter of the structure. The apex part can have (a) straight tapered, (b) convex tapered, (c) concave tapered, (d) moth eye or (e) theoretical quintic index profiles. Reprinted with permission from Ref. [22]; copyright 2010: Elsevier.

achieving a modified RI profile that improves upon the AR performance theoretically predicted by the quintic index profiles. Achieving darker materials than those already reported is always a possibility; however, imparting a broadband nature to the dark material will be a challenge.

The future may witness an advent of low-index materials that can be complementary to the efforts in AR research involving multi-films and GRIN structures. Available materials, for AR purpose, decreases with lowering of index, and they become quite rare as the index approaches 1.2 and there appears to be an index gap between n = 1.0 (air) and n = 1.3 (water) where virtually no materials can be found. New materials in this index gap can open up this field even further. Some potential candidates have been identified including porous GaN, carbon nanotubes, MgF_2 and even indium tin oxides having low RIs in the range of 1.5–1.03 due to a low material volume fraction (Fig. 4.20).

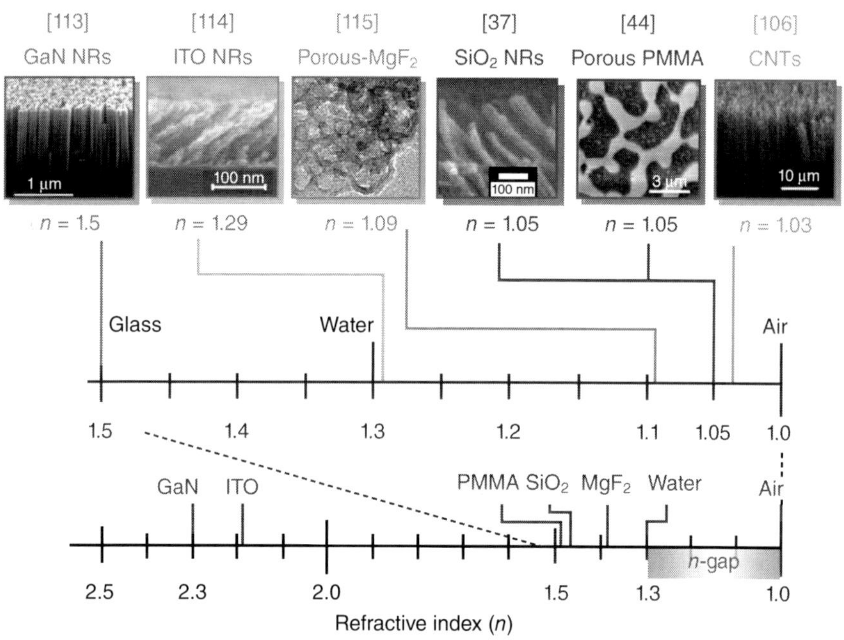

4.20 Low refractive index (RI) bulk materials on the RI scale. The RI gap (n-gap) is shown between 1.0 < n < 1.3. Representative SEM images of low index nano-materials close to and within the n gap. Representative SEM images, shown with source reference numbers (in square brackets), have been reprinted with permission from corresponding publishers: Refs. [37, 115], copyright 2007: Nature Publishing Group; Ref. [44], copyright 1999: AAAS; Ref. [113], copyright 2008: Optical Society of America; Ref. [106], copyright 2008: American Chemical Society; Ref. [114], copyright 2008: Wiley-VCH Verlag GmbH & Co. KGaA. Reprinted with permission from Ref. [22]; copyright 2010: Elsevier.

Efficient broadband AR is a theoretical goal, however, structures with high efficiency AR within selected wavelength regimes are commercially accepted and widely used. For example, in the visible range AR, surfaces are applied to many optoelectronics devices such as surface-emitting laser [116], silicon-based solar cell [117–120], GaAs/AlGaAs solar cell [121], organic light emitting diode [122–124] and light emitting diode [125–127], enhancing their efficiency. Amorphous silicon p–i–n junctions formed on quartz nanotip substrates, partially retaining the substrate morphology for AR, produced large short-circuit currents of and yielded power efficiencies of 5.9% [128]. However, much superior results were demonstrated by K. Ostrikov's group who reported a single step formation of p–n junction solar cells on vertically aligned hierarchical silicon nanotip arrays. They produced short-circuit currents of 32 mA/cm^2 and efficiencies around 11.9% [129]. This is much better than the conventional amorphous silicon thin film single junction solar cells that produced around 7% stable efficiencies. Similar nanotip structures also resulted in impressive efficiency values, 10.8%, which underline the applicability of such AR structures [130]. On the other hand, given the difficulty involved in the long wavelength range, spanning IR and THz, the progress of AR surface design is still mostly under development and optimization [131, 132]. There is also a scope for exciting research in the fields of self-cleaning AR coatings [110, 133], AR coatings on deformable substrates [134], and the use of carbon nanotubes [135]. The future of AR research lies within the scope of new materials, including low index, biomimetic or nanomaterials, and improved designs involving these.

4.6 Acknowledgements

We would like to acknowledge the financial support from the National Science Council and Ministry of Education in Taiwan, National Yang Ming University, National Taiwan University and Academia Sinica, Taiwan.

4.7 References

1. A. F. Huxley, *J. Exp. Biol.* 48 (1968) 227.
2. H. Tabata, S. Tanaka, J. Takimoto, H. Namiki and A. Yoshida, *J. Jpn. Opt.* 22 (1993) 618.
3. R. Fujii, H. Kasukawa, K. Miyaji and N. Oshima, *Zool. Sci.* 6 (1989) 477.
4. H. Ghiradella, *Appl. Opt.* 30 (1991) 3492.
5. A. C. Neville and S. Caveney, *Biol. Rev.* 44 (1969) 531.
6. S. Caveney, *Proc. Roy. Soc. Lond.* B 178 (1971) 205.
7. A. C. Neville, *Biology of Fibrous Composites: Development beyond the Cell Membrane*, Cambridge University Press, London, 1993.

8. C. H. Greenwalt, W. Brandt and D. D. Friel, *Proc. Am. Philos. Soc.* 104 (1960) 249.
9. C. H. Greenwalt, *Hummingbirds*, Doubleday & Co., New York, 1960, 183–200.
10. E. A. Monroe and S. E. Monroe, *Science* 159 (1968) 97.
11. J. V. Sanders, *Nature* 204 (1964) 1151.
12. M. J. Murray and J. V. Sanders, *Philos. Mag*. A 42 (1980) 721.
13. C. A. Murray and D. G. Grier, *Am. Sci*. 83 (1995) 238.
14. J. V. Sanders, *Philos. Mag*. A 42 (1980) 705.
15. A. R. Parker, *J. Opt. A: Pure Appl. Opt*. 2 (2000) R15.
16. M. Srinivasrao, *Chem. Rev*. 99 (1999) 1935.
17. Lord Rayleigh, *Proc. Lond. Math. Soc*. S1-11, (1879) 51.
18. R. Jacobsson, in: E. Wolf (Ed.), *Progress in Optics*, North-Holland, Amsterdam, vol. 5, (1966) pp. 247.
19. R. W. Wood, *Philos. Mag*. 47 (1899) 349.
20. R. W. Wood, *Physical Optics*, Macmillan, New York, 1934, 88.
21. H. A. Macleod, *Thin Film Optical Filters*, 3rd Edition, Taylor & Francis, London, 2001.
22. S. Chattopadhyay, Y. F. Huang, Y. J. Jen, A. Ganguly, K. H. Chen and L. C. Chen, *Mater. Sci. Engng*. R 69, 1 (2010).
23. A. Gombert, W. Glaubitt, K. Rose, J. Dreibholz, B. Bläsi, A. Heinzel, D. Sporn, W. Döll and V. Wittwer, *Thin Solid Films* 351 (1999) 73.
24. C. G. Bernhard, *Endeavour* 26 (1967) 79.
25. P. B. Clapham and M.C. Hutley, *Nature* 244 (1973) 281.
26. J. C. M. Garnett, *Phil. Trans. Roy. Soc. Lond*. 203 (1904) 385.
27. J. C. M. Garnett, *Phil. Trans. Roy. Soc. Lond*. 205 (1906) 237.
28. D. A. G. Bruggeman, *Annalen der Physik* 416 (1935) 636.
29. D. H. Raguin and G. M. Morris, *Appl. Opt*. 32 (1993) 1154.
30. B. Sheldon, J. S. Haggerty and A. G. Emslie, *J. Opt. Soc. Am*. 72 (1982) 1049.
31. W. H. Southwell, *Opt. Lett*. 8 (1983) 584.
32. E. Spiller, I. Haller, R. Feder, J. E. E. Baglin and W. N. Hammer, *Appl. Opt*. 19 (1980) 3022.
33. P. Yeh and S. Sari, *Appl. Opt*. 22 (1983) 4142.
34. P. G. Verly, J. A. Dobrowolski and R. R. Willey, *Appl. Opt*. 31 (1992) 3836.
35. E. B. Grann, M. G. Moharam and D. A. Pommet, *J. Opt. Soc. Am*. A 12 (1995) 333.
36. W. Zhou, M. Tao, L. Chen and H. Yang, *J. Appl. Phys*. 102 (2007) 103105.
37. J. Q. Xi, M. F. Schubert, J. K. Kim, E. F. Schubert, M. Chen, S. Y. Lin, W. Liu and J. A. Smart, *Nature Photonics* 1 (2007) 176.
38. D. Poitras and J. A. Dobrowolski, *Appl. Opt*. 43 (2004) 1286.
39. F. H. Nicoll, *R. C. A. Review* 6 (1942) 287.
40. F. H. Nicoll and F. E. Williams, *J. Opt. Soc. Am*. 33 (1943) 434.
41. M. J. Minot, *J. Opt. Soc. Am*. 66 (1976) 515.
42. S. F. Monaco, *J. Opt. Soc. Am*. 51 (1961) 280.
43. M. J. Minot, *J. Opt. Soc. Am*. 67 (1977) 1046.
44. S. Walheim, E. Schäffer, J. Mlynek and U. Steiner, *Science* 283 (1999) 520.
45. J. J. Steele and M. J. Brett, *J. Mater. Sci.: Mater. Electron*. 18 (2007) 367.
46. S. R. Kennedy and *M. J. Brett, Appl. Opt*. 42 (2003) 4573.
47. W. Theiß, *Surface Science Reports* 29 (1997) 91.

48. P. Menna, G. Di Francia and V. La Ferrara, *Sol. Energy Mater. Sol. Cells* 37 (1995) 13.
49. C.C. Streimer and P. M. Fauchet, *Appl. Phys. Lett.* 81 (2002) 2980.
50. J. M. Bennett and L. Mattsson, *Introduction to Surface Roughness and Scattering*, Optical Society of America, Washington D.C., 1989.
51. S. J. Wilson and M. C. Hutley, *Optica Acta* 29 (1982) 993.
52. D.G. Stavenga, S. Foletti, G. Palasantzas and K. Arikawa, *Proc. Roy. Soc. B* 273 (2006) 661.
53. A. R. Parker, Z. Hegedus and R. A. Watts, *Proc. Roy. Soc. B* 265 (1998) 811.
54. W. H. Southwell, *J. Opt. Soc. Am.* A 8 (1991) 549.
55. S. A. Boden and D. M. Bagnall, *Appl. Phys. Lett.* 93 (2008) 133108.
56. H. Ghiradella, *J. Morph.* 202 (1989) 69.
57. H. Tabata, K. Kumazawa, M. Funakawa, J. Takimoto and M. Akimoto, *Opt. Rev.* 3 (1996) 139.
58. P. Vukusic, J. R. Sambles, C. R. Lawrence and R. J. Wootton, *Nature* 410 (2001) 36.
59. S. Kinoshita, S. Yoshioka, Y. Fujii and N. Okamoto, *Forma* 17 (2002) 103.
60. P. Vukusic and J. Roy Sambles, *Nature* 424 (2003) 852.
61. K. Watanabe, T. Hoshino, K. Kanda, Y. Haruyama and S. Matsui, *Jpn. J. Appl. Phys.* 44 (2005) L48.
62. A. R. Parker and H. E. Townley, *Nature Nanotech.* 2 (2007) 347.
63. C. G. Bernhard and W. H. Miller, *Acta. Physiol. Scand.* 56 (1962) 385.
64. W. H. Miller, A. R. Moller and C. G. Bernhard, in: C. G. Bernhard (Ed.), *The Functional Organisation of the Compound Eye*, Pergamon Press, Oxford, 1966, 21.
65. S. A. Boden and D. M. Bagnall, *Proceedings of the 4th World Conference on Photovoltaic Energy Conversion*, Hawaii, 2006, 1358.
66. G. Zhang, J. Zhang, G. Xie, Z. Liu and H. Shao, *Small* 2 (2006) 1440.
67. G. Xie, G. Zhang, F. Lin, J. Zhang, Z. Liu and S.Mu, *Nanotechnology* 19 (2008) 095605.
68. A. Yoshida, *Forma* 17 (2002) 75.
69. C. H. Hsu, H. C. Lo, C. F. Chen, C. T. Wu, J. S. Hwang, D. Das, J. Tsai, L. C. Chen and K. H. Chen, *Nano Lett.* 4 (2004) 471.
70. K. H. Chen, J.-S. Hwang, H.-C. Lo, D. Das and L.-C. Chen, U. S. Patent 6,960,528 B2, Nov. 1, 2005.
71. C. H. Hsu, Y. F. Huang, L. C. Chen, S. Chattopadhyay, K. H. Chen, H. C. Lo and C. F. Chen, *J. Vac. Sci. Technol.* B 24 (2006) 308.
72. S. Chattopadhyay, L. C. Chen and K. H. Chen, *Crit. Rev. Solid Stat. Mater. Sci.* 31 (2006) 15.
73. Y. F. Huang, S. Chattopadhyay, Y. J. Jen, C. Y. Peng, T. A. Liu, Y. K. Hsu, C. L. Pan, H. C. Lo, C. H. Hsu, Y. H. Chang, C. S. Lee, K. H. Chen and L. C. Chen, *Nature Nanotech.* 2 (2007) 770.
74. Z. Yu, H. Gao, W. Wu, H. Ge and S. Y. Chou, *J. Vac. Sci. Technol.* B 21 (2003) 2874.
75. Y. Kanamori, M. Sasaki and K. Hane, *Opt. Lett.* 24 (1999) 1422.
76. J. A. Dobrowolski and S. H. C. Piotrowski, *Appl. Opt.* 21 (1982) 1502.
77. P. Lalanne and G. M. Morris, *Nanotechnology* 8 (1997) 53.
78. A. Mendoza-Galvan, K. Jarrendahl, H. Arwin, Y. F. Huang, L. C. Chen and K. H. Chen, *Appl. Opt.* 48 (2009) 4996.

79. J. A. Dobrowolski, Y. Guo, T. Tiwald, P. Ma and D. Poitras, *Appl. Opt.* 45 (2006) 1555.
80. H. G. Tompkins, *A User's Guide to Ellipsometry*, Academic Press, Inc., New York, 1999.
81. A. J. Gatesman, J. Waldman, M. Ji, C. Musante and S. Yngvesson, *IEEE Microw. Guid. Wave Lett.* 10 (2000) 264.
82. C. Brückner, B. Pradarutti, O. Stenzel, R. Steinkopf, S. Riehemann, G. Notni and A. Tünnermann, *Opt. Express* 15 (2007) 779.
83. Y. F. Huang, Y. J. Jen, K. H. Chen and L. C. Chen, *Proc. SPIE* 7039 (2008) 70390W.
84. Y.-J. Lee, D. S. Ruby, D. W. Peters, B. B. McKenzie and J. W. P. Hsu, *Nano Lett.* 8 (2008) 1501.
85. C. Lee, S. Y. Bae, S. Mobasser and H. Manohara, *Nano Lett.* 5 (2005) 2438.
86. H. Toyota, K. Takahara, M. Okano, T. Yotsuya and H. Kikuta, *Jpn. J. Appl. Phys.* 40 (2001) L747.
87. R. C. Enger and S. K. Case, *Appl. Opt.* 22 (1983) 3220.
88. Y. Kanamori, H. Kikuta and K. Hane, *Jpn. J. Appl. Phys.* 39 (2000) L735.
89. Y. F. Li, J. H. Zhang, S. J. Zhu, H. P. Dong, F. Jia, Z. H. Wang, Z. Q. Sun, L. Zhang, Y. Li, H. B. Li, W. Q. Xu and B. Yang, *Adv. Mater.* 21 (2009) 4731.
90. A. Kaless, U. Schulz, P. Munzert and N. Kaiser, *Surf. Coat. Tech.* 200 (2005) 58.
91. H. Sakata, *J. Appl. Phys.* 89 (2001) 7711.
92. P. Zhan, J. B. Liu, W. Dong, H. Dong, Z. Chen, Z. L. Wang, Y. Zhang, S. N. Zhu and N. B. Ming, *Appl. Phys. Lett.* 86 (2005) 051108.
93. J. F. DeNatale, P. J. Hood, J. F. Flintoff and A. B. Harker, *J. Appl. Phys.* 71 (1992) 1388.
94. T. V. Kononenko, V. V. Kononenko, V. I. Konov, S. M. Pimenov, S. V. Garnov, A. V. Tishchenko, A. M. Prokhorov and A. V. Khomich, *Appl. Phys.* A 68 (1999) 99.
95. R. W. Wood, *Philos. Mag.* 4 (1902) 396.
96. T. K. Gaylord, E. N. Glytsis and M. G. Moharam, *Appl. Opt.* 26 (1987) 3123.
97. N. F. Hartman and T. K. Gaylord, *Appl. Opt.* 27 (1988) 3738.
98. H. Yamaji, Y. Terada, H. Miyake and K. Hiramatsu, *Phys. Stat. Sol.* (c) 0 (2003) 2566.
99. C. H. Chiu, P. Yu, H. C. Kuo, C. C. Chen, T. C. Lu, S. C. Wang, S. H. Hsu, Y. J. Cheng and Y. C. Chang, *Opt. Express* 16 (2008) 8748.
100. C. H. Sun, B. J. Ho, B. Jiang and P. Jiang, *Opt. Lett.* 33 (2008) 2224.
101. Y. Kanamori, K. Kobayashi, H. Yugami and K. Hane, *Jpn. J. Appl. Phys.* 42 (2003) 4020.
102. C. Aydin, A. Zaslavsky, G. J. Sonek and J. Goldstein, *Appl. Phys. Lett.* 80 (2002) 2242.
103. J. I. Gittleman, E. K. Sichel, H. W. Lehmann and R. Widmer, *Appl. Phys. Lett.* 35 (1979) 742.
104. H. Sai, H. Fujii, K. Arafune, Y. Ohshita and M. Yamaguchi, *Appl. Phys. Lett.* 88 (2006) 201116.
105. K. Peng, Y. Wu, H. Fang, X. Zhong, Y. Xu and J. Zhu, *Angew. Chem. Int. Ed.* 44 (2005) 2737.

106. Z. P. Yang, L. Ci, J. A. Bur, S. Y. Lin and P. M. Ajayan, *Nano Lett.* 8 (2008) 446.
107. J. Shieh, C. H. Lin and M. C. Yang, *J. Phys. D: Appl. Phys*. 40 (2007) 2242.
108. S. Koynov, M. S. Brandt and M. Stutzmann, *Appl. Phys. Lett.* 88 (2006) 203107.
109. S.M. Yang, S. G. Jang, D. G. Choi, S. Kim and H. K. Yu, *Small* 2 (2006) 458.
110. W.-L. Min, B. Jiang and P. Jiang, *Adv. Mater.* 20 (2008) 3914.
111. C.-T. Wu, F.-H. Ko and C.-H. Lin, *Appl. Phys. Lett.* 90 (2007) 171911.
112. G. R. Lin, Y. C. Chang, E. S. Liu, H. C. Kuo and H. S. Lin, *Appl. Phys. Lett.* 90 (2007) 181923.
113. H. Y. Chen, H. W. Lin, C. Y. Wu, W. C. Chen, J. S. Chen and S. Gwo, *Opt. Express* 16 (2008) 8106.
114. J. K. Kim, S. Chhajed, M. F. Schubert, E. F. Schubert, A. J. Fischer, M. H. Crawford, J. Cho, H. Kim and C. Sone, *Adv. Mater.* 20 (2008) 801.
115. D. Grosso, C. Boissière and C. Sanchez, *Nature Mater.* 6 (2007) 572.
116. L. Vaissie, O. V. Smolski, A. Mehta and E. G. Johnson, *IEEE Photon. Technol. Lett.* 17 (2005) 732.
117. L. Tsakalakos, J. Balch, J. Fronheiser, B. A. Korevaar, O. Sulima and J. Rand, *Appl. Phys. Lett.* 91 (2007) 233117.
118. V. Sivakov, G. Andrä, A. Gawlik, A. Berger, J. Plentz, F. Falk and S. H. Christiansen, *Nano Lett.* 9 (2009) 1549.
119. Y. M. Song, J. S. Yu and Y. T. Lee, *Opt. Lett.* 35 (2010) 276.
120. E. Garnett and P. D. Yang, *Nano Lett.* 10 (2010) 1082.
121. P. C. Yu, C. H. Chang, C. H. Chiu, C. S. Yang, J. C. Yu, H. C. Kuo, S. H. Hsu and Y. C. Chang, *Adv. Mater.* 21 (2009) 1618.
122. N. K. Patel, S. Cina and J. H. Burroughes, *IEEE Sel. Top. Quantum. Electron*, 8 (2002) 346.
123. Y. H. Cheng, J. L. Wu, C. H. Cheng, K. C. Syao and M. C. M. Lee, *Appl. Phys. Lett.* 90 (2007) 091102.
124. Y. C. Kim and Y. R. Do, *Opt. Express*, 13 (2005) 1598.
125. J. K. Kim, S. Chhajed, M. F. Schubert, E. F. Schubert, A. J. Fischer, M. H. Crawford, J. Cho, H. Kim and C. Sone, *Adv. Mater.* 20, 801 (2008).
126. H. W. Huang, C. H. Lin, C. C. Yu, B. D. Lee, C. H. Chiu, C. F. Lai, H. C. Kuo, K. M. Leung, T. C. Lu and S. C. Wang, *Nanotechnology*, 19 (2008) 185301.
127. Y. M. Song, E. S. Choi, J. S. Yu and Y. T. Lee, *Opt. Express*, 17 (2009) 20991.
128. Jia Zhu, Ching-Mei Hsu, Zongfu Yu, Shanhui Fan and Yi Cui, *Nano Lett.* 10 (2010) 1979.
129. S. Xu, S. Y. Huang, I. Levchenko, H. P. Zhou, D. Y. Wei, S. Q. Xiao, L. X. Xu, W. S. Yan and K. Ostrikov, *Adv. Energy Mater.* 1 (2011) 373.
130. Yuerui Lu, Amit Lal, *Nano Lett.* 10 (2010) 4651.
131. Y. W. Chen, P. Y. Han and X. C. Zhang, *Appl. Phys. Lett.* 94, (2009) 041106.
132. C. Brückner, T. Käsebier, B. Pradarutti, S. Riehemann, G. Notni, E. B. Kley and A. Tünnermann, *Opt. Express*, 17 (2009) 3063.
133. J. Zhu, C. M. Hsu, Z. F. Yu, S. H. Fan and Y. Cui, *Nano Lett.*, Article ASAP DOI: 10.1021/nl9034237.
134. Y. L. Chueh, Z. Y. Fan, K. Takei, H. Ko, R. Kapadia, A. A. Rathore, N. Miller, K. Yu, M. Wu, E. E. Haller and A. Javey, *Nano Lett.* 10 (2010) 520.
135. K. Mizuno, J. Ishii, H. Kishida, Y. Hayamizu, S. Yasuda, D. N. Futaba, M. Yumura and K. Hata, *Proc. Natl. Acad. Sci.* 106 (2009) 6044.

4.8 Appendix: glossary of terms

AOI	angle of incidence
AR	anti-reflection
CVD	chemical vapour deposition
EMT	effective medium theory
ES	exponential sine
FAB	fast atom beam
GRIN	gradient refractive index
IR	infrared
NIR	near infrared
PSi	porous silicon
R_{eff}	effective reflectance
RI	refractive index
RIE	reactive ion etching
SEM	scanning electron microscope
SiNT	silicon nanotip
SWG	sub-wavelength grating
SWS	sub-wavelength structure
TE	transverse electric
THz	terahertz
TM	transverse magnetic
UV	ultraviolet
VIS	visible
n	refractive index of the object of interest
n_0	refractive index of the ambient
n_S	refractive index of the substrate
n_1	refractive index of the top layer in a double-layered coating (low index of refraction)
n_2	refractive index of the bottom layer in a double-layered coating (high index of refraction)
η	effective admittance at oblique AOI
λ	wavelength

5

Control of iridescence in natural photonic structures: the case of butterfly scales

S. WICKHAM, University of Sydney, Australia (currently at Dana-Farber Cancer Institute/Harvard Medical School, USA), L. POLADIAN, University of Sydney, Australia, M. C. J. LARGE, University of Sydney, Australia and Canon Information Systems Research Australia, Australia and P. VUKUSIC, Exeter University, UK

Abstract: This chapter looks at how optical microstructures in butterfly scales produce colour, as well as structural black and white. Butterflies contain some of the most diverse optical microstructures nature has to offer. The most interesting and complex of these are three-dimensional photonic crystals. The refractive index contrasts available in butterflies mean that complete band gaps (implying the structure retains its colour from all angles) are impossible. In this context, the role of chirality, composite structures, and randomness are particularly interesting. We also consider how structures in butterfly scales can exaggerate iridescence or, conversely, produce structures that either reflect or absorb across the whole visible spectrum.

Key words: photonic crystal, iridescence, structural colour.

5.1 Introduction to structural colour

There are two general types of colour found in nature. The most common is pigmentary colour, produced by chemical absorption. Less common, and far more interesting, is 'structural colour', is produced by the interaction of light with microstructures, often formed from transparent materials. Structural colour relies on optical effects such as diffraction and interference – the same sorts of processes that, in a much simpler case, produce colour from a thin film of oil. This type of colour has long been known to occur across a wide range of living organisms. While it is intrinsically permanent (unless the structures themselves are damaged) it is also usually angle dependent, or iridescent. Colour changes with angle are considered by many to be the signature of structural colour.

Work on structural colour extends back as far as Hooke (1665) and Newton (1730), who first reported the presence of structural colours in silverfish and peacock feathers. However, it was not until the end of the 19th century that the mechanisms behind these colours began to be clearly understood (Rayleigh 1887; 1888). To interact appreciably with light, these structures must have features of the order of the wavelength in size. Early work was therefore hampered by the inability of optical microscopes to resolve these sub-micron features. Nevertheless, many prominent scientists used optical microscopy to make further contributions to our understanding of this phenomenon, including: Hagen (1882), Mayer (1897), Michelson (1911), Rayleigh (1918), Onslow (1921), Rayleigh (1923), Suffert (1924), Merritt (1925), Mason (1927), and Raman (1935).

The advent of the electron microscope made it possible to resolve these optical microstructures to a much higher degree of accuracy. This spurred a large number of more detailed experimental and theoretical studies of structural colour, beginning with the work of Anderson and Richards (1942), and continuing steadily to this day. Currently, colours produced by a number of different optical effects have been identified in plants (Bone *et al.* 1985, Lee 1991, 1997, Vukusic 2003), beetles (Parker *et al.* 1998a), crustaceans (Parker 1995), damselflies (Vukusic *et al.* 2004), moths and butterflies (Ghiradella 1991), birds (Prum *et al.* 1999a, 1999b, Osorio and Ham 2002, Shawkey *et al.* 2003), spiders (Townsend and Felgenhauer 1999, Parker and Hegedus 2003), weevils (Parker *et al.* 2003), and marine organisms such as molluscs (Herring 1994), diatoms (De Stefano 2008), polychaete worms (Parker *et al.* 2001, Trecziak and Vukusic 2009), fish (McKenzie *et al.* 1995, Rowe and Denton 1997), and cephalopods (Mathger *et al.* 2009). The field has also been the subject of several reviews (Srinivasarao 1999, Parker 2000, Vukusic and Sambles 2003, Parker 2005, Parker and Martini 2006, Kinoshita 2008), which cover in more detail the different species in which structural colours are known to occur.

Remarkably, since structural colours do not degrade chemically, it has even been possible to identify some of them in the fossil record. Some examples date back half a billion years (Parker 1998). Other examples have been found in a 50 million-year-old fossilised beetle (Parker and McKenzie 2003), and in the eye of a 45 million-year-old Eocene fly preserved in amber (Parker *et al.* 1998).

5.2 Types of structural colour in butterflies

Butterflies and moths belong to the group *Lepidoptera*, a name that means 'scaly wing'. When examined under the microscope we can see that their wing membranes are covered with a tiling of tiny scales, some $75 \times 200\ \mu m$ in size (Fig. 5.1).While the majority of these scales are coloured with pigments, others produce their colour structurally. This is particularly so for blues and greens, where pigmentary colours are often either unstable or hard to produce. The diversity of microstructures that produce colour in butterfly scales is probably the greatest found anywhere in nature, and

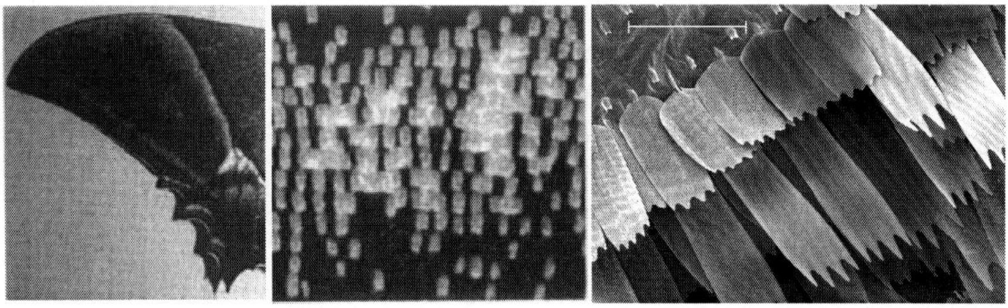

5.1 Butterfly wing scales. As viewed by: (a) digital camera, (b) light microscope, (c) scanning electron microscope. Images (a) and (b) from (Argyros *et al.* 2002).

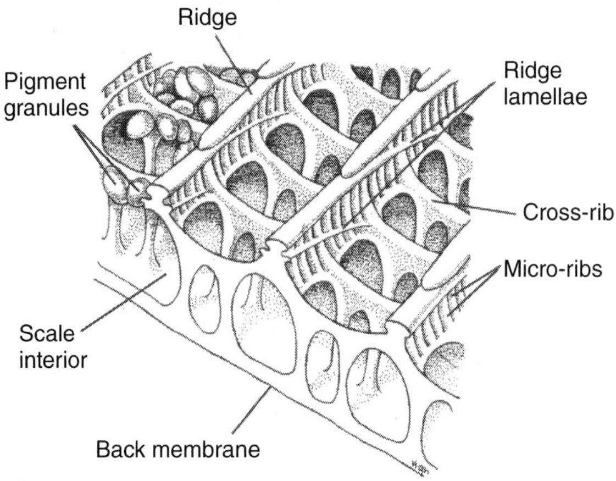

5.2 Schematic of a generic butterfly scale, from Ghiradella (1998). Key features which may be altered to produce structural colour are the ridges, micro-ribs, and the scale interior (lumen).

provides us with an exceptional database of optical designs to explore and characterise.

Figure 5.2 shows a schematic of a generic wing scale. The key features are the ridges, which run along the length of the scale, cross-ribs, which link adjacent ridges, and a back membrane of the scale which encloses the lumen, or interior region. Variations found in these key features, butterfly wing scales produce complex optical structures found in nature, and the mechanisms of these structures have been studied for over a century (Hagen 1882, Mayer 1897). Work in this field constantly highlights the broad range of microstructures responsible for different colours and optical effects.

Butterfly wing scales are composed of insect cuticle, which consists of procuticle, with a thin outer layer of epicuticle (Ghiradella 1989). The procuticle is predominantly made of the biological polymer chitin and various proteins but may also include lipids, pigments, and inorganic and organic materials (Anders 1979). The refractive index of chitin in the case of some *Morpho* butterflies has been determined to be $n = 1.56 + 0.01i$ (Vukusic *et al.* 1999). More detailed studies have revealed that the refractive index also depends on the polarisation state of the light (Berthier *et al.* 2003).

Each wing scale is assembled during metamorphosis from a single cell (Ghiradella 1989), and thus once formed and dried, remains unaltered, except for the effects of environmental wear. Butterfly structural colours can be grouped in a number of ways, and this process of classification carries an implicit statement on the relationships between the structures.

Groupings can be based on optical properties, the mechanisms which produce them, such as the degree of coherent or incoherent scattering (Prum *et al.* 2006), the physical characteristics of the structures (morphology), or a combination of all of the above. One of the most useful schemes is to categorise butterfly structural colours on the basis of both the type of reflecting and scattering elements that produce the colour and their location in the wing scale. One such scheme was developed by Vukusic *et al.* (2000a). They list three categories: Type 1 structures, produced by an elaboration of the ridges to form a thin-film stack; Type 2, where lamellae (layers of chitin parallel to the scale membrane) form within the body, or lumen, of the scale; and Type 3, where more complex scattering structures form within the body of the scale. These three groups are shown in Fig. 5.3, and are discussed in more detail in the following sections.

5.2.1 Type 1: ridge lamellae

In these species, colour is either generated by multi-layering within the ridges of the scale as a result of the elaboration of the lamellae or micro-ribs, or by diffraction from micro-ribs on the surface of the scale. Examples of Type 1 structures are shown in Fig. 5.4. As a first approximation, the first class of structures (Type 1, a, b) may be described as thin-film reflectors. That is, a series of dielectric films with alternating high and low refractive index. Such systems were first described by Rayleigh (Rayleigh 1887, 1888) and can produce strong reflections due to the constructive interference of reflections from the interfaces between the two media. The phase relationship between the waves emerging from the film gives a condition for reflection, called the Bragg condition, which is both wavelength and angle dependent. The colour reflected depends on both the spacing of the thin-film and the angle of incidence.

The most well known and understood example of this type is the brilliant blue iridescence found in the nymphalid genus *Morpho*. This type of struc-

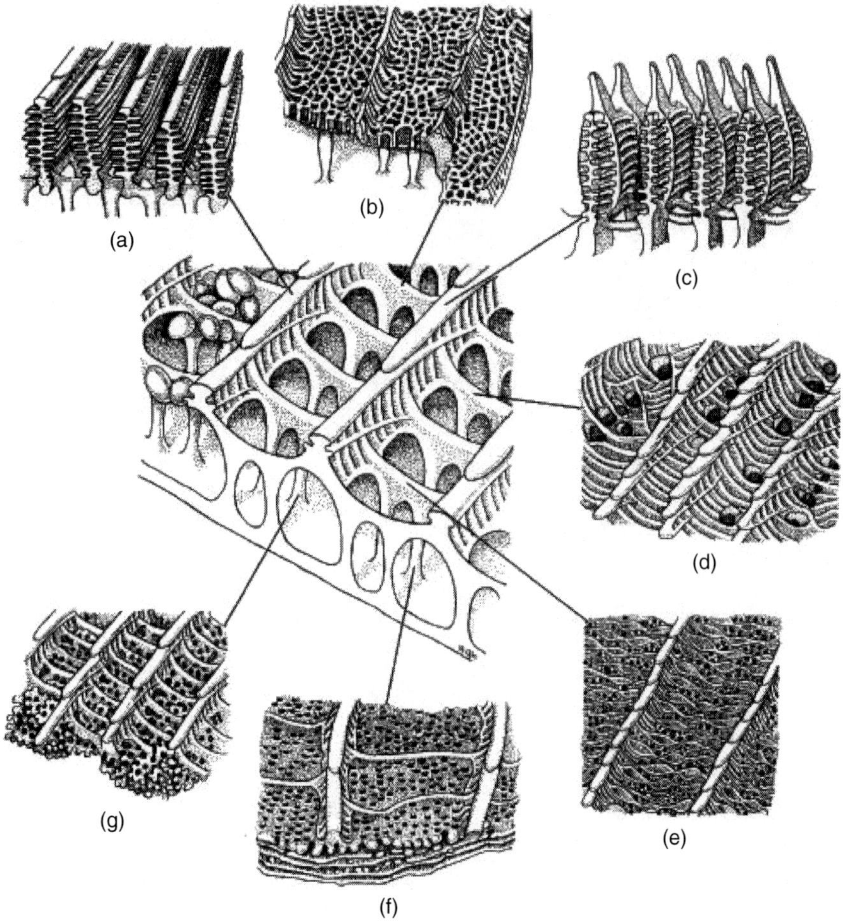

5.3 Types butterfly scale structures which produce structural colour. (a) Type 1: a thin-film stack formed by exaggeration of the ridge lamellae. (b) Type 3: inter-ridge area elaborated into alveoli that were previously thought to act as incoherent scatterers. (c) Type 1: a thin-film stack formed by the exaggeration of micro-ribs. (d) Type 1: micro-ribs form a diffraction grating-like microstructure on the scale surface. (e) Type 1: the flats between the ridges are filled with a plate and pore pattern. (f) Type 2: a thin-film stack in the body of the scale. (g) Type 3: 3D photonic crystal in the body of the scale. From Ghiradella (1998).

ture is also known to produce strong UV reflectance in *Colias*, and several other genera. Another variation of this type of structural colour is a form of high-tilt multi-layering, in which the abrupt termination of the multi-layers in this structure brings about diffraction concurrently with interference (Vukusic *et al.* 2001, 2002).

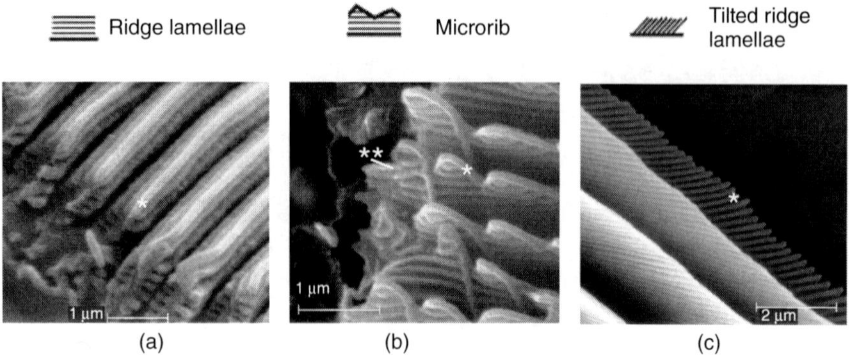

5.4 SEM images of Type 1 butterfly structural colours. (a) ridge thin-film, (b) micro-rib thin-film, (c) tilted ridge lamellae. * Indicates ridge lamellae, ** indicates micro-ribs.

5.2.2 Type 2: body lamellae

Type 2 structures, where colour is produced by multi-layering in the body of the scale, are known to be the source of a range of colours across all the butterfly families (Ghiradella 1994, Brink and Lee 1998, Tada *et al.* 1998, Vukusic *et al.* 2001). Simpler than Type 1 because they are finite in one direction rather than two, these structures are nevertheless able to produce interesting colour and polarisation effects. In particular, multi-layering found in the species *Papilio palinurus* (Vukusic *et al.* 2000b) produces two colours (additively mixed in human vision) and strong mono-colour polarisation conversion, which is attributed to curvature within the cavity containing the layering (Fig. 5.5). There are also several species in which the layers are perforated, or separated by 'spacer' beads, which are thought to be intermediate between this type of structure and Type 3, though these have not yet been the subject of detailed analysis.

5.2.3 Type 3: body scattering

In this group, colour is produced by the scattering of light by a three-dimensional (3D) structure in the body of the scale. Historically, these structures have then been further classified on the basis of the nature of that scattering, either as incoherent Tyndall or Raleigh scattering scales (Huxley 1976) (Fig. 5.6a), or as scales where colour is produced by the coherent scattering of light by highly regular 3D photonic crystals (Fig. 5.6b) (Morris 1975, Allyn and Downey 1976, Ghiradella 1989, Argyros *et al.* 2002, Biro *et al.* 2003, Balint 2005, Kertesz *et al.* 2006, Prum *et al.* 2006, Michielsen and Stavenga 2008, Stavenga 2008, Poladian *et al.* 2009, Michielsen *et al.* 2010).

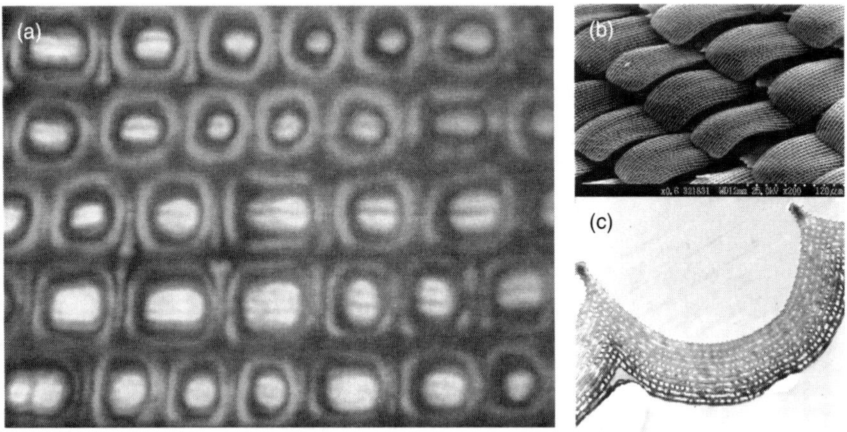

5.5 Optical (a) SEM, (b), and (c) TEM images of a Type 2 butterfly structure in *Papilio palinurus*. (b) An optical microscope image of the scale surface. Colour mixing occurs between the dipped region (which produces yellow) and steeper sides (which produce blue). After Vukusic *et al.* (2000b).

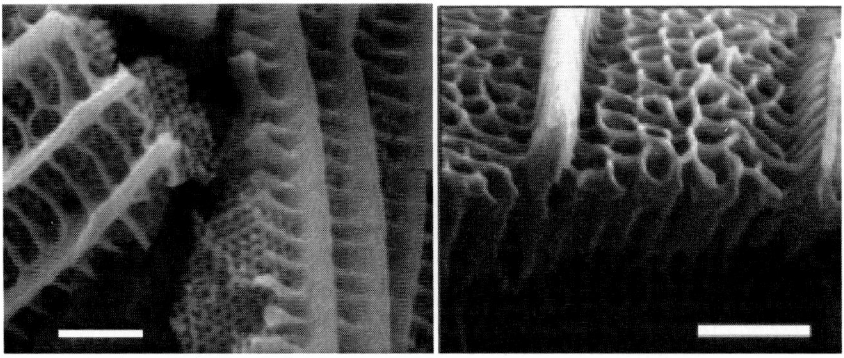

(a) (b)

5.6 SEM images of Type 3 structural colours, which have either a 3D photonic crystal in the body of the scale (a), scale bar 2 μm), or a tubular structure between the ridges (b), scale bar 1 μm).

Rayleigh scattering (Rayleigh 1871, 1899), Tyndall scattering, and Mie scattering (Mie 1908) are mechanisms of incoherent scattering by small particles. They occur when the scattering sites are randomly distributed on a length scale comparable to the size of the wavelength being scattered (Bohren and Huffman 1983), so that phase relationships between light scattered by different elements of the array can be ignored. Coherent scattering models take into account the phase of the different scattered waves, and are required when the spatial distribution of scattering sites is non-random on a length scale comparable to the wavelength (Bohren and Huffman

1983). In this case the reflected colour and whether or not it is iridescent depends strongly on the arrangement of the particles. In the limit that the scattering particles are oriented in a completely regular array, interference becomes the dominant effect. Thus Type 2 structures are theoretically a subset of Type 3.

Coherent scattering structures are also known as photonic crystals, defined as structures with a periodic variation of refractive index. Although one-dimensional photonic crystals were understood in detail as thin-film stacks by 1887 (Rayleigh 1887), the theory was not generalised to two and three dimensions for another 100 years. It was then described independently by Yablonovitch (1987) and John (1987), building on ideas from solid-state physics. In analogy to conventional crystals, photonic crystals can be thought of as a particular arrangement of material, normally described either in terms of connected struts or surfaces (analogous to the arrangement of atoms) repeated within a lattice. The strut structures associated with particular lattices are shown in Fig. 5.7. They correspond to the fundamental triply periodic (i.e. in three dimensions) minimal surfaces (Hyde 1997).

When Maxwell's equations for electromagnetic waves are solved in the presence of a periodically varying refractive index (the analogue to the potential in solid-state physics), a dispersion relation determines the existence of propagating solutions. The region for which no solutions exist is referred to as a photonic band gap. Within this band gap light cannot be propagated within the crystal and will therefore be strongly reflected. If the band gap is *complete*, there are no allowed propagating waves in the crystal in any direction for that wavelength, and so light will be uniformly reflected, resulting in a non-iridescent colour. For a *partial* band gap (or stop band), the reflected colour will change with angle as we shift from a region with no allowed propagating waves for a particular wavelength, to a region with allowed waves.

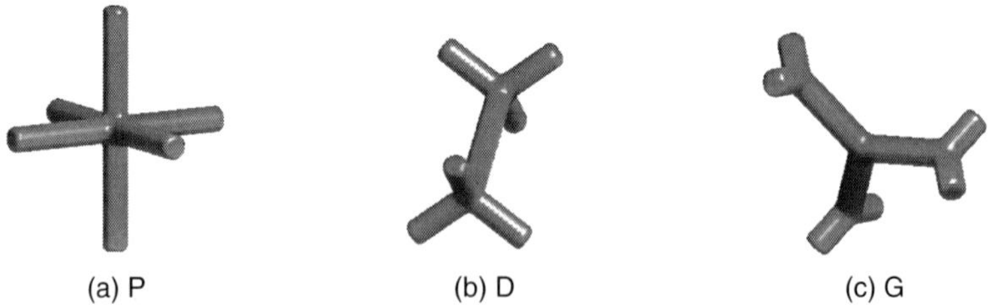

5.7 Strut structures associated with particular lattice structures: (a) six-fold coordinated P-structure in simple cubic (SC) lattice; (b) four-fold coordinated D structure in face centred cubic (FCC) lattice; (c) three-fold coordinated G-structure in body centred cubic (BCC) lattice. The letters P, D, and G refer to primitive, diamond, and gyroid respectively. After (Poladian *et al.* 2009).

Solid-state physics provides a useful tool to understand many of the optical properties of a particular crystal structure. It is a construction called a Brillouin Zone (BZ) (Kittel 1996). Each crystal structure contains a different arrangement of crystal planes, whose *spacings* ultimately determine the positions of the band gap (i.e. the centre wavelength that will be reflected), while the refractive index *contrast* will determine its width (the range of wavelengths, if it is a complete or partial band gap). The Brillouin Zone is a geometric construction in which each face represents a different crystal plane. The distance from the centre of the BZ to each point on the surface is proportional to the frequency in the centre of the band gap (i.e. the reflected colour). Thus, vertices of the BZ represent higher frequencies (shorter wavelengths) and the centres of faces (which are closer in) represent lower frequencies (longer wavelengths).

The most complete band gap will therefore be produced when the BZ is most isotropic – i.e. in 3D it should approximate a sphere. The ratio of the largest and smallest distances to the surface of the BZ (which is also the ratio of longest and shortest wavelengths) reveals how close each structure is to being isotropic. These ratios are approximately 1.7 for simple cubic (SC), 1.4 for body centred cubic (BCC), and 1.29 for face centred cubic (FCC). The most isotropic truly periodic lattice is a diamond structure (FCC) but even for this structure the widths of band gaps in each direction would need to be about 15% of the centre wavelength for the shortest and longest wavelength band gaps to overlap, (see Plate X in colour plate section between pages 96 and 97).

In nature, all of these structures are intrinsically iridescent, because complete 3D band gaps require a refractive index contrast of 2 (Joannopoulos *et al.* 1995), and chitin has a refractive index of only 1.56.

It is possible to gain some enhancements by using quasi-ordered crystals and although they do not form lattices, it is still possible to define an effective BZ (Rechtsman *et al.* 2008, Poddunby *et al.* 2010). In three dimensions, quasi-crystals with five-fold rotational symmetry are known and they have a BZ in the shape of a rhombic triacontahedron. The BZ and the iridescence map are shown in Plate XI. The ratio of longest to shortest wavelength is only 1.18 for this structure and thus 9% band gaps would be sufficient to create a complete band gap.

Photonic crystals are the most optically interesting of all the types of butterfly structural colour. Curiously though, not all the structures *look* very iridescent. Nature clearly has some optical tricks up her sleeve.

5.3 Control of iridescence

Historically it was believed that all iridescent structural colours were attributable to coherent scattering and all non-iridescent structural colours to incoherent scattering (Herring 1994). It turns out, however, that only the first of these statements is true. There are now several well-known examples of non-iridescent colours produced by coherent scattering processes (Prum *et al.* 1999a, 1999b, Osorio and Ham 2002, Wickham 2006, Prum and Torres 2003, Shawkey *et al.* 2003, Noh *et al.* 2010).

Quasi-ordered structures have been shown to produce non-iridescent colours in bird feathers (Noh *et al.* 2010) through the constructive interference of light that is scattered only once. The structures are effectively isotropic with short-range order on the length scale comparable to optical wavelength. They are non-iridescent in diffuse light (as would be found in nature) but iridescent under directional lighting.

In butterflies, some of the best-known examples of colouration attributed to Tyndall scattering, are now understood to be the result of coherent processes combined with the presence of pigments (Prum *et al.* 2006). Fourier analysis has shown that the structure found in *Papilio nireus* is quasiperiodic rather than random (Vukusic and Hooper 2005). Instead of producing colour itself, the role of the microstructure is to direct fluorescence emission from pigments that infuse the scale towards the surface. This results in much more efficient fluorescence, and is analogous to two-dimensional (2D) photonic crystals used to increase the efficiency of light-emitting diodes (LEDs) (Erchak *et al.* 2001). This example is in fact doubly instructive: not only does it anticipate the use of photonic crystals to enhance fluorescence by several millions years, but the quasi-ordered crystal that it uses is considerably more efficient than those designed for the same purpose (Lee *et al.* 2005).

Thus, nature requires some delicacy in our understanding of the words 'periodic', 'random' and even 'iridescent'. Not only are these ideas more complex than we usually think of them, the structures themselves may be too. Given a limited palette of materials, one of the emerging themes we see is the use of more than one effect at the same time, such as interference and scattering; fluorescence and photonic crystals.

5.3.1 Control of iridescence in two-dimensional structures

Reducing iridescence

Probably the best studied of all butterfly microstructures is that found in the *Morpho* butterflies (Fig. 5.3c). The scales have been extensively characterised experimentally. They can reflect up to 75% of incident blue light over an angle range of greater than 100° in one plane, and 15° in the other (Vukusic *et al.* 1999, Kinoshita *et al.* 2002). This is a far broader angular range than would be expected of a simple thin film, and understanding how this is achieved has been the focus of much study. It has been attributed to a layer of highly diffractive but colourless glass scales which lie on top of the iridescent scales (Vukusic *et al.* 1999), to the finite nature of each ridge reflector (Biro *et al.* 2003), and irregularities present within the structure (Kinoshita *et al.* 2002). The optical properties of *Morpho* structures have also been modelled extensively, and with increasing complexity, from simple multi-layer analyses to the rigorous modelling of the full 2D cross-section of the structure with lamellar grating theory (Botten *et al.* 1981, Li 1993), and finite-difference time-domain (FDTD) methods (Plattner 2004).

Plate XIV shows this variation in iridescence, with the colour of each square showing the perceived colour, based on the measured spectrum and converted to a colour using the CIE co-ordinate system.

An even greater suppression of iridescence is observed in the variant of this type in which the reflecting elements are formed by the scale micro-ribs (Fig. 5.3c, 5.4b). A comparison of an iridescence map characterised for this kind of structure is shown in Plate XV, and the even wider angular range of the colour is attributed to the increased irregularity along the length of the ridges (Wickham *et al.* 2006).

Exaggerating iridescence

For most applications, a uniform colour is most desirable, but in other cases (such as for signalling or security) it may be useful to exaggerate the iridescence. A simple variation of the *Morpho* structure achieves this very effectively (Fig. 5.4c, Fig. 5.8). The abrupt termination of the multi-layers in this structure brings about diffraction concurrently with interference.

The tilted ridge multi-layering produces bright iridescence of broad wavelength range but restricted viewing angle in the butterfly *Ancyluris meliboeus* (Vukusic *et al.* 2001, 2002). A conceptually similar structure produces grazing incidence iridescence in *Troides magellanus* (Lawrence *et al.* 2002). This subset of tilted structures have been analysed theoretically as bi-gratings (Tada *et al.* 1998, Lawrence *et al.* 2002), and occurs in a number of other species (Wickham *et al.* 2006). Plate XVIII shows the angle-

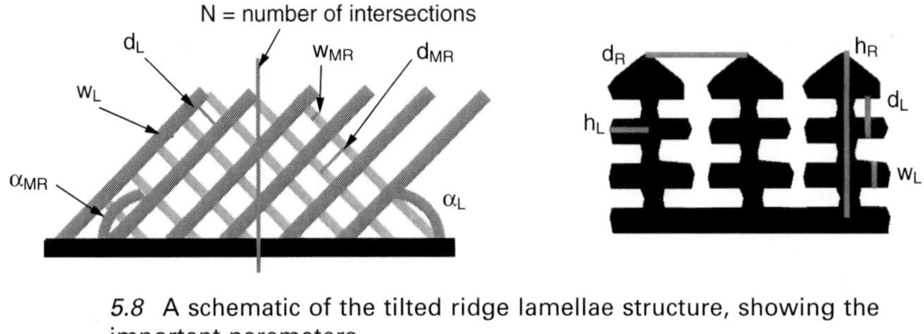

5.8 A schematic of the tilted ridge lamellae structure, showing the important parameters.

dependent spectral measurement for four tilted ridge lamellae butterflies, showing the angular range of the strongest reflection.

5.3.2 Control of iridescence in three-dimensional photonic crystals

Three-dimensional photonic crystal structures have been identified in a number of butterfly species, including the papilionids *Parides sesostris* and *Teinopalpus imperialis*, and the lycaenids *Mitoura gryneus, M. siva, Callophrys rubi* and *C. dumetorum* (Morris 1975, Allyn and Downey 1976, Ghiradella 1989, Biro *et al.* 2003). There are also a number of species which have been identified as having structures which are thought to be transitional between Types 2 and 3, where the lamellar plates in the body of the scale are separated by spacer beads and may also be porous (Ghiradella 1989).

Generally, most studies of these structures have been hampered by an inability fully to investigate the 3D nature of the lattice. Traditional scanning and transmission electron microscopy (SEM and TEM) methods are only able to give information about the surface of a sample or a 2D cross-section through it respectively. Most structural studies to date have relied on either analysis of electron micrographs of 2D sections (Michielsen and Stavenga 2008, Michielsen *et al.* 2010) or small-angle scattering X-ray (Saranathan *et al.* 2010). Electron tomography has only relatively recently been used to analyse the 3D structure of two butterfly scales: *Teinopalpus imperialis* (Argyros *et al.* 2002) and *Callophrys rubi* (Schröder-Turk *et al.* 2010).

While uncertainty as to the exact structures makes precise optical modelling impossible, there are general results which are known. All known transparent biological materials have refractive indices of less than 1.8 (McKenzie *et al.* 1995), meaning that the refractive index contrasts will be insufficient to produce a complete band gap, and hence a photonic crystal that is inherently non-iridescent. A recent study has modelled the P, D, and G structures, assuming realistic filling fractions of chitin and air (Poladian *et al.* 2009). This has confirmed that the structures should be strongly iridescent, while revealing some interesting polarisation effects.

What, then are the techniques that are used to suppress iridescence?

Randomness

The first is randomness. Figure 5.9 shows a composite of several transmission electron micrographs showing the cross-section through a complete scale of *P. sesostris*. Several distinct crystal domains can be seen. In Fig. 5.10 the domain structure of another species, *M. gryneus* is shown. The scanning

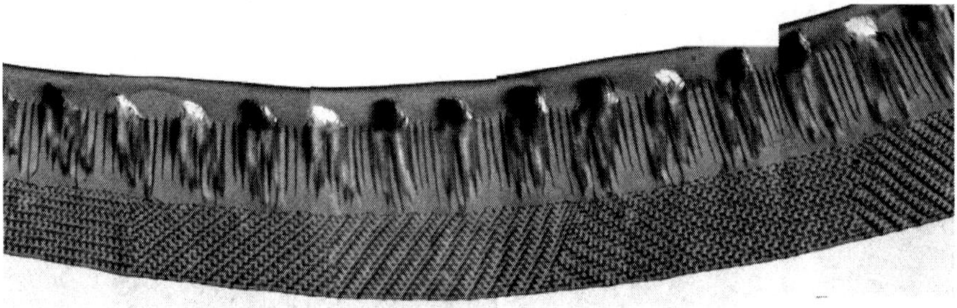

5.9 A composite of several transmission electron micrographs showing the cross-section through a complete scale of *P. sesostris* showing several crystal domains.

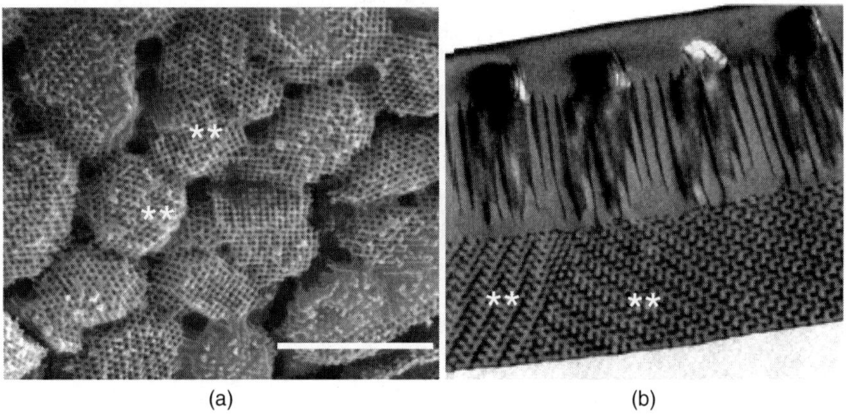

5.10 A scanning electron micrograph of 3D photonic crystal found *M. gryneus* (a) and *P. sesostris* (b), showing the different domains – two are indicated by **. Top view after removal of the honeycomb (a) and cross section (b).

electron micrograph was taken after the top surface of the scale had been peeled off using sticky tape, revealing distinctly different regions of crystal growth in the same scale.

Our studies to date indicate that these different domains have the same crystal structure, but different orientations. This variation in their orientation around that average would cause the spectral features in the band diagram to be smeared out. Indeed, under optical or confocal microscopes, the different domains in the scales appear slightly different colours (Wickham 2005). Thus, similarly to the 2D case, incorporating irregularity into the scale as a whole (as distinct from the crystal itself) could suppress its iridescence (Prum *et al.* 2006, Wickham 2006). It is not known if this represents an advantageous adaptation, or if it is an intrinsic limitation of the self-assembly process that produced them. Defect formation in synthetic self-assembled photonic crystals has proved challenging to overcome (Hatton *et al.* 2010).

Compound structures

Domain averaging (studied with optical microscopy) is not sufficient to explain the uniformity of the perceived colour in *P. sesostris* and *T. imperialis*. Rather, the optical performance seems to depend on the crystal occurring in combination with an extended network of traberculae, or 'honeycomb', that connects the ridges and scale surface to the face of the crystal in the body of the scale. This can be clearly seen in Figure 5.11. While this honeycomb structure has previously been considered to be irregular, recent studies of similar 2D structures in *P. nireus* and *P. zalmoxis*, may lead us to reconsider this assumption. Two of the species with photonic crystals that have been best studied are *T. imperialis* and *Parides sesostris*. The former is an 'at risk' species, and much of the subsequent work has focused on *P. sesostris*. The underlying crystal structure of *T. imperialis* has been determined to be triclinic (Argyros 2002) and preliminary studies (Wickham 2006) suggest that *P. sesostris* has a similar structure. The optical properties of this crystal have been modelled and show that the scale should be strongly iridescent (Poladian 2009). Although the crystal structure in *P. sesostris* has not been definitively identified, the modelling shows similar iridescent behaviour for other candidate structures.

It has been suggested that the honeycomb on the surface of the scale may act to limit the iridescence of the scale surface (Wickham *et al.* 2005, Prum *et al.* 2006). The optical function of the honeycomb structure has not been modelled, but its effect has been conclusively demonstrated (Poladian *et al.* 2009).

Plate XVII shows the scale of *P. sesostris*, viewed from above (through the honeycomb) and below at normal incidence. The scale is clearly

5.11 Close up of honeycomb on *P. sesostris*, with the honeycomb partially removed.

more reflective when viewed from below. The iridescence can only be observed when viewed from below and the scale is bent, or tilted (Plate XIX).

Thus, like the 2D examples such as the *Morpho*, the wide range of colour constancy in these species is attributable to a combination of optical effects. In this case, we may view the scale as a multi-layer structure, in which the honeycomb superstructure may collimate the incoming light before it is incident on the photonic crystal beneath.

Chirality

A final and intriguing aspect of all the butterfly photonic crystals to date is that some are chiral (i.e. they cannot be superimposed on their mirror image). *T. imperialis* was found to be triclinic (Argyros 2002) and very extensive studies of *C. rubi* (Michielsen and Stavenga 2008, Michielsen *et al.* 2010, Schroder *et al.* 2010) have shown that it is gyroid. By contrast, there are very few studies of chiral photonic crystals in the general physics literature.

Chirality has a number of implications for their optical properties. It means that they are optically active, that is, they rotate the polarisation of incoming light, and that they exhibit circular dichroism (Saba *et al.* 2010, 2011), that is, there is a difference in absorption of light between the left- and right-handed circular polarisation states. It also has implications for their band gap properties: it has been shown that such structures can result in polarisation band gaps, allowing one polarisation state to be transmitted through the crystal, while the other is reflected (Poladian *et al.* 2009). While the P and D structures result in linear polarisation effects, the gyroid will cause polarisation effects based on circularly polarised light. Detailed modelling of the structure understood to be present in *C. rubi* scales has confirmed the existence of partial band gaps for circularly polarised light and showed a polarisation stop band of width exceeding 30% (Saba *et al.* 2011) (see Fig. 5.12).

These studies of chirality in butterfly scales are at a very early stage, and raise a number of intriguing questions. Our eyes are not sensitive to polarisation, let alone circular polarisation, but the eyes of some animals are. Examples that are currently known in the biological literature are mantis shrimps (Chiou *et al.*, 2008), jewel beetles (Jewell *et al.* 2007, Sharma 2009) and the butterfly *Heliconius cydno* (Sweeney *et al.* 2003). There may therefore be a biological function to the chirality. Or perhaps it is intimately connected to the (still poorly understood) process by which the microstructure forms. Perhaps the most intriguing revelation, however, is that if we lived in a world of circularly polarised light, the band gap effects of these crystals would not be weak, despite the low index contrast.

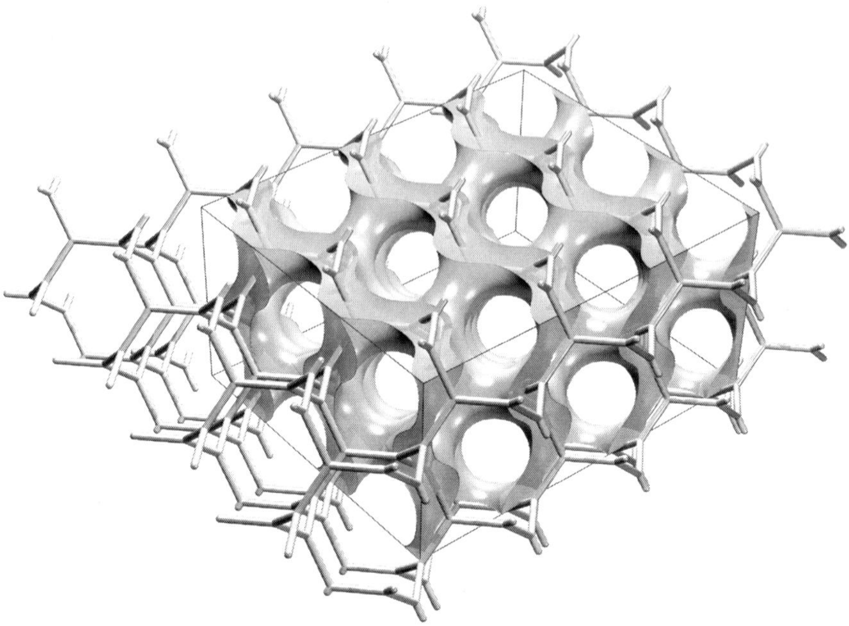

5.12 A reconstruction of the gyroid structure found in *C. rubi*. From Saba 2011.

5.3.3 Beyond colour: structural black and white

The appearance of white in any medium results from multi-wavelength optical scatter arising from the interaction of incident light with heterogeneous dispersed matter (Fox 1976). In fact, any collection of multiple oriented interfaces between non-absorbing media of sufficiently different refractive index will produce the appearance of white. For this reason it should classed as a structural colour. In animals, the many heterogeneous surfaces presented by wing scale, integumental or feather-barbule nanostructures produces white light scatter in the absence of absorbing pigment.

In addition to displaying many forms of saturated structural colour, *Lepidoptera* display a range of different whiteness appearances (Fox 1976); from matt white to micaceous and silvery whites. All these are achieved by heterogeneous cuticle-air structures found on the species' wing scales and are associated with the absence of optically absorbing melanin pigments. The white surfaces which appear more gloss-like combine scattering with broad reflections from more homeotropically aligned surfaces.

Male *Pieris rapae* butterflies, known generally as Cabbage Whites, display white wings. The structural mechanisms responsible for this appearance have been well investigated (Stavenga *et al.* 2006, Luke *et al.* 2009). They are adorned with a dense array of nanoscopic ellipsoidal beads. These beads are responsible both for enhancing optical scatter and for significant absorption at UV wavelengths. There is a positive correlation between the wing's optical scattering coefficient and mean pterin bead-array density of their scales (Luke *et al.* 2009). Detailed experimental and theoretical colour analyses of *P. rapae* wing samples with varying bead-array densities revealed that both optical scatter from the pterin bead-array and absorption by the pterin pigments themselves dictate the final colour response of the *P. rapae* wing (Stavenga *et al.* 2006, Luke *et al.* 2009).

A species of scarab beetle, *Cyphochilus*, demonstrates how it is possible to develop brilliant whiteness from an ultra-thin system (Fig. 5.13 top) (Vukusic *et al.* 2007). Its broadband diffuse scattering appearance results from a 3D photonic solid formed within its elytra, legs and body scales. Although these scales (Fig. 5.13 bottom left) are characterised by exceptional whiteness (Ganz, 1976) their perceived brightness and their optical brilliance, they are only a few microns thick. This is a spatial scale that is at least two orders of magnitude thinner than common synthetic systems designed for equivalent quality whiteness. The scales of *Cyphochilus* are

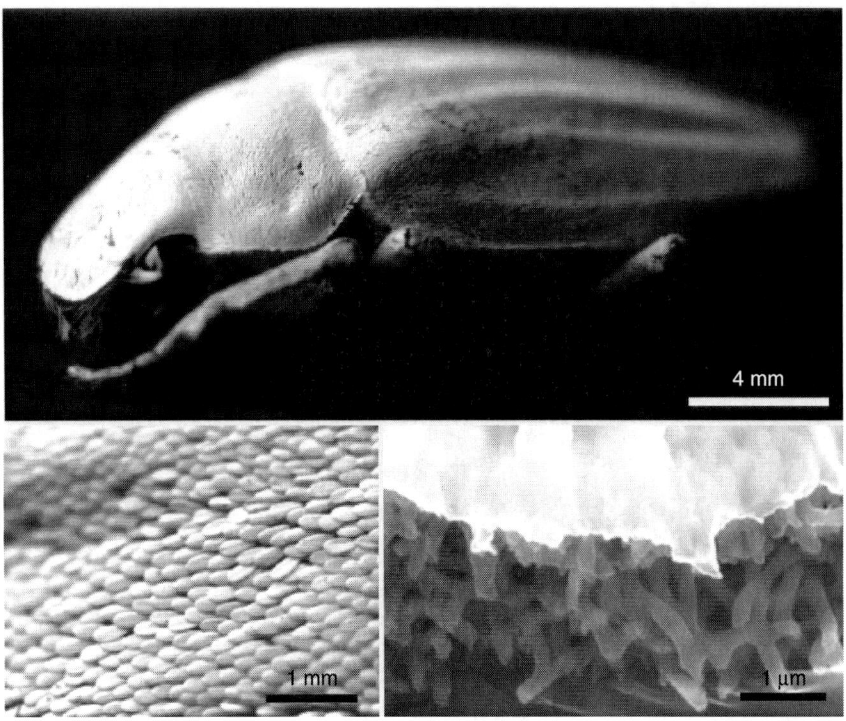

5.13 Structural whiteness in the scarab beetle, *Cyphochilus*.

approximately 5 μm thick, 250 μm long and 100 μm wide and their interiors comprise a random network of interconnecting cuticular filaments of approximately 250 nm diameters (Fig. 5.13 bottom right). The broadband scattering intensity is optimised by the intra-scale cuticle filling fraction. This filling fraction is approximately 70%.

The beetle also appears to have maximised the scattering centre number density while avoiding significant unfavourable optical crowding where the scattering centres are formed too close to one another. The relatively high void fraction in the *Cyphochilus* beetle appears to be a vital part of the system's ability to scatter light. It is this, as well as the system's aperiodicity and index contrast of approximately 0.56, that create such intense optical whiteness for very limited thickness. The quality of the beetle's whiteness and brightness (Ganz 1976) has been measured at 60 and 65 respectively, quantitatively confirming the efficient multi-wavelength scatter for the scales' limited 5 μm thickness. Work is in progress to incorporate the distributed scattering centre design evolved by the *Cyphochilus* beetle into paper manufacture processes (Hallam *et al.* 2009).

Several other coleopteran samples exhibit conspicuous whiteness too. One alternative system design has been discovered as the mechanism responsible for the whiteness of the beetle *Calothyrza margaritifera* (Luke *et al.* 2010). It has a different scale shape and structure to that of *Cyphochilus*, comprising aggregations of co-joined sphere-like particulates instead of filaments. It is the inherent disorder associated with the concurrent size and positioning of these particulates that create the broadband scatter and hence the scales' white appearance. In these two unrelated beetles, the functional requirement for the appearance of whiteness, understood to be for camouflage, has been achieved using contrasting designs.

Surfaces which appear black or deep brown rely on strong optical absorption. This appearance and optical effect is extremely important to animals that use the colour black or brown for thermoregulation, cryptic or communication purposes. However, it is also crucial if the animals' appearances comprise components of saturated structural colour. This is because of the enhancement it offers to the saturation, or colour purity, of their structural colour (Fox 1976). Also the black bordering or framing with which animals, particularly insects, often surround their colour centres, can increase the species' visibility and related conspicuousness.

Various shades of black and brown in *Lepidoptera* and most other animal orders are attributable to the effect of melanin in different forms and different concentrations (Fox 1976, Nijhout 1991). In *Lepidoptera* specifically, it appears to be present within most scale structures, rather than, say, in pigment sacs between the ridges of each scale. It has been shown that even the multi-layering of iridescent scales comprise cuticle that is infused with optically absorbing material, assumed to be melanin. This has the effect, in these scales, of reducing the overall reflectivity at peak wavelengths: for instance, *Morpho rhetenor* exhibits discretised multi-layering, comprising approximately 10 cuticle layers each spaced by a layer of air. In theory, given the refractive index contrast between cuticle and air, this should yield very close to 100% reflectivity at peak wavelength. In fact, as a result of the optical absorption associated with the presence of melanin, this peak wavelength reflectivity is nearer 85% (Vukusic *et al.* 1999). Although this might be a disadvantage in terms of overall brightness, the corresponding advantage rests with the added colour saturation of the blue reflectivity, calculated to be approximately 80% (P. Vukusic, unpublished data).

The deepest black colours in certain butterflies and other animals had been assumed to be solely due to very high concentrations of melanin. However, detailed characterisation work revealed a role played by microstructure in enhancing the blackness (Vukusic *et al.* 2004). This microstructure is in the form of irregularly shaped hollow cavities, crossribs and trebaculae within the scale. Results indicate that these act as scattering centres, which increase the scattering cross-section of the incident light and subsequently render it more efficiently absorbed by the diffusely present melanin (Vukusic *et al.* 2004).

5.4 Perspectives on butterfly biomimetics

Microstructures in butterflies offer a variety of biomimetic lessons. Indeed, a number of biomimetic studies have aimed to reproduce and adapt the butterfly's microstructure through different fabrication techniques. These range from direct bio-templating, where the material of the actual scale is replaced with another substance, such as ZnO or phosphor compounds (Silver *et al.* 2005, Li *et al.* 2006, Zhang *et al.* 2006), to methods which take inspiration from the way the butterflies suppress iridescence through a combination of regularity and irregularity (Watanabe *et al.* 2005). We believe, however, that the real value of studying microstructures is less for direct imitation than for suggesting new approaches.

One key aspect of this has been to see how biology has coped with having a very limited palate of refractive index, compared to the inorganic systems. The use of randomness/quasi periodicity, compound structures and chirality offers solutions. These solutions are technologically relevant. Polymers are much more easily processed and shaped than most other materials, but (being organic) have a similar range of refractive index to the materials found in butterfly scales.

Structural black and white are both in their way somewhat surprising. Both of course combine a fairly uniform and extremely broadband response that is unusual in the world of optical microstructures. Clearly black requires pigmentation, but the idea that microstructures could actually enhance absorption across the visible (rather than for example, cause scattering or diffraction) is not obvious. Structural white is of course less surprising but it is impressive to see that brilliant whiteness can be obtained in an extremely thin film. These results offer new approaches to improving several technologies, such as solar collectors and displays.

5.5 References

Allyn, A. C. and Downey, J. (1976). Diffraction structures in the wing scales of callophrys (mitoura) siva siva (lycaenidae), *Bulletin of the Allyn Museum* (40): 1–6. 20, 24.

Anders, S. (1979). Biochemistry of insect cuticle, *Annual Review of Insect Entomology* 24: 29–61. 16.

Anderson, T. F. and Richards, A. G. (1942). An electron microscope study of some structural colour in insects, *Journal of Applied Physics* (13): 748–758. 14.

Argyros, A., Manos, S., Large, M. C. J., McKenzie, D. R., Cox, G. C. and Dwarte, D. M. (2002). Electron tomography and computer visualisation of a three-dimensional photonic crystal in a butterfly wing-scale, *Micron* 33: 483–487. 15, 20, 24, 32, 39, 40, 53, 54, 92.

Balint, Z., Vertesy, Z. and Biro, L. (2005). Microstructures and nanostructures of high Andean penaincisalia lycaenid butterfly scales (Lepidoptera:Lycaenidae): descriptions and interpretations, *Journal of Natural History* 31(39): 2935–2952. 20, 24, 29, 38.

Berthier, S., Charron, E. and Silva, A. D. (2003). Determination of the cuticle index of the scales of the iridescent butterfly morpho menelaus, *Optics Communications* (228): 349–356. 19, 89.

Biro, L., Balint, Z., Kertesz, K., Vertesy, Z., Markand, G., Horvath, Z., Balazs, J., Mehn, D., Kiricsi, I., Lousse, V. and Vigneron, J. (2003). Role of photonic-crystal-type structures in the thermal regulation of a lycaenid butterfly sister species pair, *Physics Review*, E 67: 252–258. 12, 18, 20, 24, 29.

Bohren, C. F. and Huffman, D. R. (1983). *Absorption and Scattering of Light by Small Particles*, John Wiley and Sons, New York, 21.

Bone, R. A., Lee, D. W. and Norman, J. M. (1985). Epidermal cells functioning as lenses in leaves of tropical rain-forest shade plants, *Applied Optics* (24): 1408–1412. 15.

Botten, L., Craig, M., McPhedran, R., Adams, J. and Andrewartha., J. (1981). The finitely conducting lamellar diffraction grating, *Optica Acta* 28: 1087–1102. 18.

Brink, D. J. and Lee, M. E. (1998). Thin-film biological reflectors: optical characterization of the chrysiridia croesus moth, *Applied Optics* 19(37): 4213–4217. 19, 20.

Chiou, T.-H. *et al.* (2008). Circular Polarization Vision in a Stomatopod Crustacean, *Current Biology* 18, 429 – 434.

De Stefano, L., Maddalena, P., Moretti, L., Rea, I., Rendina, I., De Tommasi, E., Mocella, V. and De Stefano, M. (2008). Nano-biosilica from marine diatoms: A brand new material for photonic applications. *2nd International Conference on Nano-structures Self-Assembling (NanoSeA2008)*, Univ Rome Tor Vergata, Rome, Italy, Superlattices and Microstructures Volume: 46 Issue: 1–2 pp 84–89.

Erchak, A. A., Fan, D. J. R. S., Rakich, P., Joannopoulos, J. D., Ippen, E. P., Petrich, G. S. and Kolodziejski, L. A. (2001). Enhanced coupling to vertical radiation using a two-dimensional photonic crystal in a semiconductor light-emitting diode, *Applied Physics Letters* 78(5): 563–565. 22.

Fox, D. L. (1976). *Animal biochromes and structural colours*. Univ California Press, Berkeley.

Ganz, E. (1976). Whiteness: photometric specification and colorimetric evaluation. *Applied Optics,* 15, 2039-2058.

Ghiradella, H. (1989). Structure and development of iridescent butterfly scales: Lattices and laminae, *Journal of Morphology*. (202): 69–88. 16, 20, 24, 29, 54.

Ghiradella, H. (1991). Light and color on the wing: structural colors in butterflies and moths, *Applied Optics* 24(30): 4392–3500. 15, 16, 28.

Ghiradella, H. (1994). Structure of butterfly scales: Patterning in an insect cuticle, *Micro Research Technology* (27): 429–438. 16, 20.

Ghiradella, H. (1998). Microscopic anatomy of invertebrates, *Insecta*, volume 11A, Wiley, London, pp. 257–287.

Hagen, H. A. (1882). On the colour and pattern of insects, *Proceedings of the American Academy of Arts and Science* (17): 234–267. 14, 15.

Hallam, B. T., Hiorns, A. G. and Vukusic, P. (2009). Developing optical efficiency through optimized coating structure: biomimetic inspiration from white beetles, *Applied Optics* 48: 3243–3247.

Hatton, B., Mishchenko, L., Davis, S., Sandhage, K. H. and Aizenberg, J. (2010). Assembly of large-area, highly ordered, crack-free inverse opal films, *Proceedings of the National Academy of Sciences*, Volume: 107 Issue: 23 Pages: 10354–10359.

Herring, P. J. (1994). Reflective systems in aquatic animals, *Comparative Biochemistry and Physiology*, A (109): 513–546. 15, 21.

Hooke, R. (1665). *Micrographia*, Martyn and Allestry, London, 14.

Huxley, J. (1976). The colouration of papilio zalmoxis and p. antimachus and the discovery of Tyndall blue in butterflies, *Proceedings of the Royal Society, B* (193): 441–453. 20.

Hyde, S., Andersson, S., Larsson, K., Blum, Z., Landh, T., Lidin, S. and Ninham, B. W. (1997). *The Language of Shape. The Role of Curvature in Condensed Matter: Physics, chemistry, and biology*, Elsevier, Amsterdam, The Netherlands: Elsevier.

Jewell, S. A., Vukusic, P. and Roberts, N. W. (2007). Circularly polarised colour reflection from helicoidal structures in the beetle *Plusiotis boucardi*, *New Journal of Physics*, 9 (4): 99.

Joannopoulos, J., Meade, R. and Winn, J. (1995). *Photonic Crystals: Moulding the Flow of Light*, Princeton University Press, Princeton, p. 12.

John, S. (1987). Strong localization of photons in certain disordered dielectric superlattices, *Phys. Rev. Lett.* 23(58): 2486–2489. 22.

Johnson, S. G. and Joannopoulos, J. D. (2001). Block-iterative frequency-domain methods for Maxwell's equations in a planewave basis, *Optics Express* 8(3): 173–190. 24, 38.

Kertesz, K., Balint, Z., Vertesy, Z., Mark, G. I., Lousse, V., Vigneron, J. P. and Biro, L. P. (2006). Photonic crystal type structures of biological origin: Structural and spectral characterization, *Current Applied Physics* 6: 252–258. 20, 29.

Kinoshita, S. (2008). *Structural Colors in the Realm of Nature*, World Scientific Publishing Company, SBN-10: 9812707832.

Kinoshita, S., Yoshioka, S. and Kawagoe, K. (2002). Mechanisms of structural colour in the morpho butterfly: cooperation of regularity and irregularity in an iridescent scale., *Proceedings of the Royal Society*, B, . 269: 1417–1421. 18.

Kittel, C. (1996). *Introduction to Solid State Physics*, John Wiley and Sons, New York.

Lawrence, C. R., Vukusic, P. and Sambles, J. R. (2002). Grazing-incidence iridescence from a butterfly wing, *Applied Optics* 3(41): 437–441. 19, 34.

Lee, D. W. (1991). Ultrastructural basis and function of iridescent blue colour of fruits in elaecarpus, *Nature* (349): 260–262. 15.

Lee, D. W. (1997). Iridescent blue plants, *American Scientist*. (85): 56–63. 15.

Lee, T. D. M., Parker, G. J., Zoorob, M. E., Cox, S. J. and Charlton, M. D. B. (2005). Design and simulation of highly symmetric photonic quasi-crystals, *Nanotechnology* 16: 2703–2706. 22, 23.

Li, L. (1993). A modal analysis of lamellae diffraction gratings in conical mountings, *Journal of Modern Optics*. 40: 553–573. 18.

Li, B., Zhou, J., Zong, R., Fu, M., Bai, Y., Li, L. and Li, Q. (2006). Ordered ceramic microstructures from butterfly bio-template. *Journal of American Ceramic Society*. 87: 2298–2300. 19.

Luke, S. M., Vukusic, P. and Hallam, B. (2009). Measuring and modelling optical scattering and the colour quality of white *Pierid* butterfly scales, *Optics Express*, 17, 14729–14743.

Luke, S. M., Hallam, B. T., and Vukusic, P. (2010). Structural optimization for broad-band scattering in several ultra-thin white beetle scales, *Applied Optics*, 49, (22), 4246–4254.

Mason, C. W. (1927). Structural colour in insects. II. iridescent colours, *Journal of Physical Chemistry* (21): 321–354. 14.

Mathger, L. M., Denton, E. J., Marshall, N. J. and Hanlon, R. T. (2009). Mechanisms and behavioural functions of structural coloration in cephalopods, *Journal of the Royal Society Interface*, 6 S149–S163 Supplement.

Mayer, A. G. (1897). On the colour and colour-patterns of moths and butterflies, *Bulletin of the Museum of Comparative Zoology* (30): 169–254. 14, 15.

McKenzie, D. R., Yin, Y. and McFall, W. D. (1995). Silvery fish skin as an example of a chaotic reflector, *Proceedings of the Royal Society A* 451(1943): 579–584. 12, 24.

Merritt, E. (1925). A spectrophotometric study of certain cases of structural colour, *Review of Scientific Instruments*. (11): 93–98. 14.

Michelson, A. A. (1911). On metallic colouring in birds and insects, *Philosophical Magazine* (21): 554–566. 14.

Michielsen, K. and Stavenga, D. (2008). Gyroid cuticular structures in butterfly wing scales: biological photonic crystals. *J. R. Soc. Interface* 5, 85–94.

Michielsen, K., De Raedt, H. and Stavenga, D. (2010). Reflectivity of the gyroid biophotonic crystals in the ventral wing scales of the green hairstreak butterfly, callophrys rubi. *Journal of the Royal Society Interface* 7, 765–771.

Mie, G. (1908). Beiträge zur Optik trüber Medien, speziell koiloidaler Metallösungen, *Ann Physik* 4(25): 377–445.

Morris, R. B. (1975). Iridescence from diffraction structures in the wing scales of callophrys rubi, the green hairstreak, *Journal of Entomology* (A) 2(49): 149–154. 20, 24.

Newton, I. (1730). *Opticks*, 4 edn, Dover, New York. 14.

Nijhout, H. F. (1991). *The Development and Evolution of Butterfly Wing Patterns*, Smithsonian Institute Press, Washington.

Noh, H., Liew, S. F., Saranathan, V., Mochrie, S. G. J., Prum, R. O., Dufresne, E. R. and Cao, H. (2010). How noniridescent colors are generated by quasi-ordered structures of bird feathers, *Advanced Materials*, 22, 2871–2880.

Onslow, H. (1921). On a periodic structure in many insect scales and the cause of their iridescent colours, *Philosophical Transactions of the Royal Society of London B* (211): 1–74. 14.

Osorio, D. and Ham, A. D. (2002). Spectral reflectance and directional properties of structural colouration in bird plumage, *Journal of Experimental Biology* (205): 2017–2027. 15, 21.

Parker, A. (2005). A geological history of reflecting optics, *Journal of the Royal Society Interface* 2(2): 1–17. 15.

Parker, A. and Martini, N. (2006). Structural colour in animals – simple to complex optics, *Optics and Laser Technology* 4-6(38): 315–322. 15.

Parker, A. R. (1995). Discovery of functional iridescence and its coevolution with eyes in the phylogeny of ostracoda (crustacea), *Proceedings of the Royal Society B* 1365(262): 349– 355. 15.

Parker, A. R. (1998). Colour in burgess shale animals and the effect of light on evolution in the Cambrian, *Proceedings of the Royal Society Biological Sciences* (265): 967–972. 15.

Parker, A. R. (2000). 515 million years of structural colour, *Journal of Optics A: Pure and Applied Optics* (2): R15–R28. 15.

Parker, A. R., McPhedran, R. C., Botten, L. C. and Nicorovici, N. (2001). Aphrodite's iridescence, *Nature* 409, 36–37.

Parker, A. R. and Hegedus, Z. (2003). Diffractive optics in spiders, *Journal of Optics A*: *Pure and Applied Optics* 5: S111–S116. 15.

Parker, A. R., Hegedus, Z. and Watts, R. A. (1998b). Solar-absorber type antireflector on the eye of an Eocene fly (45ma), *Proceedings of the Royal Society Biological Sciences* 265: 811–815. 15.

Parker, A. R. and McKenzie, D. R. (2003). The cause of 50 million-year-old colour, *Proceedings of the Royal Society B* (270): S151–S153. 15.

Parker, A. R., McKenzie, D. R. and Large, M. C. J. (1998a). Multilayer reflectors in animals using green and gold beetles as contrasting examples, *Journal of Experimental Biology* (201): 1307–1313. 15.

Parker, A., Welch, V., Driver, D. and Martini, N. (2003). Structural colour – opal analogue discovered in a weevil, *Nature* 6968(426): 786–787. 15.

Plattner, L. (2004). Optical properties of the scales of morpho rhetenor butterflies: theoretical and experimental investigation of the back-scattering of light in the visible spectrum, *Journal of the Royal Society Interface* 1(1): 49–59. 18, 33, 36, 37, 38, 61, 62, 64, 66, 81.

Poddubny, A. N. and Ivchenko, E. L. (2010). Photonic quasicrystalline and aperiodic structures, *Physica E*, 42(7), 1871–1895.

Poladian, L., Wickham, S., Lee, K. and Large, M.C. (2009). Iridescence from photonic crystals and its suppression in butterfly scales. *Journal of the Royal Society* Interface 6, S233–S242.

Prum, R. O. and Torres, R. (2003). Structural colouration of avian skin: convergent evolution of coherently scattering dermal collagen arrays, *Journal of Experimental Biology* 14(206): 2409–2429. 21.

Prum, R. O., Torres, R., Korvach, C.,Williamson, S. and Goodman, S. M. (1999a). Coherent light scattering by nanostructured collagen arrays in the caruncles of the Malagasy asities (Eurylaimidae: Aves), *Journal of Experimental Biology* 24(202): 3507–3522. 15, 21.

Prum, R. O., Torres, R., Williamson, S. and Dyck, J. (1999b). Two-dimensional Fourier analysis of the spongy medullary keratin of structurally coloured feather barbs, *Proceedings of the Royal Society B* (266): 13–22. 15, 21.

Prum, R., Quinn, T. and Torres, R. (2006). Anatomically diverse butterfly scales all produce structural colours by coherent scattering, *Journal of Experimental Biology* 4(209): 748–765. 17, 20, 21, 22, 24, 25, 26, 29, 30, 38, 54, 92.

Raman, C. V. (1935). The origin of colours in the plumage of birds, *Proceedings of the Indian Academy of Sciences, Section A* 1–7. 14.

Rayleigh, J. (1887). On the maintenance of vibrations by forces of double frequency, and on the propagation of waves through a medium endowed with a periodic structure, *Philosophical Magazine.* 24(145–159). 14, 18, 22.

Rayleigh, J. (1871). On the light from the sky, its polarization and colour, *Philosophical Magazine.* XLI(4): 274, 447. 21.

Rayleigh, J. W. S. (1888). On the remarkable phenomenon of crystalline reflection described Prof. Stokes, *Philosophical Magazine* (26): 256–265. 14, 18.

Rayleigh, J. W. S. (1899). *Philosophical Magazine* XLVII(5): 375. 21.

Rayleigh, J. W. S. (1918). On the optical character of some brilliant animal colours, *Philosophical Magazine* (37): 98–111. 14.

Rayleigh, J. W. S. (1923). Studies of iridescent colour, and the structure producing it. IV. iridescent beetles, *Proceedings of the Royal Society of London* A 721(103): 233–239. 14.

Rechtsman, M. C. and Jeong, Hyeong-Chai and Chaikin, P. M. and Torquato, S. and Steinhardt, P. J. (2008) Optimized structures for photonic quasicrystals, *Physical Review Letters*. 101(7) 073902.

Rowe, D. M. and Denton, E. J. (1997). The physical basis for reflective communication between fish, with special reference to the horse mackerel, *Trachurus trachurus* Philosophical Transactions: Biological Sciences 352(1353) 531–549.

Saba, M., Thiel, M., Turner, M., Hyde, S., Gu, M., Grosse-Brauckmann, K., Neshev, D., Mecke, K. and Schröder-Turk, G. (2011). Circular dichroism in biomimetic cubic chiral nets. *Physical Review Letters*, 106(10). DOI: 10-1103/PhysRevLett.106.103902.

Saranathan, V., Osuji, C. O., Mochrie, S. G. J., Noh, H., Narayanan, S., Sandy, A., Dufresne, E. R. and Prum, R. O. (2010). Structure, function, and selfassembly of single network gyroid (I4132) photonic crystals in butterfly wing scales. *Proceedings of the National Academy of Sciences* 107, 11676–11681.

Schroder, G. E. (2005). *Skeletons in the Labyrinth, Medial Representations and Packing Properties of Bicontinuous Space Partitions*, PhD thesis, Australian National University, Canberra. 42, 54.

Schroder, G. *et al.* (2010) Electron tomography evidence for a cubic Gyroid structure in chitin microstructure of *C. rubi* butterfly wings, to be published.

Schröder Turk, G. E., Wickham, S., Averdunk, H., Fitz Gerald, J. D., Paladian, L., Large, M. C. J. and Hyde, S. T. (2011). The chiral structure of porous chitin within the wing-scales of *Callophrys rubi*, Journal of Structural Biology 174(2): 290–295.

Sharma, V., Crne, M., Park, J. and Srinivasarao, M. (2009). Structural origin of circularly polarized iridescence in jeweled beetles. *Science* 325, 449–451.

Shawkey, M. D., Estes, A. M., Siefferman, L. M. and Hill, G. E. (2003). Nanostructure predicts intraspecific variation in ultraviolet-blue plumage colour, *Proceedings of the Royal Society B* (270): 1455–1460. 15, 21.

Silver, J., Withnall, R., Ireland, T. G. and Fern, G. R. (2005). Novel nano-structured phosphor materials cast from natural morpho butterfly scales, *Journal of Modern Optics* 52(7): 999–1007. 19.

Srinivasarao, M. (1999). Nano-optics in the biological world: Beetles, butterflies, birds, and moths, *Chemical Reviews* (99): 1935–1961. 15.

Stavenga, D. G., Giraldo, M. A. and Hoenders, B. J. (2006). Reflectance and transmittance of light scattering scales stacked on the wings of pierid butterflies, *Optics Express* 14(11): 4880–4890. 33.

Suffert, F. (1924). Morphologie und optik der schmetterlingschuppen inbesondere die schillerfarben der schmeterlinge, *Zeitschrift für Morphologie und Ökologie der Tiere*. (1): 171–308. 14.

Sweeney, A., Jiggins, C. and Johnsen, S. (2003). Insect communication: Polarized light as a butterfly mating signal. *Nature* 423, 31–32.

Tada, H., Mann, S. E., Miaoulis, I. N. and Wong, P. Y. (1998). Effects of a butterfly scale microstructure on the iridescent colour observed at different angles, *Applied Optics* 9(37): 1579–1584. 19, 20, 35.

Townsend V. R. and Felgenhauer B. E. (1999). Ultrastructure of the cuticular scales of lynx spiders (aramaeae, oxyopidae) and jumping spiders (araneae, salticidae), *Journal of Morphology*. 240: 77–92. 15.

Trecziak, T. and Vukusic, P. (2009). Photonic crystal fiber in the polychaete worm Pherusa sp. *Physical Review* E. 80, 061908.

Vukusic, P., Sambles, J. R., Lawrence, C. R. and Wootton, R. J. (1999). Quantified interference and diffraction in single Morpho butterfly scales, *Proceedings of the Royal Society of London B* 266, 1403–141.

Vukusic, P. (2003). *Optical Interference Coatings*, Springer, New York. 15.

Vukusic, P., Hallam, D. and Noyes, J. (2007). Brilliant whiteness in ultra-thin beetle scales, *Science* 315 (5810): 348–348.

Vukusic, P. and Hooper, I. (2005). Directionally controlled fluorescence emission in butterflies, *Science* 310: 1151. 21, 22, 23.

Vukusic, P., Sambles, J. and Ghiradella., H. (2000a). Optical classification of microstructures in butterfly wing-scales, *Photonic Science News* 6(1): 61–66. 17.

Vukusic, P. and Sambles, J. R. (2003). Photonic structures in biology, *Nature* (424): 852–855. 15.

Vukusic, P., Sambles, J. R., Lawrence, C. R. and Wootton, R. J. (2001). Now you see it and now you don't, *Nature* (410): 36. 19.

Vukusic, P., Sambles, J. R., Lawrence, C. R. and Wootton, R. J. (2002). Limited-view iridescence in the butterfly Ancyluris meliboeus, *Proceedings of the Royal Society of London B* (269): 7–14. 19.

Vukusic, P., Sambles, R. and Lawrence, C. (2000b). Colour mixing in wing scales of a butterfly, *Nature* (404): 457. 20.

Vukusic, P., Sambles, R., Lawrence, C. and Wakely, G. (2001). Sculpted-multilayer optical effects in two species of papilio butterfly, *Applied Optics* 7(40): 1116–1125. 20.

Vukusic, P., Wootton, R. J. and Sambles, J. R. (2004). Remarkable iridescence in the hindwings of the damselfly Neurobasis chinensis (Linnaeus) (zygoptera: Calopterygidae), *Proceedings of the Royal Society of London* B 271: 595–601. 15, 33.

Vukusic, P., Sambles, J. R. and Lawrence ,C. R. (2004). Structurally assisted blackness in butterfly scales, *Proceedings of the Royal Society,* B., 271: S237–S239.

Watanabe, K., Hoshino, T., Kanda, K., Haruyama, Y. and Matsui, S. (2005). Brilliant blue observation from a morpho-butterfly-scale quasi-structure, *Japanese Journal of Applied Physics* 44(1): L48–L50. 20, 38.

Wickham, S., Large, M. and Poladian, L. (2005). High resolution single-scale investigation of a 3-dimensional butterfly photonic crystal, ACOLS 2005: *Proceedings of 7th Australasian Conference on Optics, Lasers and Spectroscopy*, Rotorua, New Zealand, December 5–9th. 25, 26.

Wickham, S., Large, M., Poladian, L. and Jermiin, L. (2006). Exaggeration and suppression of iridescence: the evolution of two-dimensional butterfly structural colours, *Journal of the Royal Society Interface* 6(3): 99–108. 18, 19, 38.

Wickham, S. (2006). *Three-dimensional Butterfly Photonic Crystals: A Structural and Optical Investigation*, Master's Thesis, University of Sydney.

Yablonovitch, E. (1987). Inhibited spontaneous emission in solid-state physics and electronics, *Physical Review Letters* 20(58): 2059–2062. 22.

Zhang, W., Zhang, D., Fan, T., Ding, J., Guo, Q. and Ogawa, H. (2006). Fabrication of ZnO microtubes with adjustable nanopores on the walls by the templating of butterfly wing scales, *Nanotechnology* (17): 840–844. 19.

6

Fabrication of nanostructures using natural synthesis: optical materials using silk

H. TAO, D. L. KAPLAN and
F. G. OMENETTO, Tufts University, USA

Abstract: Biopolymers are promising building blocks for a new generation of green devices. Silk proteins represent a unique family of biopolymers due to their novel structural and biological properties, which serve as a broad inspiration to develop biological foundries for technological applications that leverage nature's materials as their main constituents. This chapter focuses on the opportunities offered by the material as a promising biopolymer platform for high technology applications in optics and photonics, electronics and optoelectronics.

Key words: silk, biopolymer, photonics, electronics, sustainable materials.

6.1 Introduction

The possibility of adopting sustainable materials as alternatives to contemporary technological materials whose fabrication is based on non-renewable resource consumption such as oil, coal or natural gas is a global challenge. Naturally occurring materials provide a compelling template to reinterpret and, in some cases, simplify modern manufacturing while at the same time rendering it sustainable.

A particular and long-standing challenge is to identify, among the options available, new materials that, while being environmentally sustainable, also maintain adequate physical and material properties to meet the performance requirements and the fabrication tolerances needed successfully to interface with current technology. Further, such materials have to be widely available and be cost-competitive within the global commodity supply chain. A conjunction of practical and technological elements is needed to make naturally occurring materials a credible alternative to current plastic, semiconductors and inorganic substrates, and provide alternatives to reroute present-day materials towards more environmentally sustainable options.

Biopolymers are promising building blocks for a new generation of green devices. This class of polymers, produced and modified by living organisms, are the mainstay of naturally occurring, self-assembling, structurally hierarchical micro- and nanoscale systems, from the polysaccharide chitin in butterfly wings and beetle exoskeletons, to the collagen in lens arrays or dermal iridescences, to the keratin in peacock feathers, among many others. The direct utilization and re-engineering of these biological materials into technological material platforms is already underway in many laboratories and offers a path forward for a green revolution (Vukusic, Sambles *et al.* 1999; Sarkar and Mallick 2000; Vukusic, Sambles *et al.* 2000; Aizenberg, Tkachenko *et al.* 2001; Sundar, Yablon *et al.* 2003; Aizenberg 2004; Payne, Yi *et al.* 2005; Hooper, Vukusic *et al.* 2006; Steckl, Hagen *et al.* 2006; Steckl 2007; Yu, Li *et al.* 2007; Arockiados, Xavier *et al.* 2008; Cremona, Legnani *et al.* 2008; Oksman, Mathew *et al.* 2009; Payne, Meyer *et al.* 2009; Vukusic, Kelly *et al.* 2009; Singh, Sariciftci *et al.* 2010; Irimia-Vladu, Sariciftci *et al.* 2011).

We will focus on the opportunities offered by silk proteins as a promising biopolymer platform for high technology applications (Vollrath and Knight 2001; Shao, Vollrath *et al.* 2003; Shao, Yang *et al.* 2005; Shao, Chen *et al.* 2006; Omenetto and Kaplan 2008; Vollrath, Liu *et al.* 2008; Young, Brookes *et al.* 2008). In this chapter, we will address the use of silk proteins in optics and photonics, electronics and optoelectronics. These options represent additional developments for this technology platform that compound the broad utility and impact of this material for medical needs that have been recently described in the literature (Cannas, Santin *et al.* 1999; Motta, Migliaresi *et al.* 2002; Kaplan, Altman *et al.* 2003; Kaplan, Meinel *et al.* 2005; Scheibel 2005; Kaplan, Wang *et al.* 2006; Kaplan and Vepari 2007; Kaplan, Kluge *et al.* 2008; Kundu, Dash *et al.* 2008; Chen, Zhou *et al.* 2009). The favorable properties of the material serve as a broad inspiration to develop biological foundries for technological applications that leverage nature's materials as their main constituents.

Silk fibers have historically been highly favored in the textile industry for a few thousand years owing to their extraordinary mechanical strength plus their exceptionally smooth and gleaming appearance. In addition to the popular use in clothing, silk has recently found its applications as a promising biomaterial because of several desirable properties. In fact, silk has been used as a suture material for centuries. From a materials science perspective, silks spun by spiders and silkworms represent the strongest and toughest natural fibers known and offer unlimited opportunities for functionalization, processing, and biological integration. As shown in Table 6.1, the toughness of silk protein fibers, for example spider silks and silkworm silks, is greater than most synthetic materials, including widely used high-performance Kevlar fibers (Vepari and Kaplan 2007). In terms of strength, silk fibers (0.6–1.1 GPa) are in orders of magnitudes higher than that of commonly used polymeric biomaterials such as poly(L-lactic acid) (PLA) (28–50 MPa) and collagen (0.9–7.4 MPa) (Omenetto and Kaplan 2010). Silks are produced by several kinds of insects (over tens of thousand species)

Table 6.1 Mechanical properties of spider and silkworm silks and other materials

Material	Strength (GPa)	Toughness (MJ · m^{-3})	Extensibility
Araneus MA silk	1.1	160	0.27
Araneus viscid silk	0.5	150	2.7
Bombyx mori silk	0.6	70	0.18
Tendon	0.15	7.5	0.12
Bone	0.16	4	0.03
Elastin	0.002	2	1.5
Resilin	0.003	4	1.9
Nylon fiber	0.95	80	0.18
Kevalar 49 fiber	3.6	50	0.027
High-tensile steel	1.5	6	0.008

that include, for example, spiders from the class Arachnida, mites, butterflies and moths from the order *Lepidoptera*. Silk fibers generated from silkworms, known as the *Bombyx mori*, are of particular interest because of the feasibility of large-scale cultivation (thus cost-effective) and slightly lower but comparable properties to spider silks.

The *B. mori* silk fibroin fibers are typically about 10–25 μm in diameter, consisting of two proteins in a 1:1 ratio, i.e. a light chain (~26 kDa) and a heavy chain (~390 kDa), which are linked by a single disulfide bond and coated with adhesive and hydrophilic sericin proteins (20–310 kDa) (Kaplan, Altman *et al.* 2003). Silk fibroin is a block copolymer rich in hydrophobic beta sheet forming protein blocks that self-assemble to form strong and resilient materials with high mechanical strength and toughness. In addition to the notable mechanical properties, silk fibroin is also a degradable and resorbable material. *B. mori* silk fibroin is reported to be biocompatible and to induce minimal inflammatory responses when implanted (Kaplan, Altman *et al.* 2003). However, the presence of sericin may cause immunological and allergic reactions in humans, and the removal of sericin from the raw silk cocoons is essential for further medical applications, which can be done by boiling silk cocoons in an alkaline solution, the so-called de-gumming process. Once the sericin proteins are removed, the fibroin fibers can be dissolved into an aqueous solution of pure silk fibroin protein, which can be further processed into different materials ranging from gels and sponges to blocks and films (Rockwood, Preda *et al.* 2011). Silk films are of particular interest for optics and electronics applications because of their optical transparency (~92% across the visible range with the film thickness between 40 μm to 100 μm) and surface smoothness (with a root-mean-square surface roughness of just a few nanometers) (Omenetto and

Kaplan 2008). The refractive index of silk (n = 1.54 @ 633 nm) is similar to that of glass (n = 1.52) and higher than that of water (n = 1.33), which implies the possibilities for bio-optics applications, for instance, guiding light within a water-based biological environment with silk waveguides (Omenetto, Parker *et al.* 2009).

Silk proteins represent a unique family of biopolymers due to their novel structural and biological properties. Recent progress has been made into the relationship between structure and processing. This includes the role of self-assembly (Kaplan, Valluzzi *et al.* 2002; Braun and Viney 2003; Kundu, Khire *et al.* 2010), the role of water in assembly and processing (Jin, Park *et al.* 2005; Lee, Min *et al.* 2006; Vollrath and Porter 2009; Omenetto, Tsioris *et al.* 2011), and various options to modify the native proteins (Mori and Tsukada 2000; Goldsmith, Shimada *et al.* 2005). This has led to the transformation of this ancient and commodity material, in particular silkworm silk, into a variety of new material formats including, hydrogels, ultrathin films, thick films, conformal coatings, three-dimensional (3D) porous or solid matrices, fibers with diameters spanning the nano- to the macroscale, and many related material formats (Fig. 6.1). In addition, silk is processed in an all water-based, room-temperature, neutral pH environment, is mechanically stable, edible, biocompatible, and implantable in the human body.

What particularly distinguishes silk from other biopolymers for high technology applications are the robust mechanical properties, the facile control of materials properties through the control of water content during processing, the programmable/controllable (from instantaneous to years) degradation lifetime, and the unique optical and electronic properties of

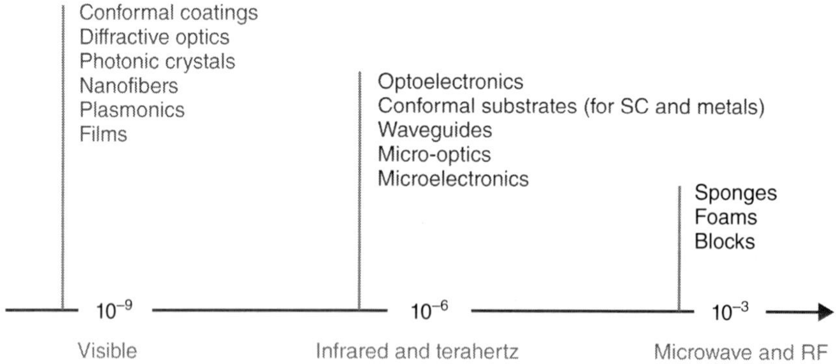

6.1 Various silk-based materials formats and devices cover a multitude of material scales with sizes relevant to applications covering a wide range of the electromagnetic spectrum from radio frequency (RF) to optical regimes for novel biomedical, optoelectronic, and photonic applications.

the material. Further the ambient environment during silk processing allows for the incorporation of labile biological components without loss of function and with retention of bioactivity over extended time frames (Kaplan, Lu *et al.* 2009; Kaplan, Lu *et al.* 2010).

This feature, in particular, provides the opportunity to embed biochemical functions within the biopolymer-based material system, and enables the direct incorporation of biological activity in a bulk material substrate. The latter offers a paradigm shift from traditional optoelectronic and electromagnetic devices that are commonly manufactured on biologically inert substrates, opening opportunities for sustainable and 'environmentally interactive' devices.

Silk-based materials appear to be suited to ideally interface with both technological and natural worlds. One of the unique traits of these materials is the ability to store functional compounds, including both organic and inorganic. Since the novel block copolymer structure of silk protein chains provide the template to maintain and preserve biochemical function of such labile biological components (e.g. enzymes, antibodies) within the material structure, new opportunities become available to affect device behavior and responses. The coexistence of technological devices and biological components to generate 'living photonic' components (or more generally, 'living materials') through the combination of nanofabrication, diffractive optics, and biological doping provides innovative venues for functional devices that are not easily (if at all) attainable with more traditional approaches. This confluence of material properties and functions is the driver for using silk for applications in photonics, electronics, and optoelectronics.

6.2 Silk optics and photonics

Silk has recently emerged as a highly promising material platform for optics and photonics applications because of its excellence on mechanical robustness, optical transparency, and surface flatness. Realization of silk devices starts from production of the silk fibroin solution that is regenerated and purified from natural silk cocoons. The clear water-like silk fibroin solution can be activated, biochemically and/or physically, by simply mixing with various organic (such as cells, proteins, and enzymes) and/or inorganic (such as quantum dots, laser dyes, and metallic nanoparticles) dopants into the solution. The either undoped or doped solution can be deposited on appropriate substrates (flat or patterned), which crystallizes through protein self-assembly upon exposure to air, without the need to resort to exogenous cross-linking reactions or post processing cross-linking for stabilization. This offers the opportunity for numerous fabrication strategies that yield a class of optical elements or a mechanically robust, biocompatible and bioresorbable substrate for thin film photonic and electronic devices (Fig. 6.2).

Silk fibroin can be easily formed into mechanically robust films of thermodynamically stable beta sheets, with the ability to control thicknesses from just below ten nanometers to hundreds of micrometers or more (Fig. 6.3).

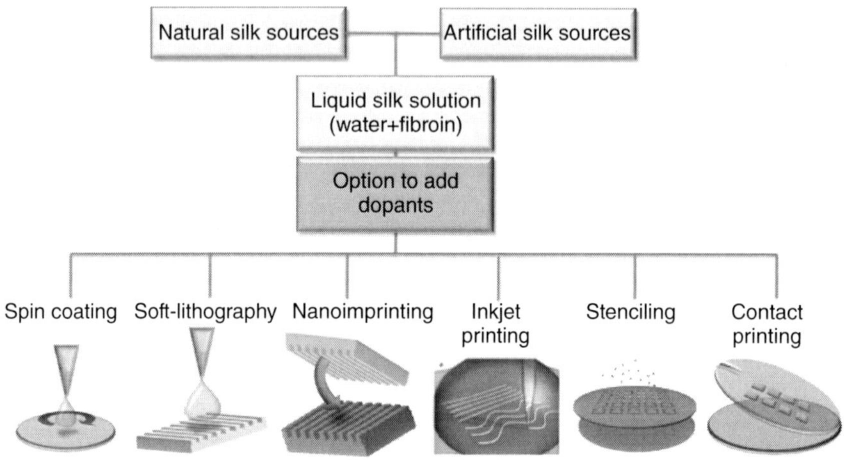

6.2 Different fabrication strategies for silk devices using silk solution derived from natural silk sources (such as cocoons, webs, and fibers) as the starting material.

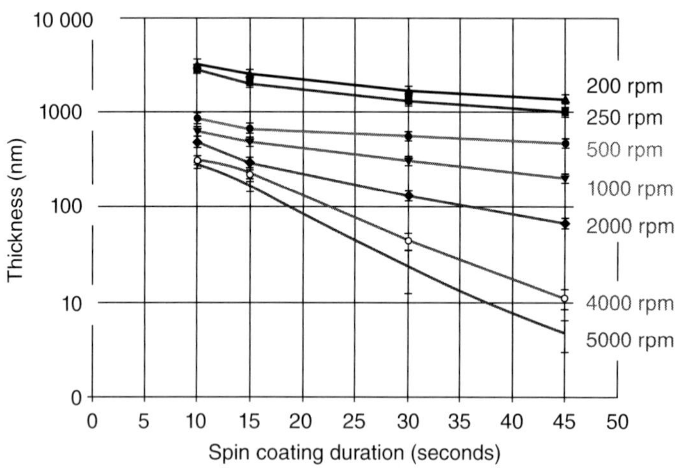

6.3 Control of the thickness of spin-coated silk film through the adjustment of the silk concentration (6 wt% is shown) and spin rate allowing for ultimate film thicknesses down to less than 10 nanometers.

The resulting hardened silk has mechanical properties, surface quality (surface roughness root mean square (rms) <5 nm) and transparency (over 90% transmission across the visible range) that are ideally suited for its use as an optical substrate (Kaplan, Lawrence *et al.* 2009). Owing to the processing and self-assembly attributes, silks are suitable for soft-lithography. By casting the aqueous solution on patterned surfaces the topographies can be replicated down to feature sizes less than 10 nanometers in the hardened, free-standing silk film. Using this approach in combination with the optical transparency of the films, a variety of photonic structures can be fabricated. These materials and devices include one-dimensional (1D) and two-dimensional (2D) diffractive structures, holograms, prisms, photonic lattices, and microlens arrays (Fig. 6.4).

While soft-lithography is effective, nanoimprinting of silk films is possible by adjusting their glass transition temperature, which depends on the water content and absorbed moisture of the films themselves (Agarwal, Hoagland *et al.* 1997). For films prepared at ambient humidity (e.g. ~35%) the glass

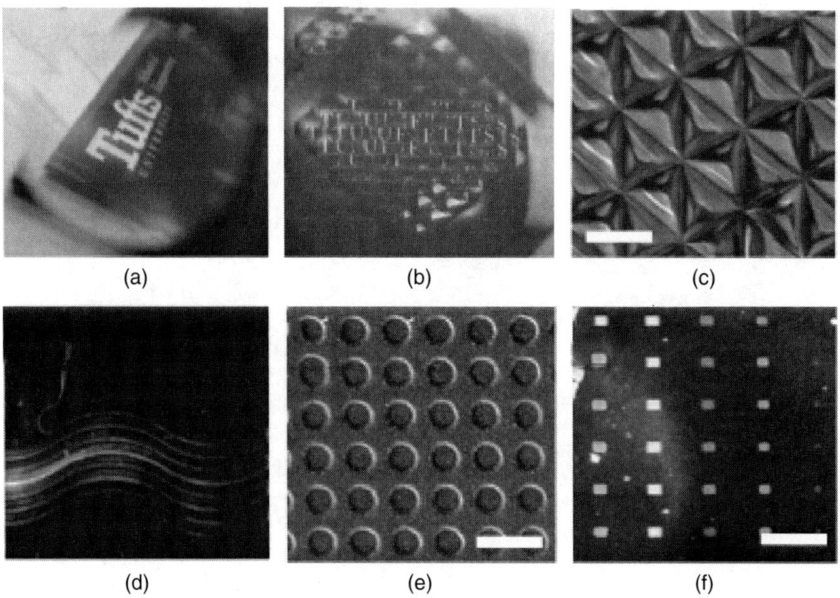

6.4 Examples of silk-based optical and photonic elements and devices: (a) an image through a 1-cm-diameter lens; (b) an image through a 6 × 6 microlens array; (c) image of a silk microprism array (scalebar = 25 micrometers); (d) 2 cm long inkjet printed silk waveguide; (e) a scanning electron microscope (SEM) image of a periodic nanohole array with lattice constant equal to 400 nm (scalebar = 500 nm); (f) periodic nanoarrays imprinted in silk (such as in (e)) under dark-field illumination showing different colors associated to the different lattice constants.

transition temperature is ~100°C. This allows nanoimprinting by using a hot embossing process in which a silk film is pressed (~30–50 psi) onto a heated (100°C) master with micro- or nanoscale patterns for a few seconds and after a brief (~1 min) cooling step can then be mechanically removed from the master (Mondia, Amsden *et al.* 2010). In contrast, progressive water saturation of the films decreases the glass transition temperature to room temperature.

This allows the additional opportunity to carry out the nanoimprinting process at room temperature by adding a small amount of water on the film surface before the master is pressed onto the silk film (Omenetto, Amsden *et al.* 2010) (Fig. 6.5). Once imprinted, the films can be annealed by exposure to water vapor or methanol to ensure their insolubility in water (Cao, Lv *et al.* 2005; Jin, Park *et al.* 2005; Wen, Xu *et al.* 2007). Upon exposure to methanol, the secondary structure of the silk fibroin protein changes from (primarily) random coil to (primarily) β-sheet, resulting in an increase in the glass transition temperature (Cebe, Hu *et al.* 2006). After annealing, the imprinted films are stable and can last for years under ambient conditions. The ability to adopt nanoimprinting techniques increases the quality and throughput of nanostructured silk materials in comparison to casting

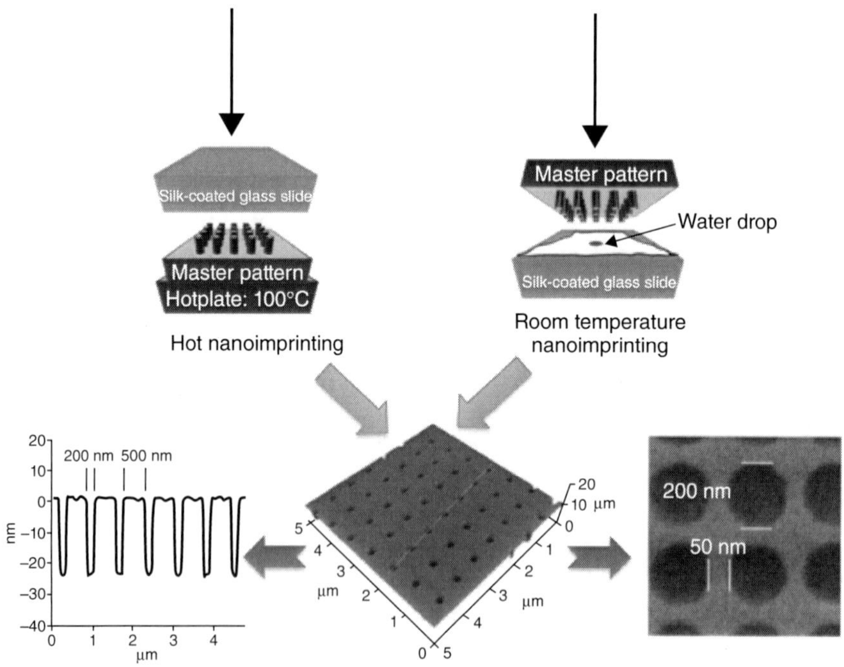

6.5 Scheme illustrating rapid nanoimprinting in silk and resulting images corresponding to the imprinting of a periodic nanohole array.

methods, as well as eliminating mechanical stresses when separating hardened silk from masters (Omenetto, Perry *et al.* 2008). In addition to the speed and fidelity of production of such films, as well as their mechanical and optical properties, these silk optical devices possess optical features and performance that exceed those of other biopolymers and biocompatible polymers such as chitosan (Park, Cheng *et al.* 2007) and polylactide (PLA) (Mills, Navarro *et al.* 2006).

6.2.1 Silk photonic lattices

The effectiveness of both soft-lithography and nanoimprinting processes replication geometries and topologies down to tens of nanometers allows for more sophisticated nanoscale photonic components to be generated from silk. Of particular interest is the ability to define ordered arrays of deterministic nanoscale features which confer predetermined scattering properties to these protein films, ultimately resulting in color localization when illuminated by white light. Light scattering in periodic systems has been investigated for decades in optics and photonics (Elachi 1976). Their classical description relies on Bragg scattering phenomena, which give rise to constructive interference at specific wavelengths along well-defined propagation directions, depending on illumination conditions, structural periodicity, and the refractive index of the surrounding medium (Baryshev, Khanikaev *et al.* 2009).

Structural color based on periodic nanopatterned two-dimensional (2D) lattices in silk protein films were demonstrated by imprinting periodic lattices of nanoscale holes (200 nm diameter, 40 nm in depth) which display different colors as a function of varying lattice spacing under white light illumination (Amsden, Perry *et al.* 2009; Omenetto, Amsden *et al.* 2010) as shown in Plate XVI (see color plate section between pages 96 and 97). The variation of the index of refraction contrast between the imprinted nanopatterned lattices and the surrounding environment causes a shift in the spectral response as a function of the perturbation.

This phenomenon can be exploited by applying liquids on the lattices to monitor their optical properties through the spectral responses (Altug, Adato *et al.* 2009). Such approaches have been used as the optical transduction mechanisms upon which to build colorimetric sensors and have been demonstrated in a range of materials (Nuzzo, Yao *et al.* 2010). Specifically for silk, this response was observed by using a nanopatterned lattice and exposing it to a variety of glucose concentrations (Amsden, Perry *et al.* 2009). Measurement of the spectral responses offered a demonstration of the adequacy of silk fibroin photonic lattices to act as an optical sensing substrate. The unique attribute of biological stabilization is a technologically enabling feature of the silk matrix, which can be coupled to the nanopatterned and optical features of the silk to function as transducers, monitors or enhancers of the entrained biological functions. This level of functionality is not commonly available in current technological materials without complex chemistries or modifications.

An example of this type of device was demonstrated by transforming a simple optical component into an interactive optical component. For these experiments, human hemoglobin was mixed with silk solution and then either cast (Omenetto, Lawrence *et al.* 2008) or imprinted (Omenetto, Amsden *et al.* 2010) to form a blood-based diffraction grating (Fig. 6.6). The hemoglobin contained in the free-standing silk film remained active, responding to the oxygenation state of the external environment. This silk-blood-grating was used as the building block of a simple spectrometer, using the diffractive properties of the doped silk as the transduction mechanism for the oxygenation levels of the entrained hemoglobin through its spectral response or, in other words, the grating senses its material constituent. The hemoglobin remains active in the silk-grating system after the device is stored for several months at ambient conditions (Omenetto, Domachuk *et al.* 2009). The ability to preserve the function of biological dopants within silk films was also demonstrated with the entrainment of enzymes such as glucose oxidase, lipase and peroxidase, where activity was preserved for over a year even at 37°C (Kaplan, Lu *et al.* 2009; Kaplan, Lu *et al.* 2010).

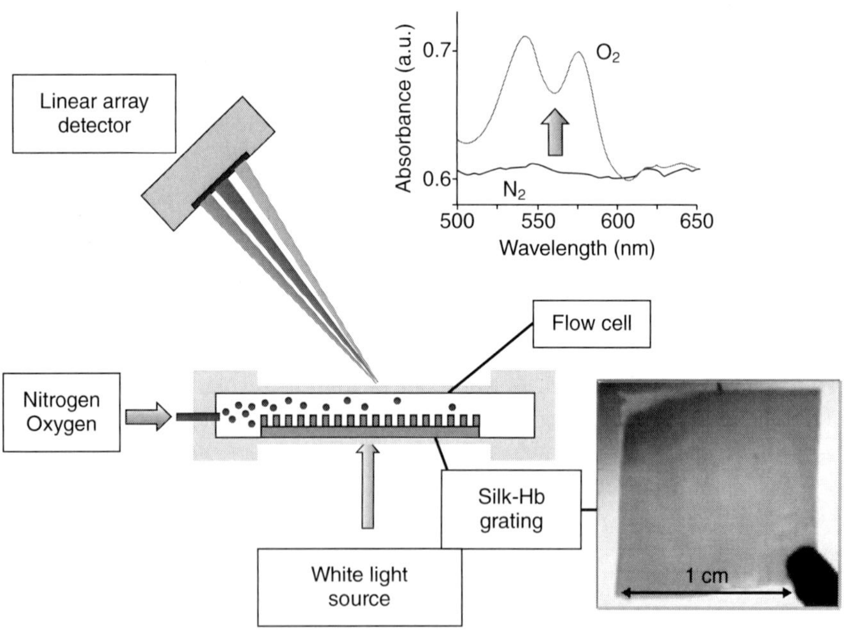

6.6 Hemoglobin-doped silk grating acting as a bio-active oxygen sensor.

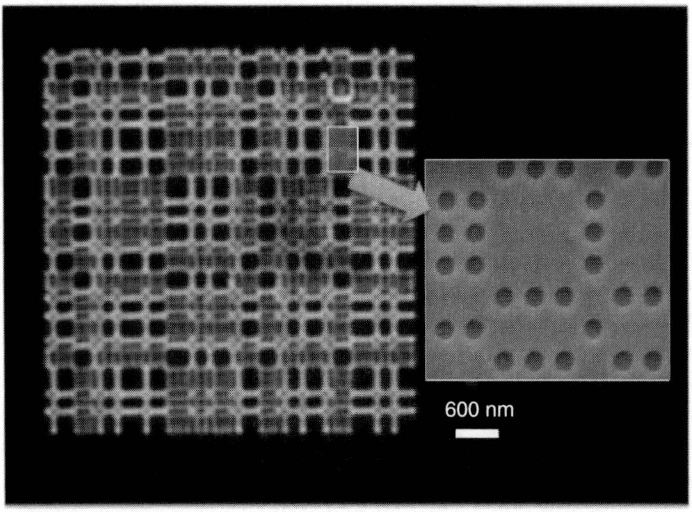

6.7 A dark-field image of an aperiodic lattice of nanoholes imprinted in silk to obtain complex colorimetric signatures. The inset shows an SEM image of the nanoholes in silk. The entire lattice covers a 50 μm × 50 μm area on a silk film.

This outcome is in contrast to stabilization of these enzymes only for a few days in similar conditions outside of the silk matrix (Rao, Gouda *et al.* 2003). To achieve increased sensitivity in the optical response of the silk nanostructured devices, different lattices can be engineered by nanoimprinting aperiodic structures (Fig. 6.7). The colorimetric fingerprints of these naturally inspired geometries result from multiple scattering of the various spectral components of white light interacting with the nanostructured silk film surfaces in combination with the broadband scattering characteristic of aperiodic systems (Boriskina, Lee *et al.* 2010; Omenetto, Lee *et al.* 2010). These unique scattering characteristics of deterministic aperiodic scattering systems and the sensitivity of the associated colorimetric fingerprints to morphological and refractive index surface variations suggest a very appealing path towards the engineering of novel sensing platforms for label-free optical detection of thin molecular layers, all based on this unique pure protein system. Silk films imprinted with such geometries generate complex signatures that are typical of light localization, underscoring the effectiveness of this biopolymer as a technological substrate for advanced photonics.

6.2.2 Silk-based optical fibers

The same silk solution that is used as the starting material to manufacture the optical films described above, can also be used as a material for microfluidic printing. Given the favorable material properties, the use of silk-based inks can be important for a variety of controlled biomaterial fabrication on the microscale. Among the possible applications of this approach is the opportunity to directly print optical devices. This was recently explored by using direct inkjet writing to fabricate silk optical waveguides (Omenetto, Parker *et al.* 2009). In the demonstration, silk fibroin inks were concentrated (aqueous solution of 28–30 wt% silk, compared to the 6–8 wt% concentration used for films in soft-lithography and nanoimprinting) to obtain the viscosity needed for high-resolution and high-quality structures. The printed silk waveguides retained rod-like morphology by crystallization in a methanol-rich coagulation reservoir. Both straight and curved silk waveguides were generated and guided light, with measured losses ranging between 0.25 dB/cm and 0.81 dB/cm for the straight and curved waveguides, respectively. In this case as well, the utility of this method can be enhanced by leveraging the ease of functionalization offered by the aqueous silk solution. Doped waveguides were generated by mixing a laser dye (Rhodamine 6G) into the silk solution prior to printing. These doped waveguides exhibited significant fluorescence while guiding $\lambda = 532$ nm light and were measured to have less losses (~<0.1 dB/cm) than their undoped counterparts. The devices realized in this fashion are equally amenable to nanoimprinting, which could prove to be valuable for facile generation of distributed feedback lasers or fiber-based biophotonic components.

6.2.3 Fluorescent silk with localized emission enhancement

The utility of nanoimprinting processes to generate silk films with complex photonic lattices can be further enhanced by similarly simple ways to functionalize the silk solution through the addition of organic and inorganic dopants that are easily mixed into the silk solution before forming the photonic elements. This opens the option to seamlessly combine nanoimprinted lattices on doped silk films, thereby consenting to design the interactions between the bulk optical properties (induced by the dopants) and the imprinted optical structures.

This was demonstrated by imprinting photonic lattices on fluorescent silk substrates (Mondia, Amsden *et al.* 2010). Directional and wavelength-specific fluorescent enhancement was obtained from silk films doped with fluorophores when nanoimprinted with periodic lattices that were designed to have suitable lattice constants and, consequently, suitable optical properties to match the spectral response of the compounds embedded in the bulk film (see Plate VII). The aqueous environment afforded by the silk solution makes it possible to easily incorporate organic and/or inorganic dyes used in biomedical applications. Such materials have the potential to be used as sensors, spectral filters, or light emitting devices and open new opportunities in research for biologically based or biologically active photonic crystals or optical resonators.

6.2.4 Silk-based microfluidics and optofluidics

The ability to reform the material into thick films enables the use of silk as a substrate for microfluidic devices. This exemplifies another potential offered by this biopolymer. Constituting such devices out of a material that is biocompatible and interfaces favorably with a biological environment presents additional opportunities over traditional substrates used in microfluidic devices such as siloxane-based elastomers or glass (Borenstein, Bettinger *et al.* 2007).

The functionality of silk can be increased by chemical modifications. This is generally achieved by exploiting the chemical bonds available within the fibroin structure, namely the acid side chains (~1% of the total amino acids) or the tyrosine residues (>5% of the total amino acids) (Kaplan, Wang *et al.* 2008). A typical modification used generally relies on azo-benzene molecules. These molecules have been widely used in a variety of applications ranging from cell culture applications (Kaplan, Wang *et al.* 2008), chemical dyes, pH indicators in a fluid environment (Carofiglio, Brigo *et al.* 2008), for optical sensors (Carofiglio, Fregonese *et al.* 2006), and in combination with Raman spectroscopy (Wang and Callahan 1998). In addition, azo-benzene has been studied extensively in materials science and engineering as an optically addressable substrate for holographic data storage (Psaltis and Burr 1998) exploiting the cis-trans isomerization of the molecule (Barrett, Mamiya *et al.* 2007).

With this in mind, chemically activated microfluidic devices can be manufactured uniting the utility of facile chemistry to the convenience of a fluidic platform that has garnered much attention for convenience of use in diagnostics. Hybrid devices combining the azo-benzene functionality within modified silk films and PDMS microfluidics were shown to generate pH-reactive platforms (Tsioris, Tilburey *et al.* 2010).

6.3 Silk electronics and optoelectronics

The use of silk for *in vivo* electronic device applications builds on an extensive history of biocompatibility based on biomedical and tissue engineering studies. This history starts with eons of use of silk as medical sutures, which are FDA approved, and has progressed in recent years to utility in the formation of many human tissues and *in vivo* studies with silks in various forms, from fibers and films to gels and sponges (Kaplan, Altman *et al.* 2003; Kaplan and Vepari 2007).

6.3.1 Silk as an electronic material

While the mechanical robustness, transparency, and surface flatness are compelling features for photonics applications, such material properties are also desirable for metal deposition for microelectronic applications. Silk films are insulating and could provide a sustainable alternative to thin dielectric layers used in microelectronics or as supports in microelectronics.

Silk biomaterials have the convenience of possessing material attributes that make them ideal constituents of technological devices such as planar microelectronic structures. Recently, there have been reports of using silk as a low-cost e-paper device (Hwang, Wang *et al.* 2011). Silk can be used as a replacement for traditional inorganic oxide layers such as SiO_2 or poly(methylmethacrylate) (PMMA) (Hwang, Wang *et al.* 2011; Muller, Hamedi *et al.* 2011; Omenetto, Capelli *et al.* 2011; Wen, Yao *et al.* 2011). Organic thin film transistors (OTFTs) have been manufactured on a glass/ITO substrate by stacking 400 nm thick silk layers as the gate dielectric, n- and p-type organic semiconductors (N,N'-ditridecylperylene-3,4,9,10-tetracarboxylic diimide (P13), and α,ω-dihexyl-quaterthiophene (DH4T) respectively), and gold gate and source electrodes (Fig. 6.8). Both P13 and DH4T are well-characterized benchmark materials for organic electronics production and provide a suitable test for the electronic performance of silk.

The reported silk dielectric constant for this device configuration is $\varepsilon_s = 6$ with charge mobility values of $\mu_n = 4 \times 10^{-2}$ cm^2/Vs for the silk-P13

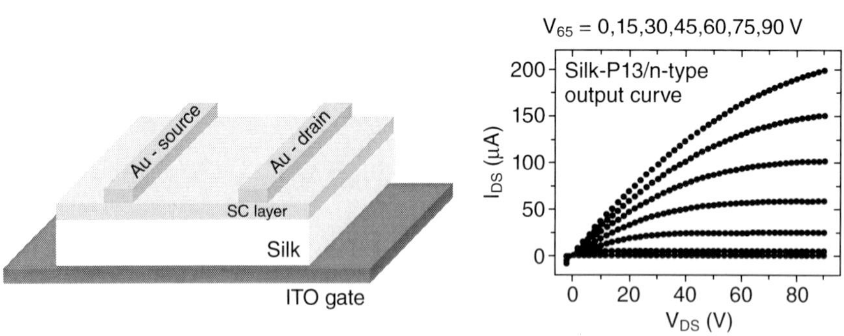

6.8 Silk-based organic field effect transistor and corresponding electrical measurement when using an n-type semiconducting material.

and $\mu_p = 1.3 \times 10^{-2}$ cm^2/Vs for the silk-DH4T. The comparison of the parameters of silk-based devices and the equivalent PMMA and SiO_2-based devices reveals that measured voltage thresholds for the silk-P13 transistor is one order of magnitude lower, namely 1.8 V at 20 V, than the respective standard organic field effect transistor (OFET). Additionally, the on/off switch ratio was found to be ~10^4, a value that matches the highest value reported for P13 and DH4T-based thin film organic transistors.

6.3.2 Silk-based electromagnetic metamaterials

Patterning metals on silk has been recently demonstrated by a shadow masking deposition method (Zhang, Tao *et al.* 2010). Micro-fabricated stencils are used to deposit metals onto silk films in a dry, chemical-free environment, preventing chemical contamination that might be involved in other photolithography-based metal patterning methods such as lift-off processes and wet etching. With this approach, deposition of microscale resonators was demonstrated on silk films. Additionally, this approach maintains the integrity and biocompatibility of the silk films making the resulting materials suitable for incorporation into biological environments such as the human body.

Alternatively, a simpler fabrication technique where a single step transfers metal micro patterns to free-standing silk films under ambient processing conditions has been demonstrated (Omenetto, Tsioris *et al.* 2011). In this approach, standard photolithography is used to define the metal structures patterned onto pre-silanized hydrophobic silicon wafers. Once this is performed, the silk solution is poured onto the wafer and dried at ambient conditions. During the drying process the silk binds to the metallic pattern on the surface of the silicon wafer resulting in the transfer of the metal patterns from the wafer onto the silk film. This approach is particularly appealing because it allows the parallel fabrication of microstructures over a large area, free-standing and flexible silk films with high precision and eliminating the need for alignment.

The properties of silk allow for the manufacturing of high-quality structures that combine the features of the biopolymer and the capacities offered by microfabrication techniques. With these approaches, large area metamaterial structures were patterned on free-standing silk substrates and exhibited strong resonance responses at Terahertz frequencies. The structures can be scaled to cover a broad area of the electromagnetic spectrum, in spite of the interest provided by the THz regime where numerous chemical and biological agents show unique 'fingerprints', which could potentially be used for identification and bio-sensing (Funk, Barber *et al.* 2005).

The canonical sub-wavelength metamaterial element is the split ring resonator (SRR) (Pendry, Holden *et al.* 1999). The electromagnetic response originates from oscillating electrons in highly conducting metals such as gold or copper, allowing for a tailored electromagnetic response. The SRR can be thought of as an LC resonator in a simple representation with a resonance frequency of $\omega_0 \sim (1/LC)^{1/2}$, where the inductance results from the current path of the SRR and capacitance is mainly determined by the split gap (O'Hara, Singh *et al.* 2008; Omenetto, Tao *et al.* 2010b). Any change in the capacitance or the inductance will result in a change in the resonant response, making metamaterials sensitive to the local environment. Therefore, metamaterials would offer an interesting option for integration with silk films for novel sensing and detection applications by providing a natural and biologically active material as the support matrix for the SRR. This was demonstrated by monitoring the evaporation of residual water in five minute intervals from the hardened silk film when the silk film was placed in a dry box, providing access to a new transduction mechanism to monitor hydration state and conformational changes of the fibroin proteins transitioning from amorphous to the crystalline state (Zhang, Tao *et al.* 2010) (Fig. 6.9). The possibility of scaling electromagnetic structures to cover different electromagnetic regimes and of silk films to be functionalized with biological components offers extensive options for biocompatible sensors, transducers and biologically integrated monitors.

An additional trait of silk films is the possibility of tuning the properties, such as solid-to-gel-to-liquid state, with stable or dissolvable features. Such tenability in material features can be attained through the control of hydration and the degree of crystallinity (beta-sheet content) in the final material. By leveraging this property, flexible electrodes that conform to curvilinear surfaces are possible (see Fig. 6.10). Electronics that are capable

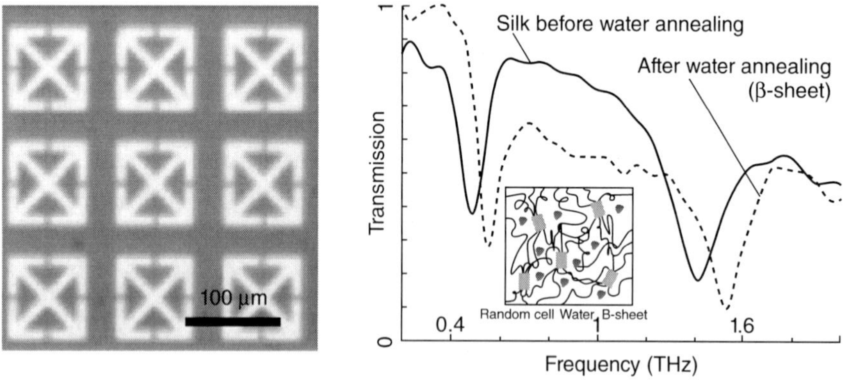

6.9 Shadow masking patterned micro split ring resonators on silk substrates showing strong resonant responses at terahertz frequencies that can be used to monitor the crystallization level of the silk protein.

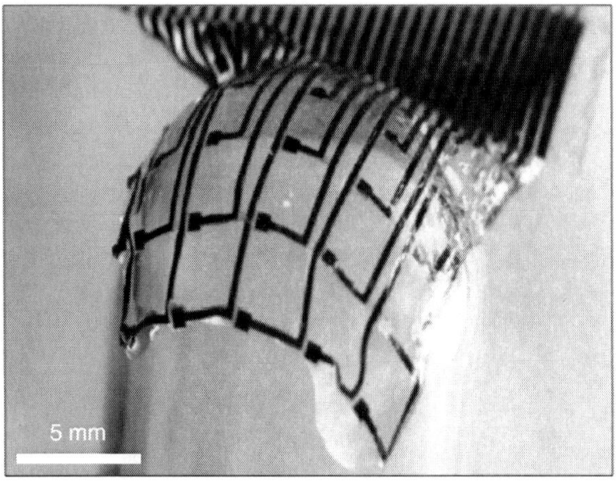

6.10 A ~1 cm × 1 cm electrode array on a dissolvable silk film conformally transferred/wrapped on a rounded surface.

of intimate, noninvasive integration with the soft, curvilinear surfaces of biological tissues offer important opportunities for monitoring, diagnosing and treating disease or injury, and for establishing brain/machine interconnects.

6.3.3 Silk-integrated plasmonics

In addition to the ability of being reshaped with complex nano- and microscale topologies, silk provides a suitable interface for microelectronic and photonic applications. This allows the use of silk for conformal coatings on a variety of materials (hydrophobic, hydrophilic, plastic, metal, among others) with control across various length scales, ranging from monolayers to the mesoscale. Thin layers of silk can be obtained by spin coating, dip coating or spray coating the solution. Solution concentration, additives, and application rates can be used to modulate coatings characteristics.

In early work we showed that sub-nanometer coatings of silk could be conformally deposited by dip coating, using a layer-by-layer approach (Kaplan, Wang *et al.* 2005; Kaplan, Wang *et al.* 2007). The concentration of silk in aqueous solution and the presence of salts were used to regulate layer thickness with precision. Similar variables were used to control spin coating of silk solution to generate precisely tunable layers of conformal coating on a variety of surfaces.

This degree of control allows the application of silk layers to specifically patterned surfaces for sensing applications. Detecting variations at the nanoscale traditionally requires sophisticated instrumentation such as atomic force or electron microscopies, imaging based on dye-assisted spectroscopic techniques, or collective resonant effects in plasmonic structures, such as sub-wavelength apertures or surface enhanced Raman scattering. The precisely tunable thicknesses allow the convenience of interfacing protein films with resonant electromagnetic structures for enhanced detection approaches.

Two demonstrations of this capacity were demonstrated in the infrared (Altug, Adato *et al.* 2009) and optical regions of the electromagnetic spectrum (Omenetto, Lee *et al.* 2010) by spin coating silk protein monolayers on engineered nanoscale surfaces. In the first case, a silk monolayer was spin coated on plasmonic nanoantenna arrays to leverage the capacity to probe optical signatures from proteins in the mid-IR regime, where the vibrational fingerprint of the protein molecular structure was accessible. By tailoring the geometry of individual nanoantennas to form resonant structures that match the molecular vibrational mode there was enhancement at four orders of magnitude of the otherwise small IR signal. This approach allows measurement of the vibrational spectra from silk films at atto-mole levels because of the enhanced absorption signals from a small amount of molecules in the vicinity of the antenna nanotips. This enhanced response is further demonstrated by coating the nanoantennas with increasing (monolayer by monolayer) protein thicknesses.

In the optical regime, silk monolayers are spin coated on aperiodic photonic lattices consisting of the ordered arrangement of 200 nanometer diameter chromium nanoparticles on a quartz substrate. The silk protein monolayer was assessed for structural color in the visible range using conventional dark-field microscopy. This detection approach is label-free and relies on the confluence of photonics and biopolymer engineering by combining the nanoscale silk layers with nanopatterned aperiodic surfaces with controlled light scattering signatures.

The structure was sensitive to local changes in refractive index, giving rise to distinctive changes in the spatial and spectral color patterns observable under white light illumination (Plate XXI). When the slide is placed under a dark-field microscope, the white light from the condenser is scattered and spectrally rearranged into a structural color pattern. Additionally the multiple, deterministic spectral components encoded in the nano-quilt (angular spectra, scattering intensity, correlation patterns) yield distinct scattering responses which define a multi-parametric sensing platform for real-time nanoscale detection of biological materials in the visible spectral range.

Moreover, the recently demonstrated ability of the silk films to stabilize biochemical dopants as described earlier allows for expanded detection schemes exploiting the selectivity and responses of these biological components as active components in these engineered colorimetric chips. This approach could ultimately yield sensitivity limits comparable to surface plasmon sensors within a simpler, cost-effective approach with microfluidics, bio-assays, and label-free detection of biological materials on the nanoscale all accessible.

6.3.4 Bioresorbable electronics

These properties have been exploited towards devices that rely on ultrathin electronics supported by bioresorbable substrates of silk. When these devices were placed on living tissue, controlled hydration causes the silk to dissolve and resorb, initiating spontaneous, conformal, wrapping of tissue driven by capillary forces at the tissue interface. A combination of experimental and theoretical studies of the materials and underlying mechanics explored the transferring of a conformal electrode fabricated on the silk substrate (with an initial Young's modulus of E_{silk} = 2.8 GPa), onto the surface of a living feline brain for EEG neural mapping experiments (Rogers, Kim *et al.* 2009; Kim, Viventi *et al.* 2010). These concepts provide capabilities for implantable or surgical devices that lie outside of wafer-based technologies or known forms of flexible electronics (Fig. 6.10). Further, as outlined earlier, the rate of silk film resorption or degradation can be tailored, from minutes to years, depending on the mode of processing.

Similar to the conformal electrode work described earlier, silk materials are suitable candidates for advanced implantable electronic devices that can interface with the body and establish conformal contact with the curvilinear surfaces of various organs. Achieving biocompatibility can be challenging, due to the complex nature of the biological response to many organic and inorganic materials. Electronic devices should be constructed out of materials that are soluble an/or biodegradable. Alternatively, a large fraction of the device can be designed to resorb, such that only small, safe, components remain, minimizing adverse biological responses. With this in mind, silk was used as the principal constituent of largely resorbable devices constituted of thin-film silicon electronics, to yield flexible devices largely resorbed *in vivo* (Rogers, Kim *et al.* 2009). The use of silicon provides high performance, good reliability, and robust operation. In this case, as described earler, silk is attractive compared to other biodegradable polymers such as poly(glycolic acid) (PGA), PLA and collagen, because of its robust mechanical properties, the ability to tailor the dissolution and/or biodegradation rates from hours to years, the generation of non-inflammatory amino acid degradation products, and the option to prepare the materials at ambient conditions to preserve sensitive electronic functions (Kaplan and Vepari 2007; Kaplan, Murphy *et al.* 2008).

6.3.5 Nanoparticle-doped silk

The use of silk doped with bioactive components can also be useful to add more latitude to the applications described above. As described earlier, the silk solution can be easily functionalized by mixing in suitable dopants to functionalize the devices, similar to what is done for silk-based drug delivery applications. Many instances of doping silks have been reported, from the inclusion of metals (Omenetto, Tao *et al.* 2010a), inorganic components such as silica (Kaplan, Foo *et al.* 2006) and hydroxyapatite (Yao, Liu *et al.* 2008), silk particles (Wang, Rajkhowa *et al.* 2010), fluorescent dyes (Mondia, Amsden *et al.* 2010; Han, Tansil *et al.* 2011), and drugs (Meinel, Hofmann *et al.* 2006; Boison, Wilz *et al.* 2008; Pritchard, Wilz *et al.* 2009), to other polymers for dynamic devices responsive to temperature (Kaplan, Gil *et al.* 2010).

Recently, silk films doped with gold nanoparticles (Au-NPs) have been used as light-activated heating elements that were interfaced with commercial thermo-electronic components to wirelessly power micro devices. In these experiments (Omenetto, Tao *et al.* 2010a) Au-NP doped silk films were interfaced to thermally available thermoelectric chips by casting the solution on the active area of the chip. Using a continuous wave (CW) green laser with an output power up to 450 mW/mm^2 at 532 nm to match the absorption of the Au-NPs induces a temperature increase in the silk layer, which in turn is used to generate 20 mW of power. This use of silk solution underscores the ease of interfacing a doped protein with planar technologies. Power requirements of the illuminating source can be further optimized by either increasing the concentration of the Au-NPs and/or the thickness of the Au-NP doped silk film.

6.4 Conclusion

These outcomes are a promising prelude to opportunities that seamlessly combine technological function with biological function by leveraging the favorable processing environment and material features of silk proteins.

Silk fibroin films offer transparency, ability to nanopattern, and water-processing enabling the confluence of optical physics with biomaterials and the biomedical sciences. The ability to embed and maintain biological activity within a robust micro/nanostructured optical material is particularly unusual. Free-standing optical elements can be engineered to contain compounds across varying 'biological scales', from small organics, to complex proteins such as hemoglobin, and enzymes, while stored at room temperature.

Additionally, an important feature of these devices is their full biodegradability and biocompatibility, which expand the application space for silk-based devices in the environmental and life sciences where these attributes are paramount. The refined optical, electronic, and electro-optical functions obtainable through the various methods and approaches described herein would then be available for a class of new devices that could unobtrusively monitor and be part of a natural environment, including the human body. With this material versatility, devices for implantation *in vivo* become feasible, eliminating the need to retrieve the system at a later point. Such controlled degradation and biocompatibility would allow for silk-based technological devices to be dispersed in the environment as well, opening new opportunities for distributed sensing and disposable detection systems.

The availability of a free-standing biological matrix that has the material toughness to withstand room temperature use under ordinary environmental conditions while simultaneously exhibiting high technological quality and biological activity is unique. This platform could open opportunities for a new class of devices by weaving together the technological and biological worlds with a silk thread.

6.5 References

Agarwal, N., D. A. Hoagland, *et al.* (1997). 'Effect of moisture absorption on the thermal properties of Bombyx mori silk fibroin films.' *Journal of Applied Polymer Science* **63**(3): 401–410.

Aizenberg, J. (2004). 'Crystallization in patterns: A bio-inspired approach.' *Advanced Materials* **16**(15): 1295–1302.

Aizenberg, J., A. Tkachenko, *et al.* (2001). 'Calcitic microlenses as part of the photoreceptor system in brittlestars.' *Nature* **412**(6849): 819–822.

Altug, H., R. Adato, *et al.* (2009). 'Ultra-sensitive vibrational spectroscopy of protein monolayers with plasmonic nanoantenna arrays.' *Proceedings of the National Academy of Sciences of the United States of America* **106**(46): 19227–19232.

Amsden, J. J., H. Perry, *et al.* (2009). 'Spectral analysis of induced color change on periodically nanopatterned silk films.' *Optics Express* **17**(23): 21271–21279.

Arockiados, T., F. P. Xavier, *et al.* (2008). 'Isolation and characterization of biologically metal-doped protein as semiconducting biopolymer.' *Materials Chemistry and Physics* **111**(2–3): 517–523.

Barrett, C. J., J. I. Mamiya, *et al.* (2007). 'Photo-mechanical effects in azobenzene-containing soft materials.' *Soft Matter* **3**(10): 1249–1261.

Baryshev, A. V., A. B. Khanikaev, *et al.* (2009). 'Diffraction processes in 3D photonic crystals based on thin opal films.' *Journal of Materials Science-Materials in Electronics* **20**: 416–420.

Boison, D., A. Wilz, *et al.* (2008). 'Silk polymer-based adenosine release: Therapeutic potential for epilepsy.' *Biomaterials* **29**(26): 3609–3616.

Borenstein, J. T., C. J. Bettinger, *et al.* (2007). 'Silk fibroin microfluidic devices.' *Advanced Materials* **19**(19): 2847–2850.

Boriskina, S. V., S. Y. K. Lee, *et al.* (2010). 'Formation of colorimetric fingerprints on nano-patterned deterministic aperiodic surfaces.' *Optics Express* **18**(14): 14568–14576.

Braun, F. N. and C. Viney (2003). 'Modelling self assembly of natural silk solutions.' *International Journal of Biological Macromolecules* **32**(3–5): 59–65.
Cannas, M., M. Santin, *et al.* (1999). '*In vitro* evaluation of the inflammatory potential of the silk fibroin.' *Journal of Biomedical Materials Research* **46**(3): 382–389.
Cao, C. B., Q. Lv, *et al.* (2005). 'Preparation of insoluble fibroin films without methanol treatment.' *Journal of Applied Polymer Science* **96**(6): 2168–2173.
Carofiglio, T., L. Brigo, *et al.* (2008). 'An optical sensor for pH supported onto tentagel resin beads.' *Sensors and Actuators B – Chemical* **130**(1): 477–482.
Carofiglio, T., C. Fregonese, *et al.* (2006). 'Optical sensor arrays: one-pot, multiparallel synthesis and cellulose immobilization of pH and metal ion sensitive azo-dyes.' *Tetrahedron* **62**(7): 1502–1507.
Cebe, P., X. Hu, *et al.* (2006). 'Determining beta-sheet crystallinity in fibrous proteins by thermal analysis and infrared spectroscopy.' *Macromolecules* **39**(18): 6161–6170.
Chen, X., G. Q. Zhou, *et al.* (2009). 'Silk fibers extruded artificially from aqueous solutions of regenerated Bombyx mori silk fibroin are tougher than their natural counterparts.' *Advanced Materials* **21**(3): 366–370.
Cremona, M., C. Legnani, *et al.* (2008). 'Bacterial cellulose membrane as flexible substrate for organic light emitting devices.' *Thin Solid Films* **517**(3): 1016–1020.
Elachi, C. (1976). 'Waves in active and passive periodic structures – review.' *Proceedings of the IEEE* **64**(12): 1666–1698.
Funk, D. J., J. Barber, *et al.* (2005). 'Temperature-dependent far-infrared spectra of single crystals of high explosives using terahertz time-domain spectroscopy.' *Journal of Physical Chemistry A* **109**(15): 3501–3505.
Goldsmith, M. R., T. Shimada, *et al.* (2005). 'The genetics and genomics of the silkworm, Bombyx mori.' *Annual Review of Entomology* **50**: 71–100.
Han, M. Y., N. C. Tansil, *et al.* (2011). 'Intrinsically colored and luminescent silk.' *Advanced Materials* **23**(12): 1463–1466.
Hooper, I. R., P. Vukusic, *et al.* (2006). 'Detailed optical study of the transparent wing membranes of the dragonfly Aeshna cyanea.' *Optics Express* **14**(11): 4891–4897.
Hwang, J. C., C. H. Wang, *et al.* (2011). 'Flexible organic thin-film transistors with silk fibroin as the gate dielectric.' *Advanced Materials* **23**(14): 1630–1634.
Irimia-Vladu, M., N. S. Sariciftci, *et al.* (2011). 'Exotic materials for bio-organic electronics.' *Journal of Materials Chemistry* **21**(5): 1350–1361.
Jin, H. J., J. Park, *et al.* (2005). 'Water-stable silk films with reduced beta-sheet content.' *Advanced Functional Materials* **15**(8): 1241–1247.
Kaplan, D. L., G. H. Altman, *et al.* (2003). 'Silk-based biomaterials.' *Biomaterials* **24**(3): 401–416.
Kaplan, D. L., C. W. P. Foo, *et al.* (2006). 'Novel nanocomposites from spider silk-silica fusion (chimeric) proteins.' *Proceedings of the National Academy of Sciences of the United States of America* **103**(25): 9428–9433.
Kaplan, D. L., E. S. Gil, *et al.* (2010). 'Mechanically robust, rapidly actuating, and biologically functionalized macroporous poly(N-isopropylacrylamide)/silk hybrid hydrogels.' *Langmuir* **26**(19): 15614–15624.
Kaplan, D. L., J. A. Kluge, *et al.* (2008). 'Spider silks and their applications.' *Trends in Biotechnology* **26**(5): 244–251.
Kaplan, D. L., B. D. Lawrence, *et al.* (2009). 'Silk film biomaterials for cornea tissue engineering.' *Biomaterials* **30**(7): 1299–1308.

Kaplan, D. L., Q. Lu, *et al.* (2010). 'Stabilization and release of enzymes from silk films.' *Macromolecular Bioscience* **10**(4): 359–368.

Kaplan, D. L., S. Z. Lu, *et al.* (2009). 'Stabilization of enzymes in silk films.' *Biomacromolecules* **10**(5): 1032–1042.

Kaplan, D. L., L. Meinel, *et al.* (2005). 'The inflammatory responses to silk films *in vitro* and *in vivo*.' *Biomaterials* **26**(2): 147–155.

Kaplan, D. L., A. R. Murphy, *et al.* (2008). 'Modification of silk fibroin using diazonium coupling chemistry and the effects on hMSC proliferation and differentiation (vol 29, pg 2829, 2008).' *Biomaterials* **29**(31): 4260–4260.

Kaplan, D. L., R. Valluzzi, *et al.* (2002). 'Silk: molecular organization and control of assembly.' *Philosophical Transactions of the Royal Society of London Series B – Biological Sciences* **357**(1418): 165–167.

Kaplan, D. L. and C. Vepari (2007). 'Silk as a biomaterial.' *Progress in Polymer Science* **32**(8–9): 991–1007.

Kaplan, D. L., X. Wang, *et al.* (2008). 'Controlled release from multilayer silk biomaterial coatings to modulate vascular cell responses.' *Biomaterials* **29**(7): 894–903.

Kaplan, D. L., X. Y. Wang, *et al.* (2007). 'Nanolayer biomaterial coatings of silk fibroin for controlled release.' *Journal of Controlled Release* **121**(3): 190–199.

Kaplan, D. L., X. Y. Wang, *et al.* (2005). 'Biomaterial coatings by stepwise deposition of silk fibroin.' *Langmuir* **21**(24): 11335–11341.

Kaplan, D. L., Y. Z. Wang, *et al.* (2006). 'Stem cell-based tissue engineering with silk biomaterials.' *Biomaterials* **27**(36): 6064–6082.

Kim, D. H., J. Viventi, *et al.* (2010). 'Dissolvable films of silk fibroin for ultrathin conformal bio-integrated electronics.' *Nature Materials* **9**(6): 511–517.

Kundu, S. C., B. C. Dash, *et al.* (2008). 'Natural protective glue protein, sericin bioengineered by silkworms: Potential for biomedical and biotechnological applications.' *Progress in Polymer Science* **33**(10): 998–1012.

Kundu, S. C., T. S. Khire, *et al.* (2010). 'The fractal self-assembly of the silk protein sericin.' *Soft Matter* **6**(9): 2066–2071.

Lee, K. Y., B. M. Min, *et al.* (2006). 'Regenerated silk fibroin nanofibers: Water vapor-induced structural changes and their effects on the behavior of normal human cells.' *Macromolecular Bioscience* **6**(4): 285–292.

Meinel, L., S. Hofmann, *et al.* (2006). 'Silk fibroin as an organic polymer for controlled drug delivery.' *Journal of Controlled Release* **111**(1–2): 219–227.

Mills, C. A., M. Navarro, *et al.* (2006). 'Transparent micro- and nanopatterned poly(lactic acid) for biomedical applications.' *Journal of Biomedical Materials Research Part A* **76A**(4): 781–787.

Mondia, J. P., J. J. Amsden, *et al.* (2010). 'Rapid nanoimprinting of doped silk films for enhanced fluorescent emission.' *Advanced Materials* **22**(41): 4596–4599.

Mori, H. and M. Tsukada (2000). 'New silk protein: modification of silk protein by gene engineering for production of biomaterials.' *Journal of Biotechnology* **74**(2): 95–103.

Motta, A., C. Migliaresi, *et al.* (2002). 'Serum protein absorption on silk fibroin fibers and films: Surface opsonization and binding strength.' *Journal of Bioactive and Compatible Polymers* **17**(1): 23–35.

Muller, C., M. Hamedi, *et al.* (2011). 'Woven electrochemical transistors on silk fibers.' *Advanced Materials* **23**(7): 898–901.

Nuzzo, R. G., J. M. Yao, *et al.* (2010). 'Functional nanostructured plasmonic materials.' *Advanced Materials* **22**(10): 1102–1110.
O'Hara, J. F., R. Singh, *et al.* (2008). 'Thin-film sensing with planar terahertz metamaterials: sensitivity and limitations.' *Optics Express* **16**(3): 1786–1795.
Oksman, K., A. P. Mathew, *et al.* (2009). 'Novel bionanocomposites: processing, properties and potential applications.' *Plastics Rubber and Composites* **38**(9–10): 396–405.
Omenetto, F. G., J. J. Amsden, *et al.* (2010). 'Rapid nanoimprinting of silk fibroin films for biophotonic applications.' *Advanced Materials* **22**(15): 1746–1749.
Omenetto, F. G., R. Capelli, *et al.* (2011). 'Integration of silk protein in organic and light-emitting transistors.' *Organic Electronics* **12**(7): 1146–1151.
Omenetto, F. G., P. Domachuk, *et al.* (2009). 'Bioactive 'self-sensing' optical systems.' *Applied Physics Letters* **95**(25).
Omenetto, F. G. and D. L. Kaplan (2008). 'A new route for silk.' *Nature Photonics* **2**(11): 641–643.
Omenetto, F. G. and D. L. Kaplan (2010). 'New opportunities for an ancient material.' *Science* **329**(5991): 528–531.
Omenetto, F. G., B. D. Lawrence, *et al.* (2008). 'Bioactive silk protein biomaterial systems for optical devices.' *Biomacromolecules* **9**(4): 1214–1220.
Omenetto, F. G., S. Y. Lee, *et al.* (2010). 'Spatial and spectral detection of protein monolayers with deterministic aperiodic arrays of metal nanoparticles.' *Proceedings of the National Academy of Sciences of the United States of America* **107**(27): 12086–12090.
Omenetto, F. G., S. T. Parker, *et al.* (2009). 'Biocompatible silk printed optical waveguides.' *Advanced Materials* **21**(23): 2411–2415.
Omenetto, F. G., H. Perry, *et al.* (2008). 'Nano- and micropatterning of optically transparent, mechanically robust, biocompatible silk fibroin films.' *Advanced Materials* **20**(16): 3070–3072.
Omenetto, F. G., H. Tao, *et al.* (2010a). 'Gold nanoparticle-doped biocompatible silk films as a path to implantable thermo-electrically wireless powering devices.' *Applied Physics Letters* **97**(12).
Omenetto, F. G., H. Tao, *et al.* (2010b). 'Performance enhancement of terahertz metamaterials on ultrathin substrates for sensing applications.' *Applied Physics Letters* **97**(26).
Omenetto, F. G., K. Tsioris, *et al.* (2011). 'Rapid transfer-based micropatterning and dry etching of silk microstructures.' *Advanced Materials* **23**(17): 2015–2019.
Park, I., J. Cheng, *et al.* (2007). 'Low temperature, low pressure nanoimprinting of chitosan as a biomaterial for bionanotechnology applications.' *Applied Physics Letters* **90**(9).
Payne, G. F., W. L. Meyer, *et al.* (2009). 'Chitosan-coated wires: conferring electrical properties to chitosan fibers.' *Biomacromolecules* **10**(4): 858–864.
Payne, G. F., H. M. Yi, *et al.* (2005). 'Biofabrication with chitosan.' *Biomacromolecules* **6**(6): 2881–2894.
Pendry, J. B., A. J. Holden, *et al.* (1999). 'Magnetism from conductors and enhanced non-linear phenomena.' *IEEE Transactions on Microwave Theory and Techniques* **47**(11): 2075–2084.

Pritchard, E. M., A. Wilz, *et al.* (2009). 'Sustained-release silk biomaterials for drug delivery and tissue engineering scaffolds.' *2009 35th Annual Northeast Bioengineering Conference*: 282–283.
Psaltis, D. and G. W. Burr (1998). 'Holographic data storage.' *Computer* **31**(2): 52–60.
Rao, A. G. A., M. D. Gouda, *et al.* (2003). 'Thermal inactivation of glucose oxidase – mechanism and stabilization using additives.' *Journal of Biological Chemistry* **278**(27): 24324–24333.
Rockwood, D. N., R. C. Preda, *et al.* (2011). 'Materials fabrication from Bombyx mori silk fibroin.' *Nature Protocols* **6**(10): 1612–1631.
Rogers, J. A., D. H. Kim, *et al.* (2009). 'Silicon electronics on silk as a path to bioresorbable, implantable devices (vol 95, 133701, 2009).' *Applied Physics Letters* **95**(26).
Sarkar, A. and H. Mallick (2000). 'An experimental investigation of electric conductivities in biopolymers.' *Bulletin of Materials Science* **23**(4): 319–324.
Scheibel, T. (2005). 'Protein fibers as performance proteins: new technologies and applications.' *Current Opinion in Biotechnology* **16**(4): 427–433.
Shao, Z. Z., X. Chen, *et al.* (2006). 'The spinning processes for spider silk.' *Soft Matter* **2**(6): 448–451.
Shao, Z. Z., F. Vollrath, *et al.* (2003). 'Structure and behavior of regenerated spider silk.' *Macromolecules* **36**(4): 1157–1161.
Shao, Z. Z., Y. Yang, *et al.* (2005). 'Toughness of spider silk at high and low temperatures.' *Advanced Materials* **17**(1): 84–88.
Singh, T. B., N. S. Sariciftci, *et al.* (2010). 'Bio-organic optoelectronic devices using DNA.' *Organic Electronics* **223**: 189–212.
Steckl, A. J. (2007). 'DNA – a new material for photonics?' *Nature Photonics* **1**(1): 3–5.
Steckl, J., J. A. Hagen, *et al.* (2006). 'Enhanced emission efficiency in organic light-emitting diodes using deoxyribonucleic acid complex as an electron blocking layer.' *Applied Physics Letters* **88**(17).
Sundar, V. C., A. D. Yablon, *et al.* (2003). 'Fibre-optical features of a glass sponge – some superior technological secrets have come to light from a deep-sea organism.' *Nature* **424**(6951): 899–900.
Tsioris, K., G. E. Tilburey, *et al.* (2010). 'Functionalized-silk-based active optofluidic devices.' *Advanced Functional Materials* **20**(7): 1083–1089.
Vepari, C. and D. L. Kaplan (2007). 'Silk as a biomaterial.' *Progress in Polymer Science* **32**(8–9): 991–1007.
Vollrath, F. and D. P. Knight (2001). 'Liquid crystalline spinning of spider silk.' *Nature* **410**(6828): 541–548.
Vollrath, F., Y. Liu, *et al.* (2008). 'Elasticity of spider silks.' *Biomacromolecules* **9**(7): 1782–1786.
Vollrath, F. and D. Porter (2009). 'Silk as a biomimetic ideal for structural polymers.' *Advanced Materials* **21**(4): 487–492.
Vukusic, P., R. Kelly, *et al.* (2009). 'A biological sub-micron thickness optical broadband reflector characterized using both light and microwaves.' *Journal of the Royal Society Interface* **6**: S193–S201.
Vukusic, P., J. R. Sambles, *et al.* (2000). 'Structural colour – colour mixing in wing scales of a butterfly.' *Nature* **404**(6777): 457–457.

Vukusic, P., J. R. Sambles, *et al.* (1999). 'Quantified interference and diffraction in single Morpho butterfly scales.' *Proceedings of the Royal Society of London Series B – Biological Sciences* **266**(1427): 1403–1411.

Wang, H. and P. M. Callahan (1998). 'Adsorption studies of azo dyes as resonance Raman spectroscopic probes at solid-liquid interfaces.' *Journal of Chromatography A* **828**(1–2): 121–134.

Wang, X. D., R. Rajkhowa, *et al.* (2010). 'Reinforcing silk scaffolds with silk particles.' *Macromolecular Bioscience* **10**(6): 599–611.

Wen, J. C., J. R. Yao, *et al.* (2011). 'Recent progress and application of non-bioactive proteins in material fields.' *Acta Polymerica Sinica*(1): 12–23.

Wen, X. J., T. Xu, *et al.* (2007). 'Modification of nanostructured materials for biomedical applications.' *Materials Science & Engineering C-Biomimetic and Supramolecular Systems* **27**(3): 579–594.

Yao, J. M., L. Liu, *et al.* (2008). 'Preparation and characterization of nano-hydroxyapatite/silk fibroin porous scaffolds.' *Journal of Biomaterials Science-Polymer Edition* **19**(3): 325–338.

Young, R. J., V. L. Brookes, *et al.* (2008). 'Deformation micromechanics of spider silk.' *Journal of Materials Science* **43**(10): 3728–3732.

Yu, Z., W. Li, *et al.* (2007). 'Photoluminescence and lasing from deoxyribonucleic acid (DNA) thin films doped with sulforhodamine.' *Applied Optics* **46**(9): 1507–1513.

Zhang, X., H. Tao, *et al.* (2010). 'Metamaterial silk composites at terahertz frequencies.' *Advanced Materials* **22**(32): 3527–3531.

7 Fluorescence control in natural green fluorescent protein (GFP)-based photonic structures of reef corals

A. SALIH, University of Western Sydney, Australia

Abstract: The green fluorescent protein (GFP) and a variety of GFP-like homologues that colour tissues of many reef organisms have revolutionized biological and biomedical research by providing the means to fluorescently tag and visualize the activity of genes and proteins in living cells. This chapter describes how we can further capitalize on what nature has produced by using the GFP group of photoactive proteins, which evolved to perform a variety of biological functions, to develop a range of biomimetic advanced biophotonic applications. The evolutionary pressures that led to the origin of GFP-like fluorescent proteins in marine organisms can be explored in designing novel biomedical sensors, solar cells, biomolecule-based materials and optoelectronic devices. As GFP-like proteins are genetically encodable, this science is posed on the brink of a new technological revolution – to create the means for interfacing biology with electronics, so that devices not only generate energy, but also diagnose diseases and detect pathogens *in vivo*.

Key words: green fluorescent protein (GFP), photoactive fluorescent protein (PAFP), solar energy, photonic, Förster resonance energy transfer (FRET), biomaterials.

7.1 Introduction

Fluorescence is a characteristic of many biological molecules and materials, such as various body fluids, including blood and urine, chitin, collagen, NADH, NADPH, FADH, flavins, phycobiliproteins, chlorophyll and numerous fluorescent dyes. The green fluorescent protein (GFP) is the most widely used fluorescent molecule in cell biology and biomedicine – a mainstay in life-science research. Its popularity is linked to the fact that it is relatively non-toxic and is the only protein in which fluorescent properties are fully genetically encoded and require no other cofactors, other than O_2, for their expression (Chalfie 1995). Thus, this molecule is a portable code that can be integrated into the genome of any organism (Fig. 7.1). GFP can be specifically fused to a protein of interest and expression in a diverse range of live cell types, tissues and organisms (Patterson 2007). The emitted fluorescence can then be imaged directly, using fluorescence techniques, such as

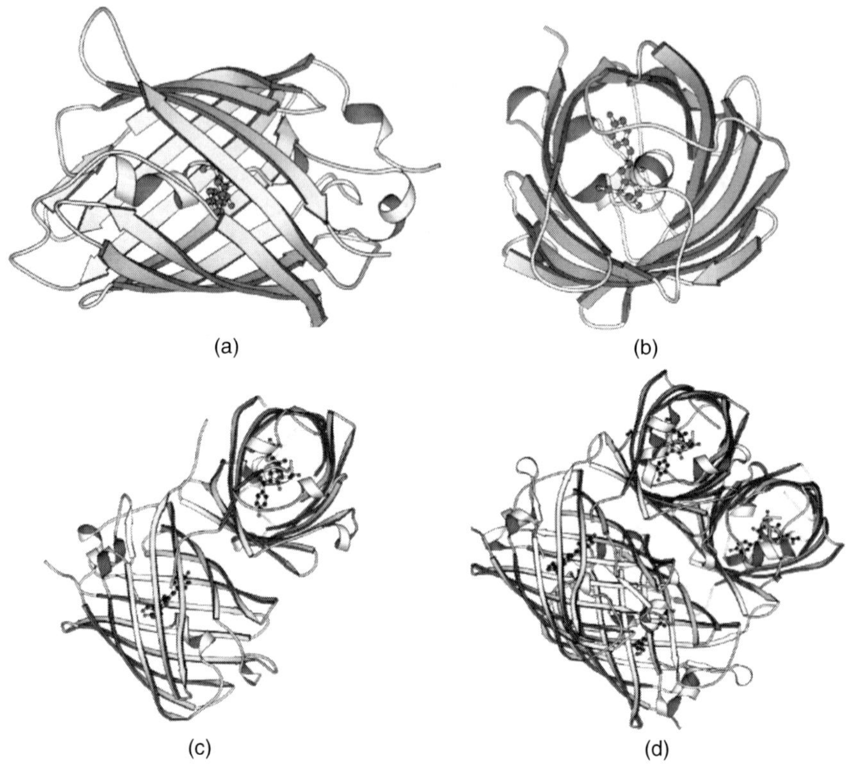

7.1 Structure of GFP-like proteins, showing the 11 β-strands (ribbons) with interconnecting loops of the β-barrel, with the chromophore suspended from an α-helix at the centre of the barrel, where it is well protected from the external environment: (a, b) monomeric form, such as wt-GFP; (c) natural or mutated dimeric form (e.g. HcRed or *Renilla* GFP); (d) and a tetrameric form of most anthozoan GFP-like proteins (e.g. DsRed, EosFP, Kaede) (adapted from Prescott and Salih 2008).

confocal microscopy, without the requirement of fixation or staining labels. In 2003, the Protein Data Bank (PDB) featured GFP as the molecule of the month, with 31 recorded GFP structures and the number rapidly increased to 266 by 2010 (Ong *et al.* 2011).

Although GFP was first isolated in the 1960s from the bioluminescent jellyfish *Aequorea victoria* (i.e. aeGFP or wild type, wtGFP) its wide-ranging utility in the life-sciences was not demonstrated until the early 1990s (Shimomura 2005). Indeed, it was first thought of as an annoying contaminant to be carefully separated away from aequorin (Shimomura *et al.* 1980; Tsien 2010). The revolution of the GFP technology started with the cloning of the wtGFP's cDNA by Prasher *et al.* (1992), followed by its heterologous expression by Chalfie *et al.* (1994). These breakthroughs were followed by the rapid explosion of GFP-based applications and culminated in the 2008 Nobel Prize in Chemistry award for the discovery and development of GFP to O. Shimomura (Marine Biological Laboratory, Woods Hole, MA), M. Chalfie (Columbia University, New York, NY) and R. Tsien (Howard Hughes Medical Institute, University of California, San Diego). GFP and related GFP-like proteins continue to be central to a vast array of studies of gene expression and protein tagging and provide novel applications for a variety of sciences.

A lesser-known fact is that GFP-like proteins are the source of colour for a majority of reef corals and many of their relatives – sea anemones, zoanthids, soft corals, sea pansies, etc. (Matz *et al.* 1999; Salih *et al.* 2000; Dove *et al.* 2001; Ando *et al.* 2002; Labas *et al.* 2002; Kelmanson and Matz 2003; Wiedenmann *et al.* 2004; Field *et al.* 2006; Alieva *et al.* 2008). The discovery of GFP-like proteins in non-bioluminescent marine organisms was made by Matz *et al.* (1999) in Lukyanov's laboratory when the group isolated a gene encoding the red fluorescent protein named DsRed from *Discosoma sp*, an anemone-like organism closely related to corals, taken from a private aquarium in Moscow. Since then, over 150 genes have been cloned from marine organisms. We now know that the striking colours, which range from bright blue, green and red fluorescent hues (see colour plate section between pages 96 and 97, Plate XX), to the deep pinks and purples of coral branch tips and anemone tentacle tips, are attributable to a super-family of proteins, variously related to the prototypical wtGFP. GFP-like proteins have also been found in evolutionary distant organisms from cnidarians – in crustaceans (copepod, *Pontellina plumata*) and even a primitive chordate, amphioxus (Shagin *et al.* 2004; Evdokimov *et al.* 2006; Bomati *et al.* 2009; Hunt *et al.* 2010). In the last decade, these new fluorescent proteins (FPs), especially those derived from reef corals and anemones, have added to the colour palette of GFP-based labels and biosensors and are overtaking in popularity the wtGFP in imaging applications.

The evolutionary pressures on coral reefs have forced this natural bio-optical system to become highly optimized. By developing an understanding of how GFPs function in coral biology, the knowledge can serve as a guide for the rational design of advanced biomimetic solutions to develop protein-based optical sensors and materials with improved functionalities. To understand how we may develop biomimetic devices by copying what nature has produced, this chapter first briefly describes the molecular structure and spectral diversity of GFP-like proteins, followed by the discussion of some of their biological functions in marine organisms. Although a full description of the wide-ranging applications of GFP-like proteins can only be scratched at here, several new directions are outlined, and future developments are discussed to set the stage for the coming generation of GFP-based advanced technologies.

7.2 Green fluorescent protein (GFP) structure and diversity

The wtGFP and other GFP-like proteins all have an 11-stranded β-barrel, with a diameter of about 24 A° (~2 nm) and a height of 42 A° (~3 nm), with an α-helix running through the β-barrel (Fig. 7.1) (Prasher *et al.* 1992; Ormo *et al.* 1996). A *p*-hydroxy-benzylideneimidazolidinone chromophore (Fig. 7.2) is located in the centre of the barrel in the majority, but not in all GFP-like proteins, and is protected from solvents by lids composed of short helices. The chromophore is formed by the cyclization of the adjacent Ser65-Tyr66-Gly67 (the number denotes the position in the intact peptide sequence) residues within this hexapeptide sequence (Cody *et al.* 1993). Aside from marine organisms, a GFP-like folding topology consisting of the 11-strand β-barrel domain, has been identified even in human extracellular matrix proteins. This protein, termed G2F (globular-2) fragment, is part of a multidomain protein and provides a binding surface for protein–protein interactions (Hopf *et al.* 2001).

Thus, the kingdom Metazoa has two gene lineages within the GFP protein superfamily: the colourless G2F lineage and the GFP-like lineage. The two branches appear to have originated from a common ancestor predating the

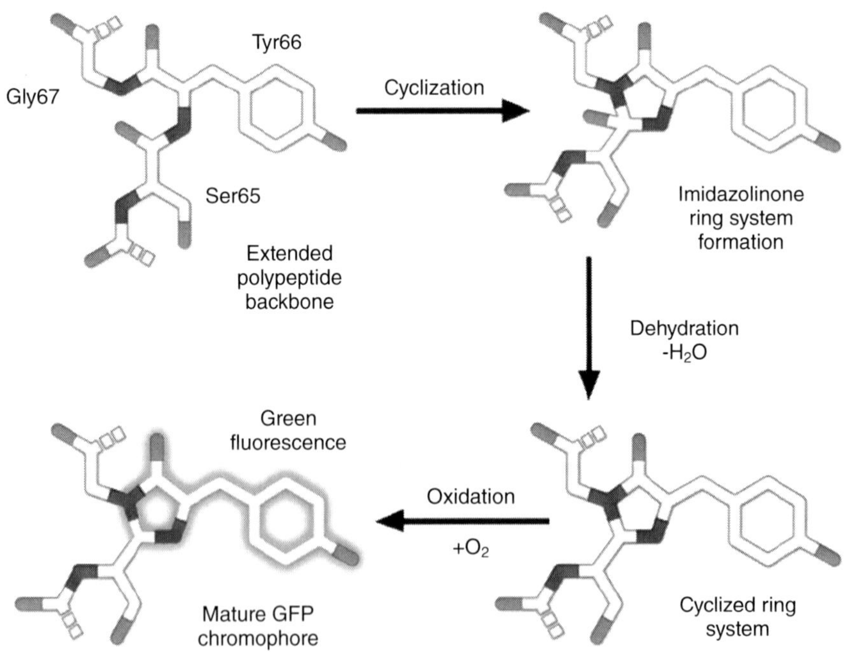

7.2 Steps in the formation of the *A. victoria* wtGFP chromophore (from Day and Davidson 2009).

divergence of Radiata (phylum Cnidaria) and Bilatera (phyla Arthropoda, Mollusca, Chordata, etc.) (Matz *et al.* 2006).

The formation of wtGFP chromophore and protein maturation has been the focus of research by several groups and can be summarized as a cyclization–oxidation–dehydration sequence of events, where the oxidation step is solely guided by main chain chemistry (Fig. 7.2) (Rosenow *et al.* 2004; Zhang *et al.* 2006; Day and Davidson 2009; Wachter *et al.* 2010). Molecular oxygen is crucial for the development of the chromophore's fluorescence and at least one oxygen molecule is required for dehydrogenation (Tsien 1998). Chromophore maturation results in a highly conjugated pi-electron resonance system responsible for the protein's spectroscopic and photophysical properties. The mechanism of maturation of the red FPs, such as DsRed, has been revised several times and is excellently reviewed by Wachter *et al.* (2010). The generation of the DsRed chromophore does not depend on light exposure and is formed by an oxidation reaction that extends the conjugated system, forming a red shifted absorbance and fluorescence (Baird *et al.* 2000; Gross *et al.* 2000; Wachter *et al.* 2010).

The similarity of the amino acid sequences of the GFP-like proteins with wtGFP from *A. victoria* ranges from 90% to less than 20%. In wtGFP and its variants the β-barrel is a cylinder, while the GFP-like proteins of anthozoans have an elliptic shape (Baird *et al.* 2000; Yarbrough *et al.* 2001). Even though these proteins are derived from completely unrelated species and sometimes have little homology, they exhibit very little variation in their tertiary structure. The cylindrical geometry of GFP appears to function in protecting the chromophore, making it remarkably photostable and bright, resistant to reactive oxygen, light, pH and many denaturing agents (Tsien 1998). The term 'light in a can' has been applied to GFP and related proteins because of these protective properties of the protein shell.

The completely autocatalytic formation of the light-emitting chromophore makes it possible for GFP to be expressed in the genetic fusions with other proteins for studying their expression, localization and function inside living cells and whole organisms. Thus, the main current applications of GFP-like proteins involve their use in imaging as fluorescent markers in single- and multicolour applications. The need simultaneously to visualize multiple proteins drives the demand for additional FP spectral variants. The wtGFP was mutated to produce variants emitting in the blue (BFP), cyan (CFP) and yellow (YFP) colours (reviewed by Tsien 1998, 2010), but orange and red mutants were not developed until GFPs were discovered in anthozoans (Matz *et al.* 1999).

The fluorescence colour repertoire of anthozoan GFPs is significantly broader than that of the bioluminescent organisms (Plate XX): cyan FPs with emissions at 470–499 nm; green FPs with emissions at 500–525 nm; yellow FPs with emissions at 526–545 nm; and orange and red FPs covering

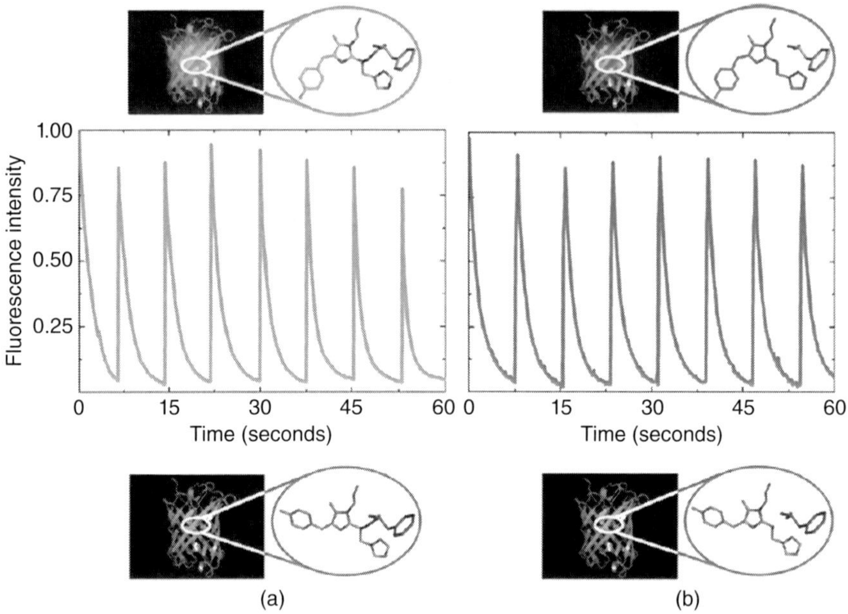

7.3 Reversible photoswitching of green and red forms of IrisFP in crystals. (a) The green form in a capillary was illuminated with 488 nm light, while 405 nm light was switched on every 7.5 s for 25 ms. (b) After green to red photoconversion of IrisFP crystal by strong irradiation at 405 nm was achieved its emissions were imaged after continuous illumination with 532 nm light and irradiation by 440 nm light switched on and off every 7.5 s for 200 ms. Inserts show representations of the β-barrel structure of IrisFP and structural models of its chromophore in its two bright and two dark states (adapted from Adam *et al.* 2010).

the rest of the visible spectrum, up to 640 nm (Matz *et al.* 1999; Ando *et al.* 2002; Kelmanson and Matz 2003; Chudakov *et al.* 2004; Wiedenmann *et al.* 2004; Alieva *et al.* 2008). DsRed alone led to the development of 14 GFP-like protein colour variants (Shaner *et al.* 2004). Far red fluorescent proteins are in high demand for *in vivo* whole body imaging; however, there has been no naturally derived colour variants above 680–700 nm discovered so far and much effort is spent on mutagenesis of red FPs, with some successful development of near-infrared variants – Katusha, eqFP650 and eqFP670 that emit at 640 to 670 nm (Shcherbo *et al.* 2007, 2010; Lukyanov *et al.* 2010).

Anthozoans also possess GFP-like chromoproteins (CPs) that are brightly coloured blue, pink or purple molecules that do not fluoresce (Dove *et al.* 1995; Takabayashi *et al.* 1995; Dove *et al.* 2001; Alieva *et al.* 2008). Several of these proteins are photoactive and can spontaneously acquire red fluorescence under green light irradiation (Lukyanov *et al.* 2000, 2005; Chudakov *et al.* 2003; Salih *et al.* 2006b; Prescott and Salih 2009) and this property is referred to as kindling (Lukyanov *et al.* 2000). Moreover, coral-derived GFPs possess other unusual photoactive characteristics, described below, that make them highly useful in bioimaging and biotechnological applications.

The majority of GFP-like proteins from non-bioluminescent organisms discovered to date display variable degrees of quaternary structure. They are composed of monomeric subunits with blue, green or red emitting chromophores (Baird *et al.* 2000). The monomers show a similar topology to GFP, but form tetrameric (Fig. 7.1(d)) or form obligate oligomeric structures and aggregates both *in vivo* and *in vitro* (Baird *et al.* 2000; Wall *et al.* 2000; Yarbrough *et al.* 2001). The monomers do not act independently from each other but are energy coupled via a Förster resonance energy transfer (FRET) mechanism (Förster 1948). When a tetrameric protein is excited by blue light irradiation, the observed red emission is not due to the direct excitation of the red chromophore, but results from the excitation transfer from the green to the red emitting fluorophore (Baird *et al.* 2000; Wall *et al.* 2000; Yarbrough *et al.* 2001). In fact, the property of tetramerization and oligomerization appears to be a unifying feature of the majority of FPs from non-bioluminescent organisms. Alternately, GFPs in bioluminescent organisms, such as *Aequorea* and *Renilla*, form hetero-tetramers with their respective photoproteins.

Two main mechanisms for protein aggregation have been proposed. One involves the non-specific interactions of 'sticky' hydrophobic patches on the molecular surface (Eaton and Hofrichter 1990). The other mechanism involves the electrostatic interactions between positively and negatively charged surfaces. Oligomerization of DsRed occurs at two chemically distinct protein interfaces to assemble the tetramer (Baird *et al.* 2000; Wall *et al.* 2000). However, it does not contain the pronounced hydrophobic areas that may cause strong interactions between tetramers. Based on computer calculation of the electrostatic potential of tetrameric DsRed, it was revealed that the protein surface is mostly negatively charged, and it was proposed that each tetramer is able to form up to four salt bridges with adjacent tetramers, resulting in the net-like polymeric structure (Yanushevich *et al.* 2002).

Oligomerization can cause many problems when FPs are used as *in vivo* imaging probes, particularly when fused to other proteins of interest, disrupting normal protein distribution and function and can critically affect cellular processes. On the other hand, the tetramerization and oligomerization of FPs presents a remarkable opportunity for the development of photostable and photoactive biopolymers or hybrid organic materials, since the quaternary structure makes the proteins highly photostable and can provide optimized means for efficient energy and electron transfer between the monomeric subunits.

7.3 Photoactive fluorescent proteins (PAFPs)

During the last decade, a remarkable group of GFP-like proteins has been discovered that can spontaneously change their optical properties in response to irradiation (Lukyanov *et al.* 2000; Ando *et al.* 2002; Patterson and Lippincott-Schwartz 2002; Chudakov *et al.* 2003; Wiedenmann *et al.* 2004; Lukyanov *et al.* 2005; Remington, 2006). They are referred to as photoactive fluorescent proteins (PAFPs), optical high-lighters, photoconverting or photoswitching proteins (Lukyanov *et al.* 2005; Patterson and Lippincott-Schwartz 2002; Lippincott-Schwartz and Patterson 2008). They all have the general characteristic that they are initially dark at the pre-activated fluorescence wavelength but upon activation display an increase in fluorescence and are 'highlighted' over a darker background, such as kindling (KFP) and Dronpa proteins from anthozoans (Chudakov *et al.* 2003; Habuchi *et al.* 2005). Other such proteins change colour upon activation, from cyan to green (Chudakov *et al.* 2004); from green to red (e.g. Ando *et al.* 2002; Wiedenmann *et al.* 2004) (see Plate XX); from red to green; and from red to yellow (Salih *et al.* 2006b). Yet others have both photoconversion and photoswitching properties, such as IrisFP (Fig. 7.3), which was developed from EosFP (Adam *et al.* 2008).

In live cell imaging PAFPs are particularly useful quantitatively to investigate the dynamics of proteins within a population in cells or tissues. They can be used directly to highlight or photoconvert a pool of molecules by the irradiation of a selected area of interest and then to follow the signal of the photoactivated population with high signal to noise ratio (e.g. Lippincott-Schwartz *et al.* 2001; Lukyanov *et al.* 2005; Prescott and Salih 2009).

The green-to-red photoconverting PAFPs appear to be especially prominent in numerous cell imaging applications and two such proteins, Kaede from the coral *Trachyphyllia geoffroyi* (Ando *et al.* 2002) and EosFP from the coral *Lobophyllia corymbosa* (Plate XX) (Wiedenmann *et al.* 2004), have become popular cellular probes. They can be photoactivated at irradiation by near UV light (~400 nm) and change colour from the pre-activated green fluorescence (emission at 516–518 nm) to a new red fluorescence (emission peak at 580–581 nm) (Plate XX). EosFP has been mutated into two dimeric forms, d1EosFP and d2EosFP, and a monomeric mEosFP (Wiedenmann *et al.* 2004). The shift in both the excitation and the emission peaks following photoconversion results in a large contrast with the background after photoactivation at more than 2000-fold increase in the red-to-green fluorescence ratio (Wiedenmann *et al.* 2004).

Kaede and EosFP, as well as other green-to-red PAFPs so far reported (e.g. IrisFP, Dendra2, KikGR, LcorFP, etc.) contain a chromophore derived from the tripeptide His-Tyr-Gly (HYG) (Mizuno *et al.* 2003). It initially emits green fluorescence in which UV wavelengths irradiation produces a cleavage between the amide nitrogen and α-carbon atoms in the histidine residue, leading to an extension of the chromophore conjugation to the histidine side chain and a spectacular transformation into bright red fluorescence (Mizuno *et al.* 2003).

Another group of PAFPs, belonging to the CP protein group described in the previous section, also contain the kindling fluorescent proteins (KFPs). These naturally occurring proteins have the basic β-barrel structure of wt-GFP and form the internal chromophore inside the β-barrel. They are less than 30% homologous to wtGFP and are non-fluorescent prior to photoactivation, but following a brief irradiation by green wavelengths spontaneously acquire bright red fluorescence (Lukyanov *et al.* 2000; Chudakov *et al.* 2003; Salih *et al.* 2006b). The first KFP was discovered in the purple tentacle tips of the sea anemone, *Anemonia sulcata* and absorbs light at 568 nm and after irradiation fluoresces at 595 nm (Lukyanov *et al.* 2000). Kindling is a reversible process as the fluorescent state gradually relaxes back to a non-fluorescent state in the dark at partial kindling (Lukyanov *et al.* 2000; Chudakov *et al.* 2003). The bright form can be instantly switched off by a flash of blue light and the switching process can be repeatedly cycled. The high-resolution crystal structure of KFP revealed that in the dark-adapted state the protein has a distorted chromophore in the *trans*-noncoplanar configuration and following light activation the chromophore undergoes cis–trans isomerization to a fluorescent state (Andresen *et al.* 2005; Quillin *et al.* 2005). By comparison, most FPs have chromophores that are *cis*-coplanar.

The diverse opportunities presented by the photoactive properties of these GFP-like proteins are only beginning to be explored and they provide an excellent foundation upon which to develop optical biomimetic technologies, some of which will be described in the sections below. In summary, the photoconversion mechanisms of PAFPs are not yet fully understood at the molecular level but it is known that they involve photo-induced isomerizations, such as cis-trans transformation, changes in protein microenvironment, other structural changes, such as bond breaking, proton and charge transfer between the protein and the chromophore.

7.3.1 Biological functions of GFP-like proteins

In bioluminescent jellyfish, wtGFP has existed for more than one hundred and sixty million years. It is localized in several hundred photo-organs in tissues on the umbrella margin, which when stimulated give off green light (Plate XX). GFP is involved in an efficient Förster resonance energy transfer (FRET) (Förster 1948) as an acceptor protein, transforming blue light (470 nm) emitted by the luminescent donor aequorin into green light

(508 nm) (Morin and Hastings 1971; Morise *et al.* 1974). However, the biological relevance of this process to the organism and in fact, within other bioluminescent organisms, is still unknown.

Indeed, the function of the enormous variety of GFP-like proteins found in the non-bioluminescent cnidarians, such as corals and other anthozoans, is only beginning to be understood (Kawaguti 1944; Catala 1959; Matz *et al.* 1999, 2006; Salih *et al.* 2000; Dove *et al.* 2001, 2008; Mazel *et al.* 2003; Wiedenmann *et al.* 2004; Field *et al.* 2006; Kao *et al.* 2007; Oswald *et al.* 2007; Gruber *et al.* 2008). As anthozoans have few predators, do not depend on mate attraction for reproduction and do not need to attract their prey, the primary function of GFPs cannot be involved in defensive light emission, reproductive signalling, prey lures and in food assimilation. One plausible hypothesis is that some tissue fluorescent patterns of symbiotic anthozoans may serve as attractants to symbionts, which have been shown to be attracted by green light (Hollingsworth *et al.* 2005). Since free-swimming symbiotic algae may be taken up from the water column throughout the host's life-history, the brightly fluorescent colours patterning the area around the mouth may serve as an attractant. Many of the known and hypothesized functions of GFP-like proteins are linked to light and many proteins are in fact, photoactive, as was discussed in the previous section of this chapter, and can spontaneously alter their optical properties in response to irradiation.

The sunscreening and light modulatory roles of GFP-like proteins are the most experimentally explored functions of these molecules in marine organisms. An early suggestion of a sunscreening by coral coloration (Kawaguti 1944) was substantiated by later research that provided the first direct evidence that GFP-like proteins function in light-optimization of coral tissue (Schlichter and Fricke 1990), and in shallow corals are photoprotective (Salih *et al.* 2000, 2001, 2006ab; Dove *et al.* 2001, 2008; Gilmore *et al.* 2003; Roth *et al.* 2010). Reef corals require effective photoprotection since they are exposed to massive doses of solar radiation in the oligotrophic clear waters of the tropical oceans in which sunlight penetrates to considerable depths (Jerlov 1976; Ackleson 2003).

While often referred to as the 'rainforest of the ocean', corals are actually animals, not plants. But they do have plant-like adaptations since one of the most fascinating aspects of their biology is the symbiotic relationship with dinoflagellate microalgae, *Symbiodinium microadriacticum* (termed zooxanthellae) that live in millions in coral's endodermal cells. Zooxanthellae carry out photosynthesis and translocate sugars into the coral's tissues, providing up to ~ 95% of coral's metabolic requirements (Muscatine *et al.* 1984). This symbiotic relationship is the reason corals require sunlight and cannot rely on solar blocking molecules, such as melanins, used by other animals. Nevertheless, they have evolved to be incredibly efficient in capturing solar energy and converting it to chemical energy via photosynthesis of their symbionts.

Even under normal environmental conditions, and even more than many photosynthetic organisms, corals are routinely exposed to very variable illumination conditions that drastically change due to diurnal variations. Underwater, wave lensing effects can create powerful light flashes of extremely high intensity (Veal *et al.* 2010). Recent studies showed that these intensities often reach above 150% of mean solar irradiance even at a depth of 20 m at a coral reef (Veal *et al.* 2010).

The presence of photosynthesizing organisms in coral tissues presents a major threat, since high levels of the resultant photo-oxidants can cause cellular damage and death. The whitening of coral tissues (referred to as coral bleaching) during conditions of adverse stress owing to high solar radiation, warming of seawater or other stressful events, is primarily caused by the loss of photosynthetic pigments and of zooxanthellae (Brown 1997). During the last two decades, coral bleaching has begun to occur on massive scales, causing many kilometres of reefs to bleach and the phenomenon has been linked to global climatic changes (Hoegh-Guldberg 1999). The mechanism of coral bleaching has been linked to the build-up of photo-oxidative stress, photoinhibition of the microalgal photosynthesis, damage to Photosystem II (PSII), increased generation of reactive oxygen species (ROS), damage to membranes, DNA and various cellular organelles in both the symbionts and the coral host (Weiss 2008). The photodamage is exacerbated at light intensities below those normally required to saturate photosynthesis (Asada 2006) when other stressful environmental or man-made factors limit the ability of zooxanthellae to acquire and assimilate carbon dioxide (CO_2). As the seawater warms due to climate change and temperatures increase above the normal summer maxima, the zooxanthellae living in symbiotic harmony with their coral hosts become damaged and expelled, leaving the starkly white, bleached coral reefs. In both the zooxanthellae and the coral tissue the various alternative sinks for electrons become turned on to prevent photo-oxidative damage (Lesser 2007; Weiss 2008).

Accumulating scientific evidence shows that GFP-like proteins have evolved to regulate the solar radiation impacting on the coral and other marine organisms, particularly those that are symbiotic with photosynthetic zooxanthellae. But even organisms that lack symbionts may experience photo-oxidative stress as a result of the fluctuating light environments encountered underwater and may rely on GFPs for light modulation to protect tissues from photodamage (Kahng and Salih 2005). In corals, experimental studies provide a strong evidence that GFP-like proteins form a highly dynamic light modulation system that can precisely alter the intracellular light environment while effectively harvesting light energy and controlling the ensuing negative consequences of living in a high-light environment. The mechanism of photoprotection by GFPs is highly sophisticated and involves light absorption, scattering and light energy transformation (Salih *et al.* 2000, 2006b; Dove *et al.* 2001; Gilmore *et al.* 2003; Cox *et al.* 2007).

Although sunscreening and light modulation are clearly important functions of GFP-like proteins, they are most certainly not the only functions. Importantly, it has been shown that FPs can directly absorb the reactive oxygen species (ROS), including H_2O_2 (Bou-Abdallah *et al.* 2006) and can function as antioxidants, protecting corals from the excessive levels of ROS generated in their cells during high-light stress (Palmer *et al.* 2009). Thus, under stressful conditions of globally changing climate, FPs can enhance the survival of pigmented corals by both sunscreening and by directly reducing the build-up of ROS by quenching reactive oxygen species (Bou-Abdallah *et al.* 2006).

Recently, a group led by K. Lukyanov made a ground-breaking discovery that upon exposure to blue light, green GFP-like proteins donate or accept electrons from biologically relevant donor or acceptor molecules and turn red (excitation 575 nm and emission 607 nm) as a result (Bogdanov *et al.* 2009). The process was referred to as redding or reddening, and occurred after only a brief irradiation. The mechanism is still not fully understood and is being intensely investigated. It was hypothesized to produce a stable radical and, since the green fluorescence disappeared faster than the red fluorescence increased, reddening was deduced to be a two-step process (Bogdanov *et al.* 2009).

Unlike electron transfer and reddening under unaerobic conditions, known to occur in GFP for some time, this newly discovered process proceeded under aeroboic conditions and revealed an exciting possibility that the primary function of GFPs is in photoreduction. The light-prompted passing of electrons by GFPs was referred to as the discovery of 'animal photosythesis' by the media. Coral FPs also undergo reddening *in vivo* in the presence of electron acceptors and their light-induced electron transfer properties are especially prominent in certain cells or organelles of coral tissues – mitochondria, cells undergoing rapid growth and, interestingly, in neurons or cells with sensory functions (Salih and Geny, unpublished). The electron transport properties appear to be key natural functions of GFP-like proteins in corals and other marine organisms.

The excited state proton transport (ESPT) is another exciting major function of the GFP-like proteins that is emerging as a result of research into protein photochemistry, but which has not yet been experimentally confirmed to occur *in vivo* in marine organisms. Proton transport is one of the most fundamental processes in biology as energy from light or respiration is used to generate a trans-membrane proton gradient. Usually, this is achieved by membrane-spanning enzymes known as 'proton pumps'. ESPT plays a key role in the biochemical processes that convert sugar to a form that cells can use – adenosine triphosphate (ATP). Cellular proton concentration changes can store this energy and the production of ATP from sunlight also involves proton transport processes.

Moreover, proton transport plays a major role in the photochemistry and activation of many receptors – the visual and the sensory rhodopsins (retinal proteins), proteorhodopsin, phytochromes and phototropin (e.g. Losi 2004; van der Horst and Hellingwerf 2004). The prerequisite for signal transduction is charge displacement within a sensory protein or protonation changes that can induce the conformational changes. The red light photoreceptors in plants, the phytochromes, for example, regulate many developmental processes such as flowering and shade avoidance (Franklin and Whitelam 2007). Another example of a proton transferring protein is the photoactive yellow protein (PYP), which is a water-soluble molecule, involved in sensing a wide range of stimuli, including light, oxygen, redox potential, and a variety of ligants (Imamoto and Kataoka 2007; Wachter and Remington 1999).

A major discovery was made when it was found that wtGFP's *p*-hydroxybenzylidene-imidazolidinone chromophore acts as a photoacid and light triggers an ultrafast ESPT (Chattoraj *et al.* 1996; Agmon 2005). The protein has two distinct absorption bands at 396 and 476 nm owing to the protonated (neutral, state A) and deprotonated (anionic, state B) forms of the chromophore (Heim *et al.* 1994; Chattoraj *et al.* 1996). While light is absorbed by both forms, fluorescence occurs from the anionic form, state B. The conversion of the neutral state A, which emits blue fluorescence, to the anionic state B, which emits green fluorescence, occurs by the excited-state ultrafast Tyr66 phenolic proton shuttling through an extensive hydrogen-bonding network (Chattoraj *et al.* 1996; Meech 2009). The change from forms A to an intermediate form I is solely a protonation change, while the change from state I to B is a conformational change (Stoner-Ma *et al.* 2005).

The X-ray structural studies of wtGFP (Yang *et al.* 1996; Ormo *et al.* 1996; van Thor *et al.* 2005) found that the active site proton wire (i.e. forming chains of water molecules and protonatable amino acids) is very extensive, originating at Tyr66 in the chromophore and beyond it to Glu5 and other groups on the protein surface (Ormo *et al.* 1996; van Thor *et al.* 2005). Ultrafast time-resolved IR spectroscopy confirmed the proton pathway and found it to be reversible, constituting a tiny 'photocycle' (Stoner-Ma *et al.* 2005). Another proton wire in GFP was also recorded (Shinobu and Agmon 2009).

Thus, in wtGFP, illumination initiates a proton transfer through a proton wire or several wires, formed by the chromophore acting as the proton donor, via a water molecule W22, Ser205 and Glu222 (the acceptor), on a picosecond time scale, with subsequent longer distance transfers occurring at nanosecond scales. Research of GFP-based proton wires is being investigated in a number of GFP-like proteins and by a variety of techniques – molecular research, X-ray protein structural determination, ultrafast fluorescence and time-resolved vibrational spectroscopy, complemented by quantum chemical calculations.

The biological function of GFP-based proton pumping and proton wires in marine organisms has only been speculated upon. In various biological systems, proton wires are made up of hydrogen bonding networks within the interior of proteins, consisting of predominantly oxygen and nitrogen atoms of amino acid side chains and the migrating protons use these atoms as stepping stones. As in other organisms, proton wires have prominent roles in trans-membrane proton transport and may perform this function in anthozoan tissues. It has been also suggested that GFPs may function as portable proton pumps within cnidarian cell cytoplasm (Agmon 2005). Thus, proton pumping properties of GFPs may perform diverse biological functions in marine organisms – photosensing, chemosensing, production of ATP (or sugar) from sunlight, calcification in corals, etc. and is a field of science that is becoming to be the focus of researchers worldwide.

7.3.2 GFP-like proteins as solar energy transfer arrays

The extraordinary adaptations for solar energy modulation and channelling by multicoloured GFP-like proteins in coral tissues present a unique biomimetic opportunity. The multitudes of anthozoan tissue patterns are attributable to diverse GFP-like proteins, many of which have different covalent structures of the protein chromophores and with different arrangements of their molecular environments. GFP-like proteins are localized in a variety of cellular and subcellular compartments in corals: in cells of the epidermal and endodermal tissue layers; in granules of fluoro-chromatophores; intracellularly, within organelles such as nucleus, mitochondria, endoplasmic reticulum, Golgi, in intracellular vesicles, cytoplasm and even within membranes (Plate XX) (Mazel 1995; Salih *et al.* 1998, 2000, 2006a; Cox and Salih 2006).

Surprisingly, it was discovered that the majority of corals usually have spectrally heterogeneous mixtures of GFP-like proteins (Plate XXe) within their cells – two or three, and often as many as four to six, and occasionally even more, can occur in a single cell. For example, four principal colours – cyan, green, red and non-fluorescent blue FPs – were cloned from the coral *Acropora millepora* and six molecularly distinct GFP-like proteins were found responsible for these colours (Cox *et al.* 2007; Alieva *et al.* 2008). The proteins formed heterogeneous mixtures within individual cells when live *A. millepora* tissues were analysed by confocal microscopy with microspectral detection (Salih *et al.* 2000, 2003, 2006a,b; Cox *et al.* 2007). They exhibited spectral characteristics of donor–acceptor pairs tuned for fluorescence energy transfer, with emissions from cyan FPs spectrally overlapping the excitation of the green emitters and the emission of greens overlapping the excitation of the reds (Plate XXe). The analysis of spectral properties of many coral species showed similar arrangements (e.g. Kelmanson and Matz 2003), with emissions of one FP precisely overlapping the excitation spectrum of the neighbouring FP, and forming progressively red shifted donor–acceptor assemblies tuned for the transfer of solar energy.

The mechanisms of energy transfer between these spectral pairs were found to be due to both the radiative and non-radiative FRET processes (Salih *et al.* 2000; Gilmore *et al.* 2003; Cox *et al.* 2006; Roth *et al.* 2010). FRET is known as the energy transfer from an excited state of a donor-chromophore to an acceptor without the emission of a photon and occurs as a result of a dipole–dipole interaction between the donor and the acceptor (Förster 1946). The fluorescent emission spectrum of the energy donor must overlap the absorption spectrum of the energy acceptor. The FRET efficiency (E) varies as the inverse sixth power of the distance between the two molecules (denoted by r):

$$E_{\mathrm{FRET}} = 1/[1 + (r/R_0)^6]$$

The distance at which energy transfer is 50% efficient, i.e. when 50% of the excited donors are deactivated by FRET, is defined by the Förster radius (R_0). The magnitude of R_0 is dependent on the spectral properties of the donor and the acceptor dyes. FRET, therefore, decreases as the inverse sixth power of the donor–acceptor distance, with typical values of the R_0 of ~2–10 nm. The probability of finding a donor and an acceptor within the Förster radius in corals can only be high within intracellular spaces in which the donor–acceptor pairs are localized within the same molecular system. The multichromophoric tetrameric and oligomeric structure of anthozoan GFP-like proteins may have evolved to optimize intramolecular FRET. However, to what degree FRET occurs among multichromophoric aggregates in coral cells is still to be determined.

One explanation for the presence of these energy channelling protein arrays in anthozoan cells is to increase the efficiency of light modulation and photoprotection of coral tissues and photosynthetic symbionts. Solar energy is transferred from protein to protein and excess energy is dissipated away from cellular targets that can be photo-oxidized (Salih 2000; Salih *et al.* 2000–2004; Gilmore *et al.* 2003; Cox and Salih 2005). FRET serves as a prominent relaxation channel, especially in multichromophoric systems (Forster 1965). The acceptor can be a fluorescent or a non-fluorescent GFP-type protein. If the acceptor is fluorescent, the transferred energy can be emitted as fluorescence characteristic of the acceptor and if the acceptor is not fluorescent, the energy can be lost through equilibration with the surrounding cytoplasm or heat dissipation via molecular vibrations. The presence of light tuning protein arrays has never been previously found in tissues of marine invertebrates other than in visual systems, certain types of non-visual retinal pigments and in photosynthetic systems. The energy transfer function of GFP-like proteins is emerging to be a true system property of non-bioluminescent marine organisms and raises significant questions regarding the structuring of GFP-like aggregates for effective excitation, energy channelling and possible, other functions, linked to intracellular photon and electron transfer.

7.3.3 Coral GFPs as improved FRET and other cellular probes

The use of GFPs as *in vivo* cellular imaging probes has been covered in many excellent publications and reviews (e.g. Tsien 1998, 2010; Lippincott-Schwartz *et al.* 2001; Remington *et al.* 2004; Verkhusha and Lukyanov 2004; Patterson 2007; Miyawaki 2009; Wu *et al.* 2011; Stepanenko *et al.* 2011) and will not be addressed here. These proteins have been successfully expressed in cell cultures and organisms, such as bacteria, yeast, plants, worms, insects and vertebrates (reviewed, Stepanenko *et al.* 2008). They have been used to study numerous diseases, such as cancer, and the activity of pharmaceutical drugs (e.g. Chishima *et al.* 1997; Hoffman 2005).

Similarly, the field of GFP-based FRET imaging probes and biosensors, used precisely to measure molecular distances between two fluorophores, is a highly advanced science, with a multitude of probes and sensors that are too numerous to be detailed here. FRET has been used to investigate and map a variety of complex biological structures, such as proteins and macromolecular assemblies including ribosomes and nucleosomes. Imaging using FRET techniques have provided unprecedented tools to quantify molecular dynamics, such as protein–protein interactions, protein–DNA interactions, protein conformational changes, concentration and movement of molecules within physiological environments, determination of their concentrations across cellular compartments, tissues and even organs (e.g. Bastiaens and Pepperkok 2000; Miyawaki and Tsien 2000; Miyawaki *et al.* 1997; Miyawaki *et al.* 2003; Miyawaki 2005; Li *et al.* 2006; Grant *et al.* 2008; Mank and Griesbeck 2008; Tsutsui *et al.* 2008).

The use of GFP-like proteins in live cell imaging has many advantages compared to fluorescent dyes and other fluorophores. They are genetically encodable, in most cases are non-toxic or have low toxicity, are very bright and photostable, are highly resistant to most proteases, organic salts, alkaline pH and high temperature and their fusion to a target protein rarely affects the function of that protein (Chudakov *et al.* 2010). Coral-derived GFP-like proteins are especially bright and often do not bleach even when exposed to high laser irradiation. Since corals live in sunlit and warm tropical seas, their GFP-like proteins are resistant to high irradiances and warm temperatures and are therefore brighter and more temperature tolerant than many (although not all) genetically unmodified GFPs from bioluminescent organisms. Their antioxidant functions in anthozoans may alter some of the cellular activity when they are expressed in cells, but this has rarely presented a problem. Importantly, these GFP-like proteins do not usually generate large quantities of toxic ROS during irradiation.

Given the natural spectral tuning for FRET of many coral GFP-like proteins discussed in the previous section, researchers searching for the best FRET probes would benefit by cloning the donor–acceptor pairs from a single coral, or even better, from a single cell of one coral. Recent research indicates that other marine organisms can be investigated for naturally FRET-tuned GFPs (e.g. Aglyamova 2011). Biomimetic approaches are likely to make major breakthroughs in GFP-based imaging applications, given the newly discovered redox active, electron and proton transferring GFPs.

7.3.4 GFP-based solar energy materials and solar cells

The global worsening environmental pollution and energy shortages have raised the awareness of the need for urgent sustainable development of alternative clean energy technologies. A multitude of green energy projects around the world are competing to design efficient methods to convert solar energy into electricity and replace the dependence on fossil fuels. Since the 1950s, the photovoltaic market has been dominated by solar cells made with wafers of crystalline silicon, known as the first-generation solar cells. However, they are expensive and are uneconomical for mass use. The second generation of solar cells, first developed in the 1970s, known as thin cells, use films of semiconductors, such as cadmium telluride (CdTe) and copper indium selenide (CIS) instead of silicon, making them cheaper to manufacture and there is a considerable effort spent to increase their durability and efficiency.

The third-generation solar cells are organic cells and cells based on biomimetics of photosynthetic light-harvesting (Gust *et al.* 2001) – the dye-sensitized solar cells (DSSCs). The highly evolved system for efficient energy channelling via GFPs in photosynthetic corals can be used as a model to develop advanced DSSCs.

Plant chlorophylls act as active light-harvesting antennae that channel light into photosynthetic centres where sugars and oxygen from carbon dioxide from air and water are formed. The process requires an electron that acts as a powerful reducing agent, reducing molecules of carbon dioxide to form carbohydrates. Only a small fraction of all chlorophylls are directly involved in the process of transformation of an electronically excited state into the primary electrochemical product (Gaffron and Wohl 1936). The majority of pigments are involved in light absorption and efficient transfer of the electronically excited states to pigment molecules where the photochemical reactions take place (Green 2003). The biological electron transfer in photosynthesis occurs over a long range very efficiently through the biomolecules (Deisenhofer 1985). Redox potential differences drive the specific energy and electron transfer events, which are also determined by the electron transfer property of molecules, especially the electron acceptors–donors, electron relay and sensitizer molecules (Karvanos 1993). Plants and certain bacterial systems achieve light-harvesting efficiencies at above 99%. As discussed earlier in the chapter, corals are incredibly efficient in capturing solar energy and converting it to chemical energy via photosynthesis of their symbionts. Although, the *in vivo* cellular generation of energy in corals by GFPs has only been hypothesized, the efficient light energy transfer among GFP multichromophoric arrays has many similarities with the photosynthetic antennae and these proteins promise to be excellent candidates for DSSC development.

The first DSSC, invented by Michael Grätzel and B. O'Regan in 1991, is known as Grätzel cell. In 2012, the Interdisciplinary Committee of the World Cultural Council awarded Michael Grätzel the Albert Einstein World Award of Science, in recognition for the major contribution of this invention to solar technologies. DSSCs belong to a photo-electrochemical solar cell type that uses two different chemical compounds to harvest light and distribute photoelectrons – microcrystals or nanoparticles of an inorganic *n*-transporter (e.g. titanium dioxide), coated by a layer of a sensitizing dye (e.g. ruthenium complex). A single coating of a fluorescent dye on the solar cells gives 30% increase in the energy conversion efficiency (η, the percentage of power converted from absorbed light to electrical energy) of the solar cell (Maruyama *et al.* 1998). Upon light absorption, the sensitizer is promoted into an electronically excited state from where it injects an electron into the conduction band of *n*-transporter. DSSCs are low-cost materials, can be produced without using toxic chemicals and can be assembled into a wide variety of sizes and shapes using simple processes similar to inkjet printing.

The constituents of the organic dye sensitizer determine to a great extent the efficiency of DSSCs (Gregg and Hanna 2003). An organic dye can be tuned to optimize its performance and combinations of dyes are used to increase the solar collecting efficiency and thereby increase electricity generation (e.g. Maruyama *et al.* 1998; Zhang and Jenekhe 2000; Hardin *et al.* 2009). The mechanism of the photocurrent generation differs from that of inorganic solar cells, since the absorption of light leads to bound electron–hole pairs, the excitons (Gregg and Hanna 2003; Hoppe and Sariciftci 2004). The efficiency of dissociation of the bound excitons is enhanced either by a stronger electric field or by FRET processes in the donor–acceptor heterojunctions, leading to a fast charge transfer across the interface (Sariciftci *et al.* 1992; Liu *et al.* 2004). The DSSC's efficiency rating η is still approximately half that of the conventional silicon photovoltaics at ~11–14%. There is a major effort, therefore, to develop organic sensitizers that will achieve a more effective harnessing of solar energy.

The outstanding photostability, brightness and FRET-based light channelling properties of GFP-like proteins, provide a real potential for an improved conversion of photons to electricity than the currently used generation of dyes. The spectral diversity and FRET properties of GFP-like proteins covering almost the full visible spectrum is relevant for the strategic undertaking to develop multi-protein arrays to enhance solar collection. We can use a similar design to photosynthetic systems where a large number of light-harvesting antenna pigments channel excitation into a small number of photochemically active pigments in designing GFP-based solar cells. The use of multiple dyes in DSSCs is already being used to increase the absorption of solar cells and increase their efficiency (e.g. Maruyama *et al.* 1998; Zhang and Jenekhe 2000). Multi-chromophoric GFP arrays, with donor–acceptor FRET pairs, will be suitably arranged to channel excitation into photochemically active layers/centres. The quaternary structure of anthozoan GFPs, which naturally form tetramers, oligomers and higher order protein assemblies, are perfectly suited for the task. The arrays can be used to absorb and re-emit light down-shifted in wavelengths that will be matched to the spectral response of the solar cells.

As certain GFP-like proteins arrays possess electron donor–acceptor capabilities (Bogdanov *et al.* 2009), these proteins can also form the constituents of the photochemically active components. Owing to their molecular properties, these proteins may exhibit structural integrity and an excellent tertiary folding architecture. The assembly of light harvesting GFPs, targeting parts of the UV and most of the visible solar spectra, together with GFPs and other suitable electroactive proteins that mimic photosynthetic reaction centre protein units, would insure a rapid energy transduction, with a high quantum yield and minimal energy losses. A significant challenge found in solar cells replicating photosynthesis is the ability to achieve long-lived charge separation and prevent charge recombination (Gust *et al.* 2001). One way to address this challenge has been to integrate successive energy gradients into covalently linked molecular dyads or molecular doublets/dimers that will quickly shuttle electrons away from the excitation site to more stable sites.

Recently, a number of studies confirmed that the electrical energy generation from coral GFPs is an effective strategy. One of the first studies to mimic photosynthesis and use GFPs to generate an electrical current was the development of a molecular-scale biophotodiode consisting of a GFP and viologen (N-allyl-N'-[3-propylamido-N",N"-di(n-octadecyl)]-4,4'-bipyridium dibromide) as a sensitizer and electron acceptor using Langmuir–Blodgett techniques (Choi *et al.* 2000).

The same team improved the system by fabricating a bio-photodiode consisting of GFP as the sensitizer and an electron acceptor with a molecular relay – a heterojunction consisting of molecular layers sandwiched between metal electrodes – to achieve faster electron transfer (Choi *et al.* 2001). This metal/insulator/metal (MIM) structured electronic device was fabricated with heterofilms made of GFP and 10 layers of viologen and TCNQ (N-docosilquinolinium tetracyano dimethan) used as the sensitizer, an electron relay and an electron acceptor, respectively, positioned between the aluminium and indium tin oxide (ITO) glass. Upon light illumination, the GFP electrons were excited from their ground to excited state (GFP*), returned to their ground state with the emission of green fluorescence at 510 nm. Since the electron acceptor was exposed to the excited sensitizer, some of GFP*'s photoexcited electrons were separated (GFP^+/viologen$^-$/TCNQ) and transferred (GFP^+/viologen/$TCNQ^-$) to the TCNQ layer via their redox potential difference, thereby generating a photo-induced electric current (Choi *et al.* 2001).

Another study used GFP as an electron sensitizer and cytochrome c as an electron acceptor in a self-assembled heterolayer (Choi and Fujihara 2004). The monolayer of thiol-modified cytochrome c was formed on Au-coated glass and GFP was molecularly tethered onto the cytochrome c surface by electrostatic attraction. A relatively efficient photo-induced current was generated with repeated on/off cycles.

One recent study combined two contemporary photovoltaic strategies – the approach of using solid-state semiconductors to produce hole–electron pairs, with the second approach represented by the Grätzel cell (Fig. 7.4) (Chirgwandi *et al.* 2008). A 30 nm anode–cathode spacing was produced on a silicon chip by electron beam lithography and self-assembled GFP was employed as the dye molecule. The improved lifetime of photo-induced hole–electron pairs in EGFP (enhanced GFP) arrays allowed for the competition of electron–hole recombination with both, the electrons from an electrode and via the de-excitation process, to generate the photo-induced current (Chirgwandi *et al.* 2008). The electron transport was found to correlate with the excitation cross-section of EGFP, the intensity of incident light, temperature, source–drain potential bias and the gate potential.

In a widely publicized, but unpublished study, the same team from Sweden's Chalmers University of Technology, headed by Z.G. Chirgwandi, generated electricity from GFP acting both as a sensitizer dye and as an electron donor (Knight 2010). A droplet of EGFP assembled itself into strands between a tiny gap of two aluminium electrodes positioned upon a silicon dioxide substrate. When exposed to UV light, EGFP acted as a dye-

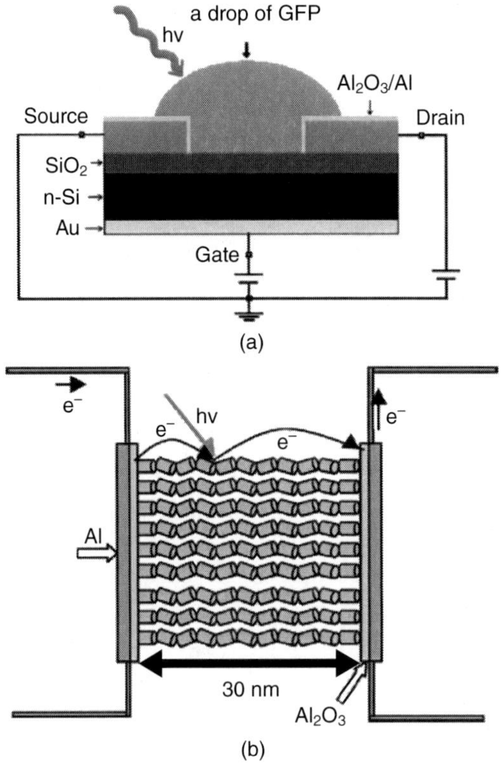

7.4 A 600 μm thick silicon wafer with an insulating SiO_2 substrate of 1000 Å thickness, a drop of GFP and aluminium (Al_2O_3) electrodes on top SiO_2 attached to a gold (Au) surface in order to produce an ohmic contact. (a) Schematic cross-section of the device and the measuring system and (b) an overall electron transfer process and current flow, but no current was observed in the absence of the GFP drop (green, represents GFP) (from Chirgwandi *et al.* 2008).

sensitized solar cell and produced electricity. Amazingly, the team even reported to have made the solar cell work in the absence of natural light, using the strategy of bioluminescent light emission from a mixture of chemicals from fireflies (magnesium and luciferase enzymes) to induce fluorescence as a light source for GFP's capacity to generate electricity. They hypothesized that since GFPs are genetically encodable, this kind of solar cell can be implanted or genetically expressed in the human body to synthesize energy from naturally occurring chemicals.

The predominant degradation of organic solar cells is attributable to the exposure to high-energy photons and the sensitivity of the photoactive layer to oxidative damage. Singlet molecular oxygen and complex photochemical reactions between the different solar cell layers, in combination with moisture, oxygen, light and heat often lead to a degradation and device failure (Brabec *et al.* 2010). The high photostability of GFPs and their antioxidant properties have the capacity to reduce the potential for photodegradation, thus further increasing their superior properties for photovoltaic technologies.

Fluorescent organic dyes are currently considered to be the most promising luminescent species for building efficient solar cells due to their high quantum efficiencies and absorption coefficients, ease of processing and relatively low cost (Klampaftis *et al.* 2009). The benefits of GFPs are that they are non-toxic, biodegradable and environmentally friendly. They can be produced cheaply by transfection and growth in bacterial, fungal or plant cultures. Availability of inexpensive and varied raw materials accompanied by an easy fabrication procedure and the ability to tune molecular properties has made GFP-based organic photovoltaics a highly attractive proposition. The truly unique property of GFP-like proteins is that they are genetically encodable and can be transfected into any organism, so that potentially, energy generation can be accomplished within living organisms.

7.3.5 Light-emitting diodes, optoelectronic and photochromic materials

Since the 1980s, both the academic and the private industries have been actively pursuing the development of all-organic materials and devices as alternatives to inorganic-based electronic and optoelectronic systems. The major advantages of organic relative to inorganic technologies are their smaller sizes, simpler fabrication through established lithography techniques, lower cost and large area electronics when coupled with conventional commercial printing technologies. The applications include organic light-emitting devices (OLEDs), lasers, photovoltaic and thin-film transistors.

Today commonly used diodes are semiconductor diodes, either solid or liquid materials, that conduct electricity at room temperature and include chemical elements and compounds such as silicon, germanium, selenium, gallium arsenide, zinc selenide and lead telluride. Light-emitting diodes (LEDs) emit visible light or infrared radiation when an electric current passes through them and are made of semiconductors mixed with substances that absorb electromagnetic radiation and re-emit it as visible light in a process known as electroluminescence. LEDs are employed in flat screen TVs, computer and other device touch screens, numerical displays of electronic digital watches and pocket calculators. An OLED is a flat light-emitting technology, made by placing a series of organic thin films between two conductors that generate bright light when an electrical current is applied. This technology was based on research initiated in the early 1960s (Kallmann and Pope 1960a, 1960b; Peter and Wolfgang 1962). OLEDs are being developed with a hope to replace crystalline/diode semiconductors and produce more efficient and cheaper LEDs.

The first application of GFPs in OLEDs was over a decade ago. A research team at the University of Southern California, headed by M. Thompson, used a biomimetic approach to generate light based on the excitation of GFP in an enzymatic chemiluminescent reaction that occurs in the jellyfish (You *et al.* 2000). Since the GFP protein scaffold adds an insulating layer around the fluorophore, that may hinder energy or electron transfer between the fluorophore and the surrounding matrix, the team prepared synthetic derivatives of GFP's imidazolidinone and oxazolone, thus eliminating the scaffold. A stronger electron-donor group was incorporated to increase the electron donating strength. Next, the GFP-like molecules were seeded onto a matrix of organic molecules, which converted energy into light, creating an OLED and eventually produced green and red types (You *et al.* 2000). Given the enormous variety of GFP-type proteins, many with different fluorophores, carrier transport and other functions, other OLEDs may be prepared, using the insight provided by the jellyfish GFP.

The exciting photoswitching properties of the photoactive fluorescent proteins (PAFPs) described earlier in this chapter can also be employed to develop advanced biomimetic solutions. PAFPs can act as nano-elements that can be switched between spectroscopically different states, enabling optical manipulation of matter and information at nanoscopic scales for optical data storage.

The limitation of the current commercial optical data storage devices (e.g. CD, DVD, Blu-ray discs, etc.) is their two-dimensional structure. To increase the storage data density, optical media using three dimensions (3D) was developed using polymers (Wang and Esener 2000; Gindre *et al.* 2006). With the development of two-photon excitation (TPE), the manipulation of optically encoded information within a 3D volume (i.e. voxel) became possible, since it enables excitation of molecules in a minute volume of the excitation beam in an opaque material (Mandzhikov *et al.* 1973).

The use of photochromic organic dyes to design rewritable optical data storage devices was first suggested ~50 years ago (Hirschberg 1956) and the science has been rapidly developing during the last decade (e.g. Berkovic *et al.* 2000; Irie 2000; Yokoyama 2000; Irie *et al.* 2002). The possibility of using photo-induced cycling between the protonated and the deprotonated forms of wtGFP for data storage at the single molecule level was first suggested by R. Tsien (Tsien 1998). Later, the photoswitching property of PAFPs for 3D data storage media was proposed for single GFP-like protein crystals (PCFPs and RSFPs) in which each protein molecule in the crystal would represent a binary encodable element, depending on whether the protein was in its 'on' (bright) or 'off' (dark) states (Hell *et al.* 2007).

PAFPs which have the capacity to alter colours following irradiation by light also provide the basis for data encoding and storage in a single crystal. This capability was recently explored by a multi-national research team from Belgium, Japan, France, USA and Germany (Adam *et al.* 2010). They pointed out that the advantages of using photoconvertible GFP-like proteins were: (i) that the two different switching/colour modes constituted alternative basis for data encoding and could be combined in a single crystal; (ii) PAFPs enable multicolour data storage; (iii) GFP-like proteins are biodegradable and provide a sustainable solution to the waste recycling of data storage devices. The researchers tested the following proteins: Kaede (Ando *et al.* 2002), Dronpa (Ando *et al.* 2004), three mutants of EosFP (Wiedenmann *et al.* 2004) – the dimeric form d1EosFP, the monomeric form mEosFP (Plate XX) (Wiedenmann *et al.* 2004), and IrisFP (Fig. 7.3) (Adam *et al.* 2008). However, they also pointed out several problems due to the biological nature of this technology, such as protein photobleaching, the need for a solvated state, sample aging and thermal instability (Adam *et al.* 2010). To overcome these problems, systematic screening of marine organisms for novel FPs was suggested, combined with random or rational mutagenesis, in order to generate more photostable and 'fatigue' resistant molecules.

The biomimetic exploitation of photoactive GFPs that photoconvert in a reversible or irreversible fashion, and between molecular states that differ in brightness and colour, creates a real possibility of bio-organic data storage devices and computing at nanoscales. Moreover, the photo-induced transformations of GFP-like proteins can be engineered to control the relative orientation and distance of donor–acceptor pairs in organic films. Nanoscale excitation that may be coherent and coupled to protein dipole shifts may generate communicative 'collective modes' within protein assemblies and provide a substrate for biological information processing (Hameroff and Schneiker 1987).

The field of photochromic materials in which photoactive GFPs may be used is too vast to be covered here. Light provides a precise and a reversible means of control without the requirement for physical contact. It is suffice to mention that photoresponsive GFP-based films, gels and polymers, with potential other biomimetic functionalities of electron or proton conductivity, of pH-, redox- and ion-responsiveness, are technologically important for a wide range of applications. These may include such devices and technologies as drug delivery and other biochemical and biomedical systems, smart microfluidic devices, microactuators and membranes in which light irradiation and optical control can be applied at micro- or nano-scale resolution and without physical contact. GFP-like protein expression in reef corals can be affected by environmental conditions and biological stresses and *in vivo* photoactivation of GFP-like proteins can be induced by the build-up of cellular ROS. This biological property can be an important source of biomimetic applications and the creation of photoactive materials that respond to biological and environmental stimuli.

7.4 Conclusion

Nature has produced a dazzling array of outstanding optical and molecular properties of GFPs and a variety of biological functions that have only become apparent in the last decade. This chapter presented several biomimetic applications of GFPs based on these properties and on their functional roles in marine organisms, and in particular, within the biophotonic system made up of the coral–algal symbiotic association.

The use of GFP and of GFP-like proteins has revolutionized the field of cell biology and biomedicine and the importance of the GFP-based technologies has been recognized with the 2008 Nobel Prize in Chemistry awarded to the pioneers in the field – Osamu Shimomura, Martin Chalfie and Roger Y. Tsien. The time has come for the field to evolve further so that biomimetic methods for improved energy production, and of biophotonic and optoelectronic developments emerge based on GFP's unique optical, molecular and biological properties.

Furthermore, GFP-like proteins are no longer considered to be inert when expressed in cells and may have a variety of functions linked to their photoactivity and to photon, electron and proton channelling properties. It may soon be possible to use GFPs not only as imaging tools or bio-sensors but as actuators with capacity to control a variety of biological functions.

During the last decade, the emergence of the use of optical methods in combination with genetic manipulations and light-driven control of cellular activity have created a new field known as 'optogenetics' (Miesenböck 2009). The ability to express GFPs in specific cellular organelles or compartments and fused to proteins of interest has become the norm. The ability to induce light-driven redox, membrane potential or electrolytic changes may provide an exciting possibility to genetically encode a light-inducible biological function and to precisely control it, using GFPs as optogenetic actuators. By further understanding the biology of GFPs in marine organisms, we can learn to use them as the basis for many new applications. It may even be feasible to develop GFP-based genetically encodable, self-powered nanodevices or even biological computers that record and transmit information, with wide-ranging applications in medicine, biological sciences and environmental monitoring.

7.5 References

Ackleson S G (2003), Light in shallow waters: a brief research review, *Limnol Oceanogr*, 48, 323–328.

Adam V, Lelimousin M, Boehme S, Desfonds G, Nienhaus K, Field M J, Wiedenmann J, McSweeney S, Nienhaus G U and Bourgeois D (2008), Structural characterization of IrisFP, an optical highlighter undergoing multiple photoinduced transformations, *Proc Natl Acad Sci USA*, 105, 18343–18348.

Adam V, Mizuno H A, Grichine A, Hotta J-I, Yamagata Y, Moeyaert B, Nienhaus G U, Miyawaki A, Bourgeois D and Hofkens J (2010), Data storage based on photochromic and photoconvertible fluorescent proteins, *J Biotechnology*, 149, 4, 289–298.

Aglyamova G F, Hunt H E, Chintan K, Modib M and Matz M V (2011), Multicolored homologs of the green fluorescent protein from hydromedusa *Obelia* sp., *Photochem & Photobiol Sci.*, 10, 1303–1309.

Agmon N (2005), Proton pathways in green fluorescence protein, *Biophys J*, 88, 2452–2461.

Alieva N O, Konzen K A, Field S F, Meleshkevitch E A, Beltran-Ramirez V, Miller D J, Salih A, Wiedenmann J and Matz M V (2008), Diversity and evolution of coral fluorescent proteins, *PLoS One*, 3, e2680.

Ando R, Hama H, Yamamoto-Hino M, Mizuno H and Miyawaki A (2002), An optical marker based on the UV-induced green-to-red-photoconversion of a fluorescent protein, *Proc Natl Acad Sci USA*, 99, 12651–12656.

Andresen M, Wahl M C, Stiel A C, Grater F, Schafer L V, Trowitzsch S, Weber G, Eggeling C, Grubmuller H, Hell S W *et al.* (2005), Structure and mechanism of the reversible photoswitch of a fluorescent protein, *Proc Natl Acad Sci USA*, 102, 13070–13074.

Asada K (2006), Update on reactive oxygen species in chloroplasts: production and scavenging of reactive oxygen species in chloroplasts and their functions, *Plant Physiology*, 141, 391–396.

Baird G S, Zacharias D A and Tsien R Y (2000), Biochemistry, mutagenesis, and oligomerization of DsRed, a red fluorescent protein from coral, *Proc Natl Acad Sci USA*, 2000, 97, 11984–11989.

Bastiaens P I H and Pepperkok R (2000), Observing proteins in their natural habitat: the living cell, *Trends in Biological Science*, 25, 631–637.

Berkovic G, Krongauz V and Weiss V (2000), Spiropyrans and spirooxazines for memories and switches, *Chem Rev*, 100, 1741–1753.

Bogdanov A M, Mishin A S, Yampolsky I V, Belousov V V, Chudakov D M, Subach F V, Verkhusha V V, Lukyanov S, Lukyanov K A (2009), Green fluorescent proteins are light-induced electron donors, *Nat Chem Biol*, 26, 5(7), 459–461.

Bomati E K, Manning G and Deheyn D D (2009), Amphioxus encodes the largest known family of green fluorescent proteins, which have diversified into distinct functional classes, *BMC Evol Biol*, 9, 77.

Bou-Abdallah F, Chasteen N D and Lesser M P (2006), Quenching of superoxide radicals by green fluorescent protein, *Biochim Biophys Acta, Gen Subj*, 1760, 1690–1695.

Brabec C J, Gowrisanker S, Halls J J M, Laird D, Jia S and Williams S P (2010), Polymer–fullerene bulk-heterojunction solar cells, *Adv Mater*, 22, 3839–3856.

Brown B E (1997), Coral bleaching: causes and consequences, *Coral Reefs*, 16 (Supplem) S129–S138.
Catala R (1959), Fluorescence effect from corals irradiated with ultra-violet rays, *Nature*, 1959, 183, 949.
Chalfie M, Tu Y, Euskirchen G, Ward W W and Prasher D C (1994), Green fluorescent protein as a marker for gene expression, *Science*, 1994, 263, 802–805.
Chalfie M (1995), Green fluorescent protein, *Photochem Photobiol*, 62, 651–656.
Chalfie M and Kain S (2006), *Green Fluorescent Protein: Properties, Applications, and Protocols*, Wiley-Interscience, Hoboken, NY, 2nd edn.
Chattoraj M, King B A, Bublitz G U and Boxer S G (1996), Ultra-fast excited state dynamics in green fluorescent protein: multiple states and proton transfer, *Proc Natl Acad Sci,* 93, 8362–8367.
Chirgwandi Z G, Panas I, Johansson L-G, Norden B, Willander M, Winkler D, Ågren H (2008), Properties of a biophotovoltaic nanodevice, *J Phys Chem Lett*, 112, 18717–18721.
Chishima T, Miyagi Y, Wang X, Yamaoka H, Shimada H, Moossa A R and Hoffman R M (1997), Cancer invasion and micrometastasis visualized in live tissue by green fluorescent protein expression, *Cancer Res*, 57, 2042–2047.
Choi J W, Nam Y S, Oh B K, Lee W H and Fujihira M (2000), Artificial molecular photodiode consisting of GFP/viologen hetero-thin film, *Synth Met*, 117, 241.
Choi J W, Nam Y S and Lee W H (2001), Rectified photocurrent of the protein-based bio-photodiode, *Appl Phys Lett*, 79, 10, 570–572.
Choi J-W and Fujihira M (2004), Molecular-scale biophotodiode consisting of a green fluorescent protein assembled heterolayer, *Appl Phys Lett*, 84, 2187–2189.
Chuang W T, Chen B S, Chen K-Y, Hsieh C-C, Chou P T (2009), Fluorescent protein red Kaede chromophore; one-step, high-yield synthesis and potential application for solar cells, *Chem Commun*, 2009, 6982–6984.
Chudakov D M, Verkhusha V V, Staroverov D B, Souslova E A, Lukyanov S, Lukyanov K A (2004), Photoswitchable cyan fluorescent protein for protein tracking, *Nat Biotechnol*, 22(11), 1435–9. Epub 2004 Oct 24.
Chudakov D M, Matz M V, Lukyanov S A and Lukyanov K A (2010), Fluorescent proteins and their applications to imaging in living cells and tissues, *Physiol Rev*, 2010, 90, 1103–1163.
Chudakov D M, Belousov V V, Zaraisky A G, Novoselov V V, Staroverov D B, Zorov D B, Lukyanov S and Lukyanov K A (2003), Kindling fluorescent proteins for precise *in vivo* photolabeling, *Nat Biotechnol*, 2003, 21, 191–194.
Cody C W, Prasher D C, Westler W M, Prendergast F G and Ward W W (1993), Chemical structure of the hexapeptide chromophore of the *Aequorea* green-fluorescent protein, *Biochemistry*, 32, 1212–1218.
Cox G and Salih A (2005), Fluorescent lifetime imaging of symbionts and fluorescent proteins in reef corals. *Multiphoton Microscopy in the Biomedical Sciences V*, *Proc SPIE*, San Jose, USA.
Cox G and Salih A (2006), Fluorescent characteristics of fluoro-chromatophores in corals. In: *Focus on Multidimensional Microscopy*. Cheng P C, Hwang P P, Wu J L, Wang G and Kim H (eds). World Scientific Publishing Co. Vol. 3. pp. 145–151.
Cox G, Matz M and Salih A (2007), Fluorescence lifetime imaging of coral fluorescent proteins, *Microscopy Res Technique*, 70, 243–251.

Day R N and Davidson M W (2009), The fluorescent protein palette: tools for cellular imaging, *Chem Soc Rev*, 2009, 38, 2887–2921.

Deisenhofer J, Epp O, Miki K, Huber R, and Michel H (1985), X-ray structure analysis of a membrane protein complex, *Nature*, 318, 618.

Dove S, Takabayashi M and Hoegh-Guldberg O (1995), Isolation and partial characterisation of the pink and blue pigments of pocilloporid and acroporid corals, *Biol Bull Mar Biol Lab*, Woods Hole 189, 288–297.

Dove S G, Hoegh-Guldberg O and Ranganathan S (2001), Major colour patterns of reef-building corals are due to a family of GFP-like proteins, *Coral Reefs,* 19, 197–204.

Dove S G, Lovell C, Fine M, Deckenback J, Hoegh-Guldberg O, Iglesias-Prieto R and Anthony K R N (2008), Host pigments: potential facilitators of photosynthesis in coral symbioses, *Plant Cell Environ*, 31, 1523–1533.

Eaton W A and Hofrichter J (1990), Sickle cell hemoglobin polymerization, *Adv Protein Chem*, 40, 63–79.

Evdokimov A G, Pokross M E, Egorov N S, Zaraisky A G, Yampolsky I V, Merzlyak E M, Shkoporov A N, Sander I, Lukyanov K A and Chudakov D M, Structural basis for the fast maturation of Anthropoda green fluorescent protein, *EMBO Rep.*, 2006, 7, 1006–1012.

Field S F, Bulina M Y, Kelmanson I V, Bielawski J P and Matz M V (2006), Adaptive evolution of multicolored fluorescent proteins in red building corals, *J Mol Evol*, 62, 332–339.

Förster T (1948), Intermolecular energy migration and fluorescence, *Ann Phys*, 2, 55–75.

Förster T (1965), In *Modern Quantum Chemistry, Istanbul Lectures*, O. Sinanoglu Academic, New York, Vol. 3, pp. 93–137.

Franklin K A and Whitelam G C (2007), Phytochrome a function in red light sensing, *Plant Signal Behav*, Sep–Oct; 2(5), 383–385.

Gaffron H and Wohl K (1936), Zur Theorie der Assimilation, *Naturwiss*, 24, 81–90.

Gilmore A M, Larkum A W D, Salih A, Itoh S, Shibata Y, Bena C, Yamasaki H, Papina M and Van Woesik R (2003), Simultaneous time-resolution of the emission spectra of fluorescent proteins and zooxanthellae chlorophyll in reef-building corals, *Photochem Photobiol*, 77, 515–523.

Gindre D, Boeglin A, Fort A, Mager L, and Dorkenoo K D, (2006), Rewritable optical data storage in azobenzene copolymers, *Opt Expr*, 14, 9896–9901.

Grant D M, Zhang W, McGhee E J, Bunney T D, Talbot C B, Kumar S, Munro, I, Dunsby C, Neil M A A, Katan M and French P M W (2008), Multiplexed FRET to image multiple signaling events in live cells, *Biophys J*, 95, L69-L71.

Green B R (2003), The evolution of light harvesting antennas. In: *Light Harvesting Antennas in Photosynthesis* (Green B R, Parson W W (eds)). Kluwer Academic Press, Dordrecht, pp. 129–168.

Gregg B A and Hanna M C (2003), Comparing organic to inorganic photovoltaic cells: Theory, experiment and simulation, *J Appl Phys*, 2003, 93, 3605.

Gross L A, Baird G S, Hoffman R C, Baldridge K K and Tsien R Y (2000), The structure of the chromophore within DsRed, a red fluorescent protein from coral, *Proc Natl Acad Sci USA*, 97, 11,990–11,995.

Gruber D F, Kao H-T, Janoschka S, Tsai J and Pieribone V (2008), Patterns of fluorescent protein expression in scleractinian corals, *Biolog Bull*, 215, 2, 143–154.

Gust D, Moore T A and Moore A L (2001), Mimicking photosynthetic solar energy transduction, *Acc Chem Res*, 34, 40–48.

Habuchi S, Ando R, Dedecker P, Verheijen W, Mizuno H, Miyawaki A and Hofkens J (2005), Reversible single-molecule photoswitching in the GFP-like fluorescent protein Dronpa, *Proc Natl Acad Sci* USA, 102, 9511–9516.

Hameroff S and Schneiker C (1987), *Ultimate Computing: Biomolecular Consciousness and Nanotechnology*, Elsevier-North Holland.

Hardin B E, Hoke E T, Armstrong P B, Yum J-H, Comte P, Torres T, Fréchet J M J, Nazeeruddin M K, Grätzel M and McGehee M D (2009), Increased light harvesting in dye-sensitized solar cells with energy relay dyes, *Nature Photonics*, 3, 406–411.

Heim R, Prasher D C and Tsien R Y (1994), Wavelength mutations and post translational autoxidation of green fluorescent protein, *Proc Natl Acad Sci*, 91, 12501–12504.

Hell S W, Jakobs S, Andresen M, Stiel A C and Eggeling C (2007), Method and apparatus for storing a three-dimensional arrangement of data bits in a solid-state body. Patents #US20070047287, DE102005040671, EP1763024.

Hirschberg Y (1956), Reversible formation and eradication of colors by irradiation at low temperatures: a photochemical memory model, *J Am Chem Soc*, 78, 2304–2312.

Hoegh-Guldberg O (1999), Climate change, coral bleaching and the future of the world's coral reefs, *Mar Freshw Res*, 50: 839–866.

Hoffman R M (2005), The multiple uses of fluorescent proteins to visualize cancer *in vivo. Nature Rev Cancer*, 5, 796–806.

Hollingsworth L L, Kinzie R A, Lewis T D, Krupp D A and Leong J A C (2005), Phototaxis of motile zooxanthellae to green light may facilitate symbiont capture by coral larvae, *Coral Reefs*, 24, 523.

Hopf M, Gohring W, Ries A, Timpl R and Hohenester E (2001), Crystal structure and mutational analysis of a perlecan-binding fragment of nidogen-1. *Nat Struct Biol*, 8, 634–640.

Hoppe H and Sariciftci N S (2004), Organic solar cells: an overview. *Mater Res*, 19, 7, 1924–1945.

Hunt M E, Scherrer M P, Ferrari F D and Matz M V (2010), Very bright green fluorescent proteins from the Pontellid copepod *Pontella mimocerami*, *PLoS One*, 5, e11517.

Imamoto Y and Kataoka M (2007), Structure and photoreaction of photoactive yellow protein, a structural prototype of the PAS domain superfamily, *Photochemi Photobiol*, 83, 1, 40–49.

Irie M (2000), Diarylethenes for memories and switches, *Chem Rev*, 100, 1685–1716.

Irie M, Fukaminato T, Sasaki T, Tamai N and Kawai T (2002), Organic chemistry: a digital fluorescent molecular photoswitch, *Nature,* 420, 759–760.

Jain V, Rajbongshi B K, Mallajosyula A T, Bhattacharjya G, KumarIyer S S and Ramanathan G (2008), Photovoltaic effect in single-layer organic solar cell devices fabricated with two new imidazolin-5-one molecules, *Solar Energy Mater Solar Cells*, 92, 1043–1046.

Jerlov N G (1976), *Marine optics*. Oxford: Elsevier Scientific Publishing Company.

Kahng S E and Salih A (2005), Localization of fluorescent pigments in a non-bioluminescent, azooxanthellate octocoral suggests a photo-protective function, *Coral Reefs*, 24, 436–437.

Kallmann H and Pope M (1960a), Positive hole injection into organic crystals, *J Chemi Physi*, 32: 300.

Kallmann H and Pope M (1960b), Bulk conductivity in organic crystals, *Nature* 186 (4718), 31.

Kao H T, Sturgis S, DeSalle R, Tsai J, Davis D, Gruber D F, Pieribone V A (2007), Dynamic regulation of fluorescent proteins from a single species of coral, *Marine Biotechnol.*, 1, 1–14.

Karvanos G J (1993), *Fundamentals of Photoinduced Electron Transfer*, VCH, New York.

Kawaguti S (1944), On the physiology of reef corals. VI. Study, of the pigments, *Palao Trop Biol Stn Stud*, 2, 617–674.

Kelmanson IV and Matz M (2003), Molecular basis and evolutionary origins of color diversity in great star coral *Montastraea cavernosa* (Scleractinia: Faviida), *Mol Biol Evol*, 20, 1125–1133.

Klampaftis E, Ross D, McIntosh K R and Richards B S (2009), Enhancing the performance of solar cells via luminescent down-shifting of the incident spectrum: A review. *Solar Energy Materi Solar Cells*, 93, 1182–1194.

Knight H (2010), Green machine: Squeezing solar juice from jellyfish. *New Scientist* http://www.newscientist.com/article/dn19416-green-machine-squeezing-solar-juice-from-jellyfish.html

Kulzer F and Orrit M (2004), Single-molecule optics. *Ann Rev Phys Chem*, 55, 585–611.

Labas Y A, Gurskaya N G, Yanushevich Y G, Fradkov A F, Lukyanov K A, Lukyanov S A and Matz M V (2002), Diversity and evolution of the green fluorescent protein family, *Proc Natl Acad Sci USA*, 99, 4256–4261.

Lesser M P (2007), Coral reefs bleaching and global climate change: Can corals survive the next generation? *Proc Natl Acad Sci*, 104, 5259–5260.

Li I T, Pham E and Truong K (2006), Protein biosensors based on the principle of fluorescence resonance energy transfer for monitoring cellular dynamics, *Biotechnol Lett,* 28, 1971–1982.

Lippincott-Schwartz J, Snapp E and Kenworthy A (2001), Studying protein dynamics in living cells, *Nature Rev Mol Cell Biol*, 2, 444–456.

Lippincott-Schwarz J (2002), Cell biology: ripping up the nuclear envelope. *Nature,* 416(6876), 31–32.

Lippincott-Schwartz J and Patterson G H (2008), Fluorescent proteins for photoactivation experiments, *Methods Cell Biology*, 85, 45–61.

Liu H R, Ramnarayanan R and Logan E B (2004), Production of electricity during wastewater treatment using a single chamber microbial fuel cell. *Environ Sci Technol*, 38(7), 2281–2285.

Losi A (2004), The bacterial counterparts of plant phototropins, *Photochem, Photobiol Sci*, 3, 566–574.

Lukyanov K A, Fradkov A F, Gurskaya N G, Matz M V, Labas Y A, Savitsky A P, Markelov M L, Zaraisky A G, Zhao X, Fang Y, Tan W and Lukyanov S A (2000),

Natural animal coloration can be determined by a nonfluorescent green fluorescent protein homolog, *J Biol Chem*, 275, 25879–25882.

Lukyanov K A, Chudakov D M, Lukyanov S, Verkhusha V V (2005), Innovation: photoactivatable fluorescent proteins. *Nat Rev Mol Cell Biol*, 6(11), 885–891.

Lukyanov K A, Loschenov V B, Luker G D and Chudakov D M (2010), Near-infrared fluorescent proteins. *Nat Methods*, 7(10), 827–829.

Mandzhikov V F, Murin V A and Barachevskii V A (1973), Nonlinear coloration of photochromic spiropyran solutions, *Sov J Quant Electron*, 3, 128–129.

Maruyama T, Shinyashiki Y, Osako S (1998), Energy conversion efficiency of solar cells coated with fluorescent coloring agent, *Solar Energy Mater Solar Cells*, 56, 1–6.

Matz M V, Fradkov A F, Labas Y A, Savitsky A P, Zaraisky A G, Markelov M L and Lukyanov S A (1999), Fluorescent proteins from nonbioluminescent Anthozoa species, *Nat Biotechnol*, 17, 969–973.

Matz M V, Labas Y A and Ugalde J (2006), Evolution of function and color in GFP-like proteins, *Methods Biochem Anal*, 47, 139–161.

Mazel C H (1995), Spectral measurements of fluorescence emission in Caribbean cnidarians, *Mar Ecol Prog Ser*, 120, 185–191.

Mazel C H, Lesser M P, Gorbunov M Y, Barry Y M, Farrell J H, Wyman K D and Falkowski P G (2003), Green fluorescent proteins in Caribbean corals, *Limnol Oceanogr*, 48, 402–411.

Mank M and Griesbeck O (2008), Genetically encoded calcium indicators, *Chem Rev*, 108, 1550–1564.

Meech S R (2009), Proton transfer – excited state reactions in fluorescent protein, *Chem Soc Rev*, 38, 2922–2934.

Miesenböck G (2009), The optogenetic catechism, *Rev Sci*, 326, 395–399.

Miyawaki A, Llopis J, Heim R, McCaffery J M, Adams J A, Ikura M and Tsien R Y (1997), Fluorescent indicators for Ca^{2+} based on green fluorescent proteins and calmodulin, *Nature*, 388, 882–887.

Miyawaki A and Tsien R Y (2000), Monitoring protein conformations and interactions by fluorescence resonance energy transfer between mutants of green fluorescent protein, *Methods in Enzymology*, 327, 472–500.

Miyawaki A, Mizuno H, Nagal T and Sawano A (2003), Development of genetically encoded fluorescent indicators for calcium, *Methods in Enzymology*, 360, 202–225.

Miyawaki A (2005), Innovations in the imaging of brain functions using fluorescent proteins, *Neuron*, 48, 189–199.

Miyawaki A (2009), Great expectations, *Science*, 326(5951), 339.

Mizuno H, Mal T K, Tong K I, Ando R, Furuta T, Ikura M and Miyawaki A (2003), Photo-induced peptide cleavage in the green-to-red conversion of a fluorescent protein, *Mol Cell*, 12, 1051–1058.

Morin J G and Hastings J W (1971), Energy transfer in a bioluminescent system, *J Cell Physiol*, 1971, 77, 313.

Morise H, Shimomura O, Johnson F H and Winant J (1974), Intermolecular energy transfer in the bioluminescent system of Aequorea, *Biochemistry*, 13(12), 2656–2662.

Muscatine L, Falkowski P, Porter J and Dijbinsky Z (1984), Fate of photosynthetically-fixed carbon in light and shade adapted colonies of the symbiotic coral Stylophora pistillata, *Proc R Sot Lond Ser B*, 222, 181–202.

Ong W J, Alvarez S, Leroux I E, Shahid R S, Samma A A, Peshkepija P, Morgan A L, Mulcahy S and Zimmer M (2011), Function and structure of GFP-like proteins in the protein data bank, *Mol Biosyst*, 7(4), 984–992.

Ormo M, Cubitt A B, Kallio K, Gross L A, Tsien R Y and Remington S J (1996), Crystal structure of the *Aequorea victoria* green fluorescent protein, *Science*, 273, 1392–1395.

Oswald F, Schmitt F, Leutenegger A, Ivanchenko S, D'Angelo C, Salih A, Maslakova S, Bulina M, Schirmbeck R, Nienhaus G U, Matz M V and Wiedenmann J (2007), Contributions of host and symbiont pigments to the coloration of reef corals, *FEBS J*, 274, 1102–1109.

Palmer C V, Modi C K and Mydlarz L D (2009), Coral fluorescent proteins as antioxidants, *PLoS ONE*, 4(10), e7298.

Patterson G H and Lippincott-Schwartz J (2002), A photoactivatable GFP for selective photolabeling of proteins and cells, *Science*, 297, 1873–1877.

Patterson G H (2007), Fluorescent proteins for cell biology. From: *Methods in Molecular Biology, vol. 411: Reporter Genes: A Practical Guide* (Anson D (ed)). © Humana Press Inc., Totowa, NJ, pp. 47–80.

Peter M and Wolfgang H (1962), Space-charge-limited currents in organic crystals, *J Appli Phys*, 33, 205.

Prasher D C, Eckenrode V K, Ward W W, Prendergast F G and Cormier M J (1992), Primary structure of the *Aequorea victoria* green-fluorescent protein, *Gene*, 111, 229–233.

Prescott M and Salih A (2009), Chapter 4: Genetically encoded fluorescent probes and their applications in the life sciences. In: *Fluorescence Applications in Biotechnology and Life Sciences* (Goldys E (ed.)), John Wiley & Sons, Inc., Hoboken, New Jersey. USA & Canada. pp. 55–77.

Quillin M L, Anstrom D M, Shu X, O'Leary S, Kallio K, Lukyanov K A, and Remington S J (2005), The kindling fluorescent protein from *Anemonia sulcata*: dark state structure at 1.38Å resolution, *Biochemistry*, 44, 5774–5787.

Remington S J, Hanson G T, Cannon M, Aggeler R, Oglesbee D, Capaldi R A and Tsien R Y (2004), Third generation GFP biosensors for real time readout of pH and redox potential in living cells, *Biomed Opt*, 5329, 1–12.

Remington G (2006), Fluorescent proteins: maturation, photochemistry and photophysics. *Current Opinion Struct Biol*, 16, 714–721.

Rosenow M A, Huffman H A, Phail M E and Wachter R M (2004), The crystal structure of the Y66L variant of green fluorescent protein supports a cyclization-oxidation-dehydration mechanism for chromophore maturation, *Biochemistry*, 43, 4464–4472.

Roth M S, Latz M I, Goericke R and Deheyn D D (2010), Green fluorescent protein regulation in the coral *Acropora yongei* during photoacclimation, *J Exp Biol*, 213, 3644–3655.

Salih A, Hoegh-Guldberg O and Cox G (1998), Photoprotection of symbiotic dinoflagellates by fluorescent pigments in reef corals, *Proc Aust Coral Reef Society 75 Anniv Conference*, Greenwood & Hall, pp. 217–230.

Salih A, Larkum A W D, Cox G, Kühl M and Hoegh-Guldberg O (2000), Fluorescent pigments in corals are photoprotective, *Nature*, 408, 850–853.

Salih A, Larkum A W D and Cox G (2001), Photoprotection from photoinhibion of symbiotic algae in corals by fluorescent pigments. *PS2001 Proceedings: 12th International Congress on Photosynthesis*, CSIRO Publish: Melbourne, Australia. http://www.publish.csiro.au/ps2001/cf/search/article.cfm?ID=4530661

Salih A, Cox G and Larkum A W D (2003), Cellular organization and spectral diversity of GFP-like proteins in coral cells studied by single and multi-photon imaging and microspectroscopy. *Multiphoton microscopy in Biomedical Sciences III*, (Pereasamy A and So P T (eds)). *Proceedings of SPIE*, California. Vol. 4963, pp. 194–200.

Salih A, Cox G, Szymczak R, Coles S, Baird A, Dunstan A, Cocco G, Mills J and Larkum A W D (2006a), The role of host-based color and fluorescent pigments in photoprotection and in reducing bleaching stress in corals. In *Proceedings of the 10th International Coral Reef Symposium*. Okinawa, Japan: ICRS. http://www.coralcoe.org.au/pub-scientific2006.html

Salih A, Wiedenmann J, Mikhail M, Larkum A W D and Cox G (2006b), Photoinduced activation of GFP-like proteins in tissues of reef corals. *Proc SPIE,* 6098.

Schlichter D and Fricke H W (1990), Coral host improves photosynthesis of endosymbiotic algae. *Naturwissenschaften*, 77, 447–450.

Shagin D A *et al.* (2004), GFP-like proteins as ubiquitous metazoan superfamily: evolution of functional features and structural complexity, *Mol Biol Evol*, 21, 841–850.

Shaner N C, Campbell R E, Steinbach P A, Giepmans B N, Palmer A E and Tsien R Y (2004), Improved monomeric red, orange and yellow fluorescent proteins derived from *Discosoma* sp. red fluorescent protein, *Nat Biotechnol*, 22, 1567–1572.

Shaner N C, Patterson G H and Davidson M W (2007), Advances in fluorescent protein technology, *J Cell Science*, 120(24), 4247–4259.

Shcherbo D, Shemiakina I I, Ryabova A V, Luker K E, Schmidt B T, Souslova E A, Gorodnicheva T V, Strukova L, Shidlovskiy K M, Britanova O V and Zaraisky A G, Fluorescent indicators for Ca^{2+} based on green fluorescent proteins and calmodulin.

Shcherbo D, Merzlyak E M, Chepurnykh T V, Fradkov A F, Ermakova G V, Solovieva E A, Lukyanov K A, Bogdanova E A, Zaraisky A G, Lukyanov S and Chudakov D M (2007), Bright far-red fluorescent protein for whole-body imaging, *Nat Methods*, 4(9), 741–746.

Shimomura O, Inoue S, Johnson F H and Haneda Y (1980), Widespread occurrence of coelenterazine in marine bioluminescence, *Comp Biochem Physiol, Part B: Biochem Mol Biol*, 65, 435–437.

Shimomura O (2005), The discovery of aequorin and green fluorescent protein, *J Microscopy*, 217, 1, 3–15.

Shinobu A and Agmon N (2009), Mapping proton wires in proteins: carbonic anhydrase and GFP chromophore biosynthesis, *J Phys Chem A*, 113, 7253–7266.

Stepanenko O V, Shcherbakova D M, Kuznetsova I M, Turoverov K K and Verkhusha V V (2011), Modern fluorescent proteins: from chromophore formation to novel intracellular applications, *BioTechniques*, 51, 313–327.

Stepanenko V, Verkhusha V V, Kuznetsova I and Uversky V N (2008), Fluorescent proteins as biomarkers and biosensors: throwing color lights on molecular and cellular processes, *Current Protein Peptide Sci*, 9, 338–69.

Stoner-Ma D, Jaye A A, Matousek P, Towrie M, Meech S R, Tonge P J (2005), Observation of excited-state proton transfer in green fluorescent protein using ultrafast vibrational spectroscopy, *J Am Chem Soc*, 127, 2864–2865.

Takabayashi M and Hoegh-Guldberg O (1995), Ecological and physiological differences between two colour morphs of the coral Pocillopora damicornis, *Mar Biol*, 123, 705–714.

Tsien R Y (1998), Green fluorescent protein: review, *Ann Rev Biochem*, 67, 509–544.

Tsien R Y (2010), Nobel lecture: constructing and exploiting the fluorescent protein paintbox, *Integr Biol*, 2, 77–93.

Tsutsui H, Karasawa S, Okamura Y and Miyawaki A (2008), Improving membrane voltage measurements using FRET with new fluorescent proteins, *Nature Methods*, 5, 683–685.

van der Horst M A and Hellingwerf K J (2004), Photoreceptor proteins, 'star actors of modern times': a review of the functional dynamics in the structure of representative members of six different photoreceptor families, *Acc Chem Res*, 37, 13–20.

van Thor J J, Georgiev G Y, Towrie M and Sage J T J (2005), Ultrafast and low barrier motions in the photoreactions of the green fluorescent protein, *Biol Chem*, 280, 33652–33659.

Veal C J, Carmi M, Dishon G, Sharon Y, Michael K, Tchernov D, Hoegh-Guldberg O and Fine M (2010), Shallow water wave lensing in coral reefs: a physical and biological case study, *J Expe Biol*, 213, 4304–4312.

Verkhusha V V and Lukyanov K A (2004), The molecular properties and applications of Anthozoa fluorescent proteins and chromoproteins, *Nat Biotechnol*, 22, 289–296.

Wachter R M and Remington S J (1999), Sensitivity of the yellow variant of green fluorescent protein to halides and nitrate, *Curr Biol,* 9, R628–R629.

Wachter R M, Watkins J L and Hanseong K (2010), Mechanistic diversity of red fluorescence acquisition by GFP-like proteins, *Biochem*, 49, 7417–7427.

Wall M A, Socolich M and Ranganathan R (2000), The structural basis for red fluorescence in the tetrameric GFP homolog DsRed, *Nat Struct Biol*, 7, 1133–1138.

Wang M M and Esener S C (2000), Three-dimensional optical data storage in a fluorescent dye-doped photopolymer, *Appl Opt*, 39, 1826–1834.

Weiss V M (2008), Cellular mechanisms of cndiarian bleaching: stress causes the collapse of symbiosis, *J Exp Biol*, 211, 3059–3066.

Wiedenmann J, Ivanchenko S, Oswald F, Schmitt F, Rocker C, Salih A, Spindler K and Nienhaus G U (2004), EosFP, a fluorescent marker protein with UV-inducible green-to-red fluorescence conversion, *Proc Natl Acad Sci USA*, 101, 15905–15910.

Wu B, Piatkevich K D, Lionnet T, Singer R H and Verkhusha V V (2011), Modern fluorescent proteins and imaging technologies to study gene expression, nuclear localization, and dynamics, *Curr Opin Cell Biol*, 23, 310–317.

Yang F, Moss L G and Phillips G N Jr (1996), The molecular structure of green fluorescent protein, *Nat Biotechnol*, 14, 1246–1251.

Yanushevich Y G, Staroverov D B, Savitsky A P, Fradkov A F, Gurskaya N G, Bulina M E, Lukyanov K A and Lukyanov S A (2002), A strategy for the generation of non-aggregating mutants of Anthozoa fluorescent proteins, *FEBS Letts*, 511(1–3), 11–14.

Yarbrough D, Wachter R M, Kallio K, Matz M and Remington S J (2001), Refined crystal structure of DsRed, a red fluorescent protein from coral, at 2.0-Å resolution, *Proc Natl Acad Sci USA*, 98, 462–467.

Yokoyama Y (2000), Fulgides for memories and switches, *Chem Rev*, 100, 1717–1739.

You Y, He Y, Burrows P E, Forrest S R, Petasis N A and Thompson M E (2000), Fluorophores related to the green fluorescent protein and their use in optoelectronic devices, *Advanced Materials*, 12(22), 1678–1681.

Zhang X and Jenekhe S A (2000), Electroluminescence of multicomponent conjugated polymers. 1. Roles of polymer/polymer interfaces in emission enhancement and voltage-tunable multicolor emission in semiconducting polymer/polymer heterojunctions, *Macromolecules*, 33(6), 2069–2082.

Zhang L, Patel H N, Lappe J W and Wachter R M (2006), Reaction progress of chromophore biogenesis in green fluorescent protein, *J Am Chem Soc*, 128, 4766–4772.

Index